VICTORIAN SCIENCE AND VICTORIAN VALUES: LITERARY PERSPECTIVES

ANNALS OF THE NEW YORK ACADEMY OF SCIENCES

Volume 360

VICTORIAN SCIENCE AND VICTORIAN VALUES: LITERARY PERSPECTIVES

Edited by James Paradis and Thomas Postlewait

The New York Academy of Sciences
New York, New York
1981

The cover picture of the Crystal Palace is from a nineteenth-century engraving owned by The Bettmann Archive, Inc.

Library of Congress Cataloging in Publication Data

Main entry under title:

Victorian science and Victorian values.

(Annals of the New York Academy of Sciences; v. 360)
1. Science — Social aspects — England — Addresses, essays, lectures. 3. Literature and science — Addresses, essays, lectures. I. Paradis, James. II. Postlewait, Thomas. III. Series: New York Academy of Sciences. Annals; v. 360. Q11.N5 vol. 360 [Q175.52.G7] 500s [304.4'83] 80-29513

CCP
Printed in the United States of America
ISBN 0-89766-109-5 (cloth)
ISBN 0-89766-110-9 (paper)

ANNALS OF THE NEW YORK ACADEMY OF SCIENCES

VOLUME 360

April 20, 1981

VICTORIAN SCIENCE AND VICTORIAN VALUES: LITERARY PERSPECTIVES

Editors
JAMES PARADIS AND THOMAS POSTLEWAIT

❧ ❧

CONTENTS

Preface

SINCE C. P. Snow first published *The Two Cultures and The Scientific Revolution* in 1959, the barriers between the sciences and humanities have seemed more formidable than ever. As Snow noted, however, we have ourselves to blame. The modern debate about science and human values is often less than sincere; it is stylized and ritualistic. Artists and humanists are likely to claim the realms of affective thought and symbol; scientists and applied scientists appear content with certain logical realms of systematic thought and fact. Set conflicts abound, including those between belief and reason, feeling and computation, figurative and literal language, the personal and the public, and so on. The result is that mind in the modern world is divided among the categories of the specialist, its dimensions and potentialities curiously parceled out among academics and other professionals.

The Victorians carried out an extended discussion of science and value in the same inevitable terms as our modern debate, but with one notable difference. Most of them — scientists, artists, and critics alike — assumed that a culture, however diversely organized, was itself an organic unity in which the values motivating the scientist could be reconciled with those of the artist and humanist. This assumption inspired a remarkable series of public discussions, literary works, and journalistic controversies that collectively weighed the cultural significance of science. The dynamics and consequences of this great Victorian discussion provide the authors of this collection of essays with their primary topics.

Victorian Science and Victorian Values: Literary Perspectives took its title from a panel convened at the 1978 meeting of the Modern Language Association in New York City. The panel was composed of scholars who were interested less in mapping out the discrete territories of the Victorian professional than in finding interdisciplinary perspectives that would recover some sense of how Victorian values bound together the activities of the scientist, artist, and humanist. These papers and discussions provided a stimulus and example for a second panel the following year on the uses of language in Victorian science and literature. The initial essays, together with a number of additional background studies produced by a general call for papers, make up this collection. We were encouraged by the amount of interest many generalists, historians of science, and historians of literature expressed in the project and regret that we were unable to include several fine essays.

vii

The essays of this collection are arranged chronologically, beginning with discussions of science and human values early in the nineteenth century and ending with the last Victorian treatment of the subject in the works of George Bernard Shaw. Four broad areas of interdisciplinary interest are represented here: (1) the intellectual backgrounds shared by Victorian science and art; (2) science biography in Victorian literature; (3) the social forces supporting Victorian science and art; and (4) problems of language and definition in Victorian science and art.

The Victorians whose work is discussed in this volume include scientists such as Herschel, Lyell, Darwin, Whewell, Lockyer, and Huxley; literary artists such as George Eliot, Tennyson, Dickens, Hopkins, and Shaw; and cultural critics such as Carlyle, Mill, Ruskin, and Arnold. All of these diverse thinkers and writers were deeply conscious that a vast scientific and technological movement was in progress, and each of them had interests that were interdisciplinary in scope. Herschel read Comte, Darwin read Alexander Bain and Joshua Reynolds, George Eliot studied Carpenter's lectures on physiology, Tennyson devoted himself to contemporary geology and astronomy, Dickens studied psychology in medical journals, and Ruskin and Hopkins studied natural history and meteorology.

We have been fortunate in having many thoughtful supporters and helpful critics and would like to thank our many colleagues in Victorian literary, historical, and interdisciplinary studies who responded to our call for papers. We would also like to express our appreciation to the School of Humanities and Social Science at M.I.T. and to the National Endowment for the Humanities for financial support during this project.

Finally, we would like to express our appreciation to The New York Academy of Sciences for its support and valuable assistance in guiding this project from its initial stages to its final form. Our special appreciation goes to Bill Boland, the Executive Editor, Joyce Hitchcock, the Managing Editor, and to editors Denis Cullinan and Justine Cullinan.

JAMES PARADIS
THOMAS POSTLEWAIT

Massachusetts Institute of Technology

Introduction

JAMES PARADIS
THOMAS POSTLEWAIT
Department of Humanities
Massachusetts Institute of Technology
Cambridge, Massachusetts 02139

[I]

IT is often observed that the Victorian era was one of increasing specialization and professionalism; yet it is also true that the outlines of disciplines, in the modern sense, were only beginning to emerge. The efforts of Victorian writers to preserve the generalist spirit of knowledge, seen against the powerful trend towards specialization, account in large part for the energy of the Victorian literary discussion of science and its cultural significance. Carried out in a variety of artistic and discursive forms by both artists and scientists, this discussion can be seen as an attempt to place the activities and results of science within the perspective of the traditional values and institutions of English culture. The bonds between science and literature forged by this discussion are the main concerns of our collection.

[II]

Nineteenth-century science became a subject of Victorian literature because it so thoroughly manifested itself throughout Victorian society. From the agricultural and industrial workers who operated the machines of the new technology, to Albert, the Prince Consort, who promoted science education and planned industrial exhibitions, few English people remained indifferent to the scientific revolution and its effect on the economic and social order. For many, science represented the essence of knowledge and thought. It took on the luster, even glamour, associated with social, commercial, and intellectual progress. But for others, like Robert Louis Stevenson, English science promised only the loss of cultural stability, carrying the intellect into "zones of speculation, where there is not habitable city for the mind of man." Nearly all social commentators, whether critics like Stevenson, novelists like George Eliot, or political economists like Karl Marx, made serious efforts to account for the forces of science. Few responsible writers felt that they could adequately interpret and reconstruct their era without understanding at least the basic outlines of scientific thought. Innumerable views of Victorian science thus survive in writings as various as George Eliot's *Middlemarch*, Charles Darwin's *Journal of Researches*, Friedrich Engels's *The Condition of the Working Class in England*, Thomas Huxley's *Lay Sermons, Addresses, and Reviews*, Charles Dickens's *Hard Times*, and Alfred Lord Tennyson's *In Memoriam*.

The Victorians were witnesses to a remarkable social revolution. Never before had science so publicly altered the ways in which individuals viewed their common world. Never before had centers of culture and education been confronted with such a vast new body of knowledge. Not only did scientific interests promote a variety of new educational institutions, academic faculties, and professional societies, but science also found public expression in the popular lecture series of scientists like Michael Faraday, William Carpenter, John Tyndall, Thomas Huxley, and many others. Such lecture series, which for young scientists like Huxley and Tyndall were important sources of income and reputation, were calculated efforts to educate the public of all classes and occupations, from artisan to academic. And the public flocked to such presentations, showing great enthusiasm for science and its applications.

Museums and public exhibitions made science visible in still other ways. New establishments like the Natural History Museum at South Kensington, London, displayed a dazzling array of unfamiliar objects assembled from every corner of the world and introduced the important, if relatively simple, innovation of providing accurate explanations and descriptions. This innovation, which transformed the museum from a private reserve of curiosities into a center of public education, made institutions like the Museum of Practical Geology on Jermyn Street in Piccadilly into valuable tools for science instruction. Highly appropriate to the growing urban demand for public entertainment, these institutions expressed the Victorian passion for practical education and self-improvement. In like manner, the public exhibition was a popular source of instruction and entertainment, often presented as extravaganza. The Great Exhibition of 1851 in Hyde Park, a personal project of Prince Albert, was a world event which drew Sunday crowds in the tens of thousands. Opened by Queen Victoria herself in Joseph Paxton's great Crystal Palace, which covered more than nineteen acres, this event linked science, technology, commerce, and government with the material and imperial aspirations of the British.

In Victorian periodicals, of which several hundred flourished in England, a public forum was provided for discussing science and its consequences — its influence on religion, its social promise, its values and traditions. This discussion was livelier and more profound, in many respects, among the Victorians than among the moderns. Newspapers such as the *Pall Mall Gazette* reported on various professional meetings of scientists, giving synopses of official dinner speeches, government commission proceedings, popular lectures, and even scientific gossip. Other periodicals such as the *Westminster Review* attempted to keep abreast of scientific developments by reviewing the best of the burgeoning number of scientific books. James Knowles's *Nineteenth Century*, one of the great Victorian periodicals, rose to prominence partially, if not largely, through Knowles's success, as impresario of Victorian controversies, in staging literary debates between the partisans and critics of science. In addition, a variety of professional journals were begun and supported by men of science, two such journals being Norman Lockyer's *Nature* and Thomas Huxley's *Natural History Review*. These journals provided centers of intellectual interest for specialists and not only set professional standards for published research and established conventions of language, but also made reputations.

At the center of much of this new intellectual and social activity was the "man of science," no longer the isolated system-builder, but now the collaborative thinker with a public life and professional status. Even Darwin, the famous recluse of Down in Kent, had a vast and stormy public life, fortunate as he was to have it managed by Huxley, the greatest popular essayist of the century. Men of science were commonly seen in public in a variety of traditional and new occupations. Some were academics like William Whewell at Cambridge, George Rolleston at Oxford, and William Thompson at Glasgow; some were wealthy and socially prominent like Sir Roderick Murchison and Sir John Lubbock; others were public employees like the Hookers at Kew Gardens and Norman Lockyer in the government's Science and Art Department. Perhaps the most visible Victorian scientists were the popular lecturers and essayists, some of whom like Huxley and Tyndall became celebrities in later years, the subjects of gossip columns. The man of science had become a noteworthy and influential public species, a symbol, for many, of the new age.

[III]

It was natural, then, for Victorian writers to explore the human dimensions of the man of science and to consider the cultural meaning of his emergence within English society. Not surprisingly, the lives, ideas, and values of this new intellectual and social class became the raw material for a variety of artistic works. Science was in the air by the mid-century, and Darwin's *On the Origin of Species* in 1859 made scientific definitions and theories central to ethical, aesthetic, and social ideas. The literary artist and social critic who wished to recreate some aspect of Victorian life, whether in fiction or poetry, in social or intellectual commentary, was destined to come face to face with the remarkable fact that science and its applications were not only transforming the structure of English society, but that they were also redefining the images and values upon which the culture as a whole had traditionally been based. Some writers were profoundly disturbed by this influence and believed that it was the source of cultural decay and disunity, as was sometimes argued in the works of Thomas Carlyle and John Ruskin. Others held that the methods and results of science were the primary avenues to the understanding and control of nature, as was often the argument of Herbert Spencer and John Stuart Mill.

Those writers who set out consciously to address the cultural influences of science often did so by demonstrating in literary works how scientific forces significantly redirected or redefined the lives of individuals. As signs of the times, scientific images and ideas became important literary metaphors. Not unlike medieval literary artists, who appropriated religious images, metaphors, and ideas in order to create a world view, Victorian artists like Tennyson, Eliot, and Hardy often appropriated the images and metaphors of science in order to reflect a contemporary sense of reality. This process, which was rarely a simple matter of importing technical material verbatim into literary texts, gave artists and social critics a vocabulary with which to represent the changing sensibilities and realities of the age. In Tennyson's *In Memoriam*, Carlyle's *Sartor Resartus*, and Eliot's *Daniel Deronda*, we find a continual, deliberate transformation of topical science into metaphors and symbols of the self and its spiritual life. Such responses, while often personal and idiosyncratic in origin, attained a general significance because they measured science by standards of human value and emotion that transcend specific disciplines and even ages.

Rarely were the representations of literature and criticism crudely unsympathetic to science, even though debate was common. Such objections as were raised to scientific ideas were generally efforts to redirect what some saw as misguided or unregulated applications of science. Because most Victorian literary artists and social critics believed unswervingly in the moral function of art and criticism, they argued vigorously about the values and cultural contributions of science. But Huxley and Arnold, debate as they would over the respective values instilled by scientific and humanistic education, found not two cultures but one, its traditions located in the accumulated wisdom of both humanism and science.

Scientists As Intellectuals:
The Early Victorians

S. S. SCHWEBER

The Martin Fischer School of Physics
Brandeis University
Waltham, Massachusetts 02154

E VEN though it is somewhat arbitrary and artificial to demarcate an age I will take "early Victorian" to refer to the two decades from 1830 to 1850 — roughly the period from the first Reform Bill to Sir Robert Peel's death and the Crystal Palace Exhibition. Similarly, to try to characterize the spirit of an age, particularly when the period was a turbulent one is a difficult task.[1] Nonetheless I will attempt to describe the atmosphere, or rather the atmospheres, of the early Victorian period, a time when the stresses created by a rapidly increasing population and by the processes of industrialization and urbanization were threatening the stability of the social order. For the atmosphere, the psychosocial environment, sets the stage upon which the social drama — of which scientific activities are a part — is enacted: a drama in which the scenery is reshaped, the play is rewritten, the various actors assume shifting roles, directors are replaced, and even the audience changes.

[I]

It is, of course, impossible within the confines of an article to encompass the complex and multifarious aspects of British science in the early Victorian period. By the 1830s, scientific activities had already become sufficiently differentiated so that many interdependent and complementary institutional spheres existed, for example, both the provincial and the national, London-based scientific societies, "Oxbridge" and the Scottish universities. Science in each of these spheres meant something different. To most members of the Manchester Literary and Philosophical Society, the cultivation of science was a mechanism for "the social legitimization of marginal men."[2] To the active members of the Geological Society of

1

London, science was a norm of Truth.[3] Similarly, a "scientific education" had one meaning to the sponsors of the mechanics institutes and another to the reformers of Cambridge University in the 1830s. In Britain the 1830s marked the emergence of new kinds of professional men: the industrially employed scientist, the civil engineer, and the civil service bureaucrat who assisted in the running of a government that had become too complex and technical to be run by the ruling class and their dependents. Not that such professionals could not be found in earlier periods: in history there are no clear beginnings. Rather it is that by the 1830s in many fields these professionals had achieved a "critical mass" and new institutional structures were being formed as a result. It was not only those engaged in scientific work who became more self-conscious about their "profession;" the same was true of the civil engineer[5] and the managerial class in the government[6] among others. What differentiated "scientists" from the other intellectuals — that is, persons performing roles in which comprehension is both means and ends — [7]was that the elite of the scientific community formed the apex of a new intellectual class. Annan has adumbrated the salient features of the intellectual aristocracy that shaped British thought during the nineteenth century.[8] For the early Victorian period Cannon has located the vital center of the intelligentsia in the Cambridge network and has perspicaciously and vividly described its membership.[9] These early Victorian intellectuals were not revolutionaries. They saw themselves as within tradition. Moreover, as has been consistently observed, British men of science "were more involved with general culture than were their counterparts in countries like France and Prussia."[10] The deep transformation of British society from the time of the French Revolution to the 1830s, and the deep transformation of natural knowledge during the same period, placed the early Victorian scientific elite in a unique position to mediate the conflicts generated by changing conceptions of the sources and bases of the social order. As educational reformers they were committed to the gradual transformation of the accepted institutions. Yet they challenged the opinions and positions of the ruling class and clergy. They were able to do so because of their mastery of both the older intellectual traditons as well as the new scientific knowledge. Their role as creators and interpreters of this new knowledge allowed them to move between the "worlds of speculation and government."[11]

What is striking about the early Victorian intellectual elite is their cultivation of polymathy, their insistence on mastering all that was known and could be known. The minimal standard of an educated man was considered by Augustus De Morgan to be represented by "a man who knows something of everything and everything of something." The description that Charles Lyell gave of Poulet Scrope, the professor of geology at Oxford during the 1820s who later became an eminent political economist:

> Scrope . . . has a most active intellectual mind — alive to everything, politics, political economy, Hume, Berkeley and Reid's metaphysics, geology, Irving, the Roweites, and progress of fanaticism"[12]

would apply equally well to the elite of the scientific intelligenstia, for example, Lyell, Whewell, and Babbage, Henslow, Sedgwick, De Morgan, John Herschel, Holland, P. M. Roget, and Mary Somerville.

Omniscience was not William Whewell's weakness; rather it was a lack of humility and graciousness: he did not bear his knowledge well. And Herschel, Peacock, Jones, De Morgan, and W. R. Hamilton all had the same craving for omniscience!

While these polymathic qualities gave the scientific elite its special characteristics and its prestige, these same qualities also helped to retard the transformation of British science into its modern cast. The ideal of the Renaissance man, if not opposed to the newer standards of professionalism and specialization, nonetheless could not be maintained in the face of the rapidly growing scientific disciplines. Many of the scientific elite held an ambiguous attitude toward the new ethos, reflected in their indecision about the relative merits of acquiring and disseminating as compared to producing knowledge. The delay in the appearance of professionalized science in Britain reflected "a national attitude towards both specialized knowledge and professional norms"[13] that placed a higher value on generalist education without necessarily ignoring the value of specialization and professionalism. Thomas Young, one of the heroes of the early Victorian scientific elite, had in 1829 defended himself against the charges by James South that he was not competent[14] to be superintendent of the *British Nautical Almanac and Astronomical Ephemeris* with the statement that for men of science

> the exclusive cultivation of any single branch of science must naturally be inclined to exaggerate the importance of the objects which engross their own attention.[15]

John Herschel declined the Lucasian Professorship in 1826 so that he would not be limited in the scope of his scientific pursuits. It is only in the second half of the nineteenth century that science and scientific insti-

tutions emerged in their modern manifestations in Great Britain: after a greater degree of specialization had taken place in the fields of medicine, chemistry, geology, physics and engineering; after Oxford, Cambridge and the Scottish universities[16] began to accept national responsibilities under the impact of various Royal Commissions; and after religious tests were fully abolished. The character of the scientific intelligentsia also changed during the second half of the century and it began to assume an anti-tradition stance.[17]

The present essay is a study of intellectuals as cultural and political activists, who, broadly speaking, were nonetheless conformists. Although in the modern context an intellectual is popularly considered to be a person whose primary interest is in the social sciences and the humanities, historically this was not the case. It was precisely their scientific accomplishments *combined* with their broad interests and capabilities that gave the early Victorian intellectuals their great influence and prestige.

The organization of the paper is as follows: Sections II and III analyze facets of the social and intellectual background of the generation that shaped early Victorian science. Section IV illustrates how the cultural background of the early Victorian period was reflected in the scientific enterprise. Section V considers more personal traits of some of the early Victorian scientists and focuses on John Herschel. An epilogue offers some conclusions.

[II]

Britain emerged from the continental wars as the richest and the most stable of the great states of the world. Her navy was the most powerful in the world and protected the largest shipping fleet carrying the most varied commerce to all parts of the world. Her population was growing at an impressive rate. The agrarian transformation of England in the eighteenth century, improved farming practices, industrial expansion, the increase in rural birth rates, the disappearance of sharp peaks in mortality and the lowering of the infant mortality rate all contributed towards a profound demographic transformation: In 1751 Great Britain's population was 7½ million people; in 1831 it was 28 million. Even more dramatic was the growth of urban centers: Manchester grew from a population of around 15,000 in 1760 to more than a quarter million in 1831. In the decade from 1821 to 1831, Liverpool's and Manchester's populations increased by more than 40 percent.[18]

The impact of these demographic changes was widely felt. Between 1812 and 1822 the number of students at Cambridge increased by considerably more than a third — "an increase much beyond what had taken place in the preceding half-century."[19] Although this particular increase also reflected the termination of the war with France, the enrollment at Cambridge and Oxford continued to grow in later decades.

Similarly, the reading public grew substantially in the decades between 1820 and 1840. The improvement in printing presses in the second decade of the nineteenth century greatly increased the amount of printed matter that could be produced and dramatically lowered its cost. By 1830 a large publishing enterprise catered to a constantly growing demand for reading materials. Vast numbers of novels, books on religion, on travel, on science, and on politics were being issued. A sizable fraction of these were critically reviewed in periodicals, whose proliferation attests to the magnitude and importance of the publication phenomenon. Thus, for example, the *Edinburgh Review*'s first issue in October 1802 consisted of 750 copies.[20] By 1818 the same journal had reached a circulation of over 13,500.

By the 1830s, the Industrial Revolution in Great Britain had created new classes with separate religious affiliations and political commitments, deeply divided in outlook. Urbanization had added a spatial dimension to the dissension. The Industrial Revolution had also reinforced the British propensity toward atomization and individualization, not only atomization as an explanatory principle and individualism as a political credo, but also both as praxis. Josiah Wedgwood had made his fortune by atomizing production into separate steps each with its particular function, by training and assigning workers to master individuated operations, and by endeavoring to "make such machines of the men as cannot err."[21] Nor was Wedgwood alone. Watts turned millwrights into engineers; Arkwright, clockmakers into machinists; Stephenson, pitmen into railway architects; and Brindley, miners into tunnelers.[22] These men believed that by emulating their own examples and by becoming more industrious, workers would transform their worlds and create schools for the children, hospitals for the ill, homes for the orphans, and libraries for everyone. All that was needed was the cooperation of the workers.

Self-help had launched the self-made industrialist into his better world. The machine technology would similarly catapult the industrious worker. The economics of machine technology thus elicited an ethic that valued self-dependence at the expense of social dependence and that argued that man's happiness was the result of his own individual efforts. It was Utilitarianism that tried to justify the suggestion that this philosophy of the new middle class was appropriate for all people.[23] Society must be reformed using individual reason to promote progress. It must be transformed into an aggregate of autonomous individuals who, by seeking their own self-interest, by increasing their pleasure and decreasing their pain, would bring about the "greatest happiness to the greatest number." Since the human mind was "the instrument of the greatest degree of happiness,"[24] education was the all-important great equalizer. Brougham, the exemplar of Utilitarianism, as leader of Whig opposition in the pre-Reform House of Commons, headed the "education-mad party." For these Benthamites, the Hartleyan and Millian epistemology, which gave education a central role,[25] complemented a practical Baconism that took the goal of science to be the enrichment of human life with new discoveries and wealth. The Industrial Revolution had indicated that science was the path to prosperity. Promote the former and you accelerate the latter. These philosophical commitments produced the numerous encyclopedias, cabinet libraries, the mechanics institutes, the Society for the Promotion of Useful Knowledge, and University College in London.[26]

It is, of course, not surprising that the new class of industrial entrepreneurs was attracted to French Enlightenment views. Except for its Deism, their shared philosophical view had many similarities with that of the *philosophes*. Being the representatives of man's expanding capacities and achievement, particularly when helped by science, they looked confidently forward. They found the *philosophes'* buoyant optimism attractive. Condorcet's utilitarianism was not very different from that of the members of the Lunar Society.[27] Progress for all of them consisted in harnessing technology to increase man's life span, in eradicating disease, ending human slavery, making all men equal in civil rights, and in raising the cultural level throughout the world.[28] The members of the Lunar Society also felt that progress was inevitable and regression impossible, because of the cumulative effect of scientific knowledge. Wedgwood, Erasmus Darwin, and Priestley were all sympathetic to the aims of the French Revolution. They supported—financially and spiritually—Godwin and the young Coleridge and Wordsworth when the latter were considered radicals.

The French Revolution was, of course, the watershed. After Robespierre and the Terror, support of the Revolution was no longer possible for the middle classes in the western European countries. The *philosophes'* ideologies having become suspect, the crucial question in the first half of the nineteenth century became: What are the new visions that will animate the hopes of the various social classes in the face of a waning Christianity? Germany, France, and Great Britain each had characteristic responses to this challenge, and the way science developed in these countries was deeply affected by the respective path taken. There was, of course, always the possibility of reversion to the former traditional Christian ways: the path advocated by Burke in England and Bonald, Chateaubriand, and de Maistre in France. But that alternative was not attractive (although it was surely partly responsible for the tenacity of Cuvierism in France and Mosaic geology in England).

Late eighteenth- and early nineteenth-century Idealism represented the German attempt to realize a new religious meaning in life. Religious sentiment was of importance to all forms of German Idealism. This reflects one of the main differences between German and other western European Enlightenment. There were few anti-religious ideas to be found in the German Enlightenment thought. There was no important expression of philosophical materialism in Germany until Feuerbach, that is, until after the death of Hegel and Schelling.[29]

In Germany during this period, in contrast to Great Britain, large entrepreneurial undertakings and high finance were rare and many of the manufactories were state-owned. Artisans and small merchants composed the largest part of the population of German cities. The German-speaking states did not have a confident, self-conscious, commercial middle class that was proud of its abilities and of its contributions to the welfare of the state. The German middle class did not seek political rights for itself. In fact, it was deeply committed to the absolutism and authoritarianism that characterized the political structure of the German States.[30] Civil servants had the greatest influence upon public life. Pastors, professors, and secondary-school teachers were civil servants because Church and schools were state institutions. Freedom of thought (*freiheitslehre*) was the political demand exacted from the state by these civil servants.[31] This is the social and political context in which German science developed in the first half of the nineteenth century.[32]

French science at the turn of the eighteenth century has received detailed study in recent years and has been the subject of outstanding presentations.[33] Thus the contrast between French and British science is better understood.[34] In Catholic France the Church was identified with the *ancien régime* and the Restoration, whereas democratic principles were the legacy of freethinkers who wanted to make the cult of Reason the religion of the state. This conflict over the relation between Church and State was a central issue in French cultural and political life throughout the nineteenth century. The growth of science in France reflected this tension, particularly in the natural sciences, where metaphysical issues were of importance.[35] Science in France also reflected the much greater degree of specialization, institutionalization, and centralization that had begun in the Napoleonic era.[36]

Great Britain had much in common with both France and Germany. With Germany it shared a deep concern for religion and religious sentiment, manifested in Britain by Methodism and Evangelicalism; hence the interest among intellectuals and theologians in German Idealist thought. But in contrast to Germany, the state in Britain was considered "the realization of legality but not of morality: the state had no absolute moral value in itself."[37] Political institutions were merely agencies and tools to implement the pragmatically moral goals of society. The patterns of educational reforms in England and Scotland reflect this basic attitude.

[III]

The generation that shaped British science during the early Victorian period received its education during the period from 1810 to 1830. Broadly speaking, there were two important influences molding its intellectual outlook: the Scottish Enlightenment and French views on the natural sciences.

Equally important, each of these traditions also furnished models of intellectual communities and norms for membership in its intellectual elite. The Scottish tradition, with its "men of parts," was well-known through the lives of David Hume, Adam Smith, and James Hutton. Until the end of the second decade of the century, John Playfair, John Leslie, and Dugald Stewart were living examplars of that tradition. Their scientific output, their influential contributions on a wide range of subjects to the *Edinburgh Review,* and their textbooks attested to their impressive polymathic capabilities.[38] The usual continental tour introduced the early Victorian scientists to the French intellectual elite in Paris and also familiarized them with French scientific institutions. The Institut and the Société d'Árceuil became paradigms for scientific productivity and creativity.

The Scottish Enlightenment was an important influence in both of its broad manifestations: the skeptical humanism of Hutcheson, Hume,

Smith, and Ferguson on the one hand, and the Common Sense school of
Thomas Reid, James Beattie, and particularly Dugald Stewart on the
other.[39] Although often opposed in their views, both schools were partially responsible for the secularization of British throught during the
eighteenth century. Hume and his intellectual descendants, by their empirical concerns, their empirical method in philosophical inquiry, their
skeptical humanism, the modesty of their claims for human intelligence,
their historicism,[40] and their materialism, were the bridge from the old to
the new *epistēmē*. All the Scottish philosophers from Hume to Stewart
and Thomas Brown agreed that one could not explain anything by
reference to ultimate principles or processes whose existence could not be
inferred from the observation of individual instances. Hume, by his
disproof of the idea of necessary connection in causal descriptions, and
Thomas Brown, by his later discussion of causality,[41] were largely
responsible for a uniformity in the British methodological approach to
both the physical and social sciences. If absolute causality was denied to
the physical sciences, then they were reduced to a level of certainty more
similar to that of the social sciences.[42]

Another important aspect of the Scottish intellectual tradition was
that its Christian God became a benevolent Deity: a God who supported
hopes of success in individual activities in this world.

The French influence on the early Victorian scientists was represented
by the views of Turgot and Condorcet and was realized in French scientific institutions. With the fall of Robespierre, an ambitious educational
program that had originally been conceived by Condorcet — a national
system based on the studies of the sciences and natural history — had
been enacted. The Constitution of the year III created the Institut National de France — the pride of the ideologues. The Institut consisted of
three classes, each subdivided according to discipline. The first and
largest class, with sixty resident members, was devoted to the sciences,
including mathematics. The second class consisted of the moral and
political sciences and had thirty-six resident members. Literature and the
fine arts made up the third class, which numbered forty-eight members.
Each class met twice every *décade* (ten days). Once a month the three
classes sat together "to deliberate on such affairs as relate to the general
interests of the Institut."[43]

The Institut stood at the summit of a vast educational enterprise which
had as its objective a system of universal education. The Law of 25 October 1795, which founded the Institut, also revived many of the institutions the Convention had abolished in 1793: the Observatoire de Paris,

the École des Ponts et Chaussées, and the Jardin du Roi (under the new name of Muséum d'Histoire Naturelle). The law also decreed the establishment of the École Polytechnique and the École Normale (which opened on January 20, 1795 but closed on May 15, 1795, and was reborn under Napoleon in 1812 by the Decree of 17 March 1808). The École Polytechnique was to become the paradigm for scientific and technical institutions throughout Europe. Its first graduates include some of the most distinguished French scientists and mathematicians of the nineteenth century: Cauchy, Coriolis, Poncelet, Poisson, Gay-Lussac, and Fresnel. Its faculty consisted of the most eminent French scientists of the day among them: Lagrange, Monge, and Laplace.

The *Mémoires de la 2ᵉ Classe de l'Institut* of the period from 1795, the date of the inception of the Deuxième Classe (the Second Class) to 1803, when Napoleon expressed his suspicion and contempt of the ideologues by reorganizing this class, constitute some of the most remarkable documents of the revolutionary period. Taken as a whole they represent an attempt to give a materialistic interpretation of all aspects of the psychological, social, and moral world of man. And since to P.J.G. Cabanis, the leading spirit of the ideologues, all the sciences were branches of one trunk, and no two sciences were more closely related than the physical and intellectual studies of man, the discussions of the Second Class in fact encompassed all knowledge, its interrelationships and modes of apperception.[44]

The political and social framework of Great Britain during the first two decades of the nineteenth century was such that it was not expedient to approve of things French. Nonetheless, French intellectual contributions and the institutional settings in which these advances were made were admired openly by the British.

During the first two decades of its publication, the *Edinburgh Review* regularly reviewed French scientific activities,[45] and particular notice was given to the *Mémoires de Physique et de Chimie de la Société d'Arceuil* when the were published in 1807, 1809, and 1817. John Leslie, reviewing the first *Mémoires* in the *Edinburgh Review* in 1809, commented that this volume was the production "of a little association, better calculated, we conceive, than the older establishments for advancing the progress of physical science," and expressed the hope that "similar associations, furnished with more ample means, will soon be formed at home."[46] Thomas Young reviewed the first two *Mémoires* for the *Quarterly Review*. He lavished praise on their contents, and also noted that

There is not uncommonly a degree of zeal and emulation attending the pursuits of a private association, which cannot always be obtained in an equal degree by any public encouragement held out to science.[47]

Both reviews manifest the strong British propensity toward individualism and the encouragement of *private* institutions.[48]

The impressive accomplishments of Cuvier, Laplace, and Lagrange and those of the younger generation (for example Malus, Biot and Gay-Lussac), during the period from 1780 to 1810 were, of course, reason enough for the British scientific community to look favorably on French science — particularly the physical sciences. But there is another reason for British admiration of French science in the first decades of the nineteenth century: many of the leading scientific practioners — for example, John Leslie and Humphry Davy — had studied, and at some stage had adopted, the Condorcet — Turgot philosophy of the physical sciences, that is, the view of the Condorcet for whom mathematics is equated with universality and certainty, who guarantees that no mathematized knowledge is ever lost, and who sees in the calculus of probabilities the vehicle to mathematize social phenomena and thus the possibility of establishing a social *science* from which predictions could be made.[49] This is not to say that the Condorcet of the *Esquisse* had no influence on the shaping of British science. I believe that it was his vision that animated Herschel, Babbage, Peacock, and the others in that group of remarkable young men who constituted the Analytical Society.[50] I would characterize them as a British realization of Condorcet's "body of progressive scientists." The members of the Analytical Society adhered to Condorcet's tenet that such a body be independent and voluntary. Whoever joined it did so freely because he believed that it would fulfill his most expansive hopes. And these hopes already in 1813 were nothing less than a transformation of British science. The Analytical Society was a collective, cooperative and elitist scientific enterprise. Herschel, Babbage, Peacock, Maule, and Bromhead banded together to form the Analytical Society with a clear vision and conception of how to transform British science. It would be necessary not only to modernize English mathematics and science, but also to reform English scientific organizations and educational institutions.[51] All would attempt to make mathematical contributions by helping one another, but the remaining tasks would be divided among them. Babbage would devote some of his special talents to organizational matters, Peacock would be responsible for the educational battle lines, and Maule — that keen and perceptive student of the political and social scene — would advise on the strategy and tempo by which change would be effected. Herschel, Babbage, and

Peacock — with Condorcet — always believed in the interdependence of the social, mathematical and physical sciences. But Condorcet's atheism and anti-Christian attitude made it impossible to acknowledge him. The French Revolution also rendered suspect Condorcet's application of probabilities to the making of moral and political decisions.[52] The vision of a state run, without debate, on the basis of applications of the calculus of probabilities recalled the ghost of revolutionary France run, without debate, on the basis of the operation of the guillotine. Thus the application of probabilities to social phenomena was foreclosed, although the idea of a mathematical social science, based on statistics, once again proved attractive in the mid 1830s. Quetelet was its main promulgator and the Statistical Section of the British Association for the Advancement of Science (BAAS) and the Statistical Society of London were the vehicles for disseminating this new sanitized vision.

Even though probability theory as applied to social phenomena was forbidden territory, Condorcet's and Laplace's view of probability as the appropriate mathematical method to deal with complex phenomena was embraced.[53] But the phenomena dealt with initially would be drawn from the physical sciences, not the moral or social ones. Herschel believed that it would be possible to abstract from sufficiently complex phenomena certain features that were universal.[55] These universal features were precisely the ones that could be addressed by a probabilistic description. The latter in turn would allow the possibility of making predictions. Such predictions "although only of a probable nature" would nonetheless "be highly valuable and useful."[54] I believe that it was this feature that attracted Herschel to meteorology and to terrestrial magnetism. After the 1830s, when the political climate in England had become more liberal, probabilities as applied to social phenomena once again became a focus of interest. Herschel's seminal and influential article, "Quetelet on Probabilities,"[56] marks a high point in that debate.

There were other important French-ideologue influences in England which manifest themselves primarily in the medical and biological sciences. Although recent scholarship is again beginning to take note of Bichat's considerable contributions to the development of the life sciences in the first half of the nineteenth century,[57] Cabanis's influence is only beginning to obtain similar attention.[58]

Cabanis, as Darwin did later, argued for an uninterrupted chain between inanimate and animate matter and developed a psychology that was based wholly on physiology. Of particular interest is Cabanis's emphasis on the substantial physiological and psychological *differences* between men and the *differences* in their reaction to the changes that occur during their life cycles. There was no *type* common to the whole human species. I suspect that Bichat and Cabanis had more influence on British thought, particularly evolutionary thought, than is commonly inferred.[59] However, the political and social context made it impossible for anyone to acknowledge such influences, for the ideologues were the embodiment of materialism and represented the first attempts to understand *all* aspects of the world without God. Finally, I believe that in the late 1820s some Saint-Simonian influence was beginning to manifest itself in British scientific circles. That J. S. Mill and Thomas Carlyle had come under this influence during this period is well known.[60] Quetelet is a prime candidate for importing such views into the Cambridge network. Babbage, in his continental travels, may also have obtained first-hand knowledge. Babbage in many ways embodies the Saint-Simonian ideal of man as "rational, activist and religious, at once mind, will and feeling."[61] He is artist, scientist, and industrialist all in one. Babbage has not yet received his due in the assessment of the development of early Victorian science.[62] He is too often seen as the "irascible genius" whose vision of an "analytic engine" was very much ahead of his time.[63] But this evaluation ignores his deep involvement in the arts, his wide scientific erudition, and his powerful analytical and logical skills. It also ignores his great organizational skills and his role in the shaping of the Analytical Society, the Astronomical Society, the British Association for the Advancement of Science, the Statistical Society, and in the reorganization of the Royal Society. His house on Dorset Street was the weekly gathering place for influential artists, scientists, bankers, industrialists and government officials.[64] His keen intuition was often enlisted by Lyell to answer questions relating to processes at high temperature and pressures. His intimate knowledge of machines and manufacturing processes and his careful analysis of their economic impact made him a respected member of the political-economy community. Babbage's *On the Economy of Machinery and Manufactures*[65] was widely read. In it he extols Adam Smith's principle of division of labor. Division of labor was a central component in Darwin's principle of the divergence of character. One of the sources of Darwin's knowledge of division of labor was undoubtedly Babbage.

[IV]

That scientific activities should reflect the social context in which they are carried out is to be expected. In the present section I shall illustrate this truism with some examples of how the social context is mirrored in the outlook and the concerns of the early Victorian scientific *community*. My intent is also to indicate how tightly coupled the scientific community was to the rest of the intellectual community.

Machines and machinery were the essential components in the transformation of Great Britain during the Industrial Revolution. Indeed, Carlyle called the Industrial Revolution the Age of Machinery. It is thus not surprising that machines should be an important source for the metaphors and models in terms of which Englishmen described and explained the world. What I find remarkable is that only fifty years separate Wedgwood's effort to "make such *machines* of the *men* as cannot err [emphasis mine]" and Andrew Ure's vision of the automated factory "which dispenses entirely with manual labor"[66] and in which production is "conducted by self-acting machines":

> . . . I conceive that this title [factory], in its strictest sense, involves the idea of a vast automaton, composed of various mechanical and intellectual organs, acting in uninterrupted concert for the production of a common object, all of them being subordinated to a self-regulated moving force.[67]

The same time span saw the adoption of similar models and metaphors of self-regulating mechanism in the sciences. O. Mayr has emphasized the importance of the steam engine in the development of Adam Smith's concept of society as a self-regulating system.[68] In the 1830s, Lyell regarded the earth as a self-regulating geological machine. It is not too much of an extrapolation to suggest that Darwin in his notebooks thinks of an organism as an automaton composed of various organs (physiological, mechanical, and intellectual) "acting in uninterrupted concert for the production of a common object" — their own reproduction — all organisms being subordinated to a self-regulated moving force, natural selection. In fact, Wallace, in explaining natural selection, stated that "the action of this principle is exactly like that of the centrifugal governor of the steam engine."[69]

The study of history was one of the characteristics of the age — a legacy of the Enlightenment. Barthold Niebuhr was a "god" at Trinity College during the 1820s. Niebuhr's historical and naturalistic interpretation of the Bible was the center of heated debates and the cause of many religious crises among the undergraduates at Cambridge.[70] Early Victorian England was the age in which Carlyle and Macaulay wrote their histories. The 1830s saw a parallel development in the study of the history of science in England.[71] Whewell's *History of the Inductive Sciences*[72] is only the most famous and wide-ranging book on the history of science during the period. David Brewster's *Life of Isaac Newton*[73] first appeared in 1831. It was later expanded into his important *Memoirs of the Life, Writings and Discoveries of Sir Isaac Newton*[74]. Baily's *Account of the Rev. John Flamsteed, the First Astronomer Royal*[75] was issued in 1835. Rigaud's *Historical Essay on the First Publication of Sir Isaac Newton's Principia*[76] appeared in 1838. All these works reflect the new standards of scholarship that historical works were attaining and were based on original manuscripts, documents, and letters that had never before been published.

Augustus De Morgan illustrates the early Victorian scientist's penchant for history at its best. An outstanding mathematician and astronomer, he was twenty-seven when elected the first professor of mathematics at the newly founded University of London. A deep commitment to education led him to write famous textbooks on arithmetic and the differential and integral calculus. This passion is also reflected in the tracts he wrote for the Society for the Diffusion of Useful Knowledge and in his extensive contribution to the *Penny Cyclopaedia*, for which he wrote more than 800 articles. Between 1831 and 1857 he contributed twenty-five historical and bibliographical papers to the *Companion to the British Almanac*, including a long essay in 1837, "Notices of English Mathematical and Astronomical Writers between the Norman Conquest and the year 1600," and in 1845 his famous *References for the History of the Mathematical Sciences* was published. His great bibliography, *Arithmetical Books from the Invention of Printing to the Present Time*[77] is a paradigm for all such undertakings. Its annotations attest to De Morgan's erudition and his mastery of the sources.

The early Victorian period was a world in which rhetoric and the eloquent speaker played a much greater role than they do today. There was a great deal of preaching, and among Dissenters, "a minister largely lived by his voice."[78] The anti-Slavery movement, the anti-Corn law movement, and the Temperance Movement all depended on oratory. Part of Whewell's, Buckland's, and Sedgwick's standing in the community surely derived from their oratorical skills. At the meeting in 1838 in Newcastle of the British Association for the Advancement of Science, Adam Sedgwick gave an outdoor public lecture to more than 3,000 "colliers and rabble" on the geology and the economy of the coal fields. John Herschel — always sensitive to the forces that mold the social world — on the occasion of dedicating the Windsor and Eton Public Library, apologized to his audience for addressing it "from a written paper."[79]

There is no better example of the function of oratory than the speeches

at the dinner honoring John Herschel on June 15, 1838, on his return
from the Cape of Good Hope. *The Athenaeum* gave the event a five-
page report.[80] Assembled for the occasion were more than 400 persons:
representatives of the Crown (the Duke of Sussex was in the chair),
bishops, members of the robe and of Parliament, the high officials in the
Admiralty and the Treasury responsible for the support of scientific
enterprises, and the elite of the scientific community. Most of the evening
was taken up by the extemporaneous speeches made by the various
dignitaries in Herschel's honor. Herschel was given a vase at the dinner
and the presentation speech was made by the Duke of Sussex (at the time
the President of the Royal Society). This speech, as reported in *The
Athenaeum*, highlights the early Victorian hierarchy of values:

> Gentlemen, if I were then to state my reasons, for the hearty welcome we
> give him, I would advert at first to his academic life and to that period
> when he carried away laurels in a contest with men of acknowledged
> talent, one of whom [Fallows] died a few years ago a victim to his arduous
> exertions, in the study of astronomy.(hear, hear). I might likewise say,
> looking around this room, that there is not an individual in it, who does
> not acknowledge the debt of gratitude we all owe him for his deep and
> unwearied research, into all the different branches of science. I might fur-
> ther advert to his Introduction to Natural Philosophy, in which he com-
> bined method and system with extensive knowledge, and brought under
> the particular notice of the public the great debt we owe to that enlightened
> man, Dr. Young; and in which I may be permitted to observe, he has
> proved he could follow up the deepest researches in science, and the
> abstraction of philosophy, without neglecting the culture of liberal
> literature. In that work he has shone as the philosopher and literary man
> (cheers). And now I come to a virtue that is of a still more interesting kind,
> because it is one of a family nature, and one in which the illustrious in-
> dividual at my right ought to be held up as a bright example: I mean that
> sense of filial duty which prompted him to follow the example of his
> distinguished relative—a relative who was protected, and honored, and
> favoured by my illustrious father (great cheering)—in adopting and
> discovering means by which we have been enabled to examine into the
> deepest recesses of the universe; and thereby I hope I may be permitted to
> say, obtaining additional means of coming nearer to the footstool of the
> Architect of the Universe (hear,hear). In making these remarks, I am not
> presumptuous; but allow me to say, that, attached as I am to science—at-
> tached as I am to religion, I am satisfied that the real philosopher is the
> most religious man; and it is in looking to the operations of nature that the
> finger of the Almighty leads us to the lesson. Gentlemen, I have given you,
> in part, the reasons why we are here assembled. The brightest ornaments of

the British Empire are all met here with one accord—to demonstrate to my excellent friend their kind feelings—and all are united upon one point, which shows what merit can produce (hear). Alas! too many have the opportunities without availing themselves to them; but here is a bright example of an individual, who comes before us, and by his own merit, his own private exertion, and the goodwill of his fellow-countrymen, unites all parties; and I will venture to say, at this moment, there is not a heart or a hand that is not united in his favour (cheers). In paying this tribute, Gentlemen, to the worth of our distinguished guest, I must be allowed, likewise, to notice the disinterestedness with which he undertook the task (cheers). Prompted by filial duty, and wishing to establish in perpetuity the fame and the character of his father—and inheritance which no revolution but nature itself can destroy—he lent himself to what he thought his duty, abandoned his country and his countrymen, left friends behind, took friends with him. He did all this, I say, for the purpose of establishing his father's fame, and recording to posterity the merit of a man whose name, I trust, never will, and never can, be forgotten (cheers). This, Gentlemen, I certainly think, adds much to the lustre of my excellent friend's character; and I am delighted to find, that from the exertion and industry of his father, and his own prudence, he had the means of doing this independent of every person's assistance; because I take this opportunity of stating, that my worthy friend went abroad, refusing, gratefully and respectfully, the offers of government. Though he refused at that time, all assistance, still he is about to publish cheaply, and to distribute generally, the discoveries he has made, and those mysteries of science which he has unveiled, and which the public is looking for with anxiety. Gentlemen, while I admire his independence, yet my excellent friend will allow me to say, in a meeting like this, that I hope it will not certainly be looked upon as a precedent, or furnish argument for refusing every man of science that aid which, under circumstances, may be indispensable.

The eulogy is for Herschel's polymathy, for his individualism, for his independence, and for his refusal of "every person's [financial] assistance," as well as the offer of governments to help defray his research costs. In accepting the honor, Herschel remarked that indeed

Much assistance was proffered to me from many quarters, both of instruments and others of a more general nature—offers in the highest degree honourable to all parties; and I should be sorry to have it thought, that in declining them, I was the less grateful for them (hear). I felt, that if they were accepted, they would compel me to extend my plan of operations, and make a larger campaign; and that in fact, *it would compel me to go, in some degree, aside from my original plan.* [italics mine]

But Herschel's stand was representative of a dying order. The Duke of

Sussex's remarks clearly indicate that the government was aware that it had an important stake in science, and that it appreciated the importance of selectively supporting scientific activities. The scientific community had been ready to accept the offer. In fact, although Herschel had refused support for his personal research, he actively promoted government support for others.

The speeches at the Herschel dinner also shed light on another important sphere of early Victorian England: religion. By the 1830s, the tide of Evangelicalism was ebbing, but evangelical Christianity had affected the frame of mind of most middle- and upper-middle-class Englishmen. Although the oration made the usual claim of natural theology that science could provide a proof of the existence of a Deity, the speeches contain no religious extrapolations. For the intellectual elite, religion was becoming grounded in ethical life rather than in metaphysical claims regarding revelation and cognition.[81] Religion was also in the process of becoming a private matter. Herschel, declining an invitation to write the Bridgewater Treatise on Astronomy, which was supposed to demonstrate the power and wisdom of God as manifested in the Creation, wrote to Gilbert

> No one, as you well know, is more deeply impressed with the great truths to be inculcated in this work; but in precisely the same proportion is the repugnance I feel to weaken the weight of my testimony in their favor by promulgation them under the direct and avowed influence of pecuniary reward.[82]

The most poignant statement of the claim that religious beliefs are a private matter was contained in Augustus De Morgan's will:

> I commend my future with hope and confidence to Almighty God; to God the Father of our Lord Jesus Christ, Whom I believe in my heart to be the Son of God, but Whom I have not confessed with my lips, because in my time such confession has always been the way up in the world.[83]

Like everyone else, the early Victorian scientists were affected by the Romanticism that polarized the outlook of that generation. Reading their "life and letters,"[84] one is struck by the stress they placed on the cultivation of their imaginative powers and the importance they accorded to emotion. Whewell, Herschel, De Morgan, and W.R. Hamilton all wrote poetry—romantic poetry—and presented their poems to one another. These early Victorian scientists also had a love for what lay outside the experience of ordinary man. There is a romantic element in the expedition that Herschel makes to the Dolomites and in Whewell's excursions on his continental trips in search of unique places. The geological trips of

Lyell, Scrope and Sedgwick, while answering scientific needs, surely also satisfied their romantic yearnings. There is no better way to obtain an insight into the Romantic values and inspiration of the early Victorian scientist than to read Darwin's diary of his voyage on the *Beagle*. Finally, I would suggest that the cultivation of polymathy is likewise an expression of the Romantic element among these intellectuals. It is not a matter merely of *sapere aude*, but of daring to know everything that can be known and immersing oneself in this challenge.

There is one other facet of early Victorian science which reflects its social character: the role of the Victorian reviews as a forum for the exchange of controversial scientific ideas. I consider this subject to be of the utmost importance and one that merits a detailed and careful analysis, but a summary will suffice here.

In contrast to France, Great Britain had a sizable literate public that supported a considerable popular scientific book-publishing enterprise and a network of unrefereed magazines,[85] of which the *Edinburgh Review*, the *Quarterly Review*, and the *Westminster Review* are best known. The anonymous articles in these Victorian magazines allowed their authors to raise and discuss controversial philosophical and scientific issues and, more important, allowed diverse and conflicting opinions to be debated.

The reforms of mathematical notations in the first decade of the century were certainly stimulated by Playfair's criticism in the *Edinburgh Review*.[86] Reform of the curricula at Oxford and Cambridge was undoubtedly accelerated by the articles that appeared in the *Edinburgh Review* and *Quarterly Review* (by Lyell and others).[87] The Declinist arguments in the early 1830s were aired in these reviews.[88] The reviews of Lyell's *Principles of Geology* were important contributions[89] to the debate over Uniformitarianism. Similarly, the reviews of Chambers' *Vestiges* contributed substantially to the debate over evolution in the mid 1840s.[90] Many other instances could be adduced.

The connection between these scientific reviews and the British tradition of the cultivation of freedom of intellectual expression merits further scrutiny. That intellectual freedom was an accepted part of intellectual activities in England by early Victorian times was apparent to contempory observers. Thus Professor Moll of Utrecht, countering Babbage's opinion that science had declined in England, noted

that much of the emolument bestowed on continental savans is held at the pleasure of the government, which takes out in submission what it advances in cash. We do not say, that it exacts servility, but silence and obedience; or, previously to the use of the gift of speech, revision by the minister. What would Mr. Herschel or Mr. Babbage say if the speeches made from the chair, on the delivery of the King's or Copley medals, were to be submitted to the approbation of the home secretary? What would Mr. Babbage's feeling be, if no one could be admitted in the Royal Society, unless his choice was approved by the Court — if members might be ejected because . . . their political opinions were objected to.[91]

It is this freedom of expression, particularly within the context of the anomymous, uncensored and unrefereed Victorian reviews, aided by a lack of centralization, that contributed to the advances of the more speculative sciences in Great Britain, for example, biology and anthropology.

[V]

The social context offers insights into the dynamics of scientific beliefs, but social forces are not the only ones shaping science. An analysis of the Victorian reviews helps clarify some aspects of the criticisms within the evolutionary debates and illuminates the linkage between Victorian accounts of natural reality and the social order. But the sociology of scientific knowledge can only offer a partial explanation of these phenomena. The actions and beliefs of individuals, although surely polarized by the social context, as well as internal factors, must also be taken into account.

In this section, I want to focus on John Herschel in order to explore aspects of Victorian scientific practice. Individual temperament, personality, beliefs, and the social context all interact to determine the actions of individuals. These in turn affect the context. But the views and actions of the leading practitioners tend to reverberate more strongly within the scientific community than those of others. John Herschel was certainly such a person. More than anyone else, Herschel represented the early Victorian ideal of the scientist. His *Preliminary Discourse on the Study of Natural Philosophy*[92] was a most successful exposition of the methodology of science, and became a bible to young persons aspiring to become scientists. The comments of Charles Darwin and Michael Faraday on the *Discourse* are well known[93] and indicative of its widespread influence. J.S. Mill learned his philosophy of science from the *Discourse*. But the *Discourse* was more than a set of guidelines for the successful pursuit of science. It was also an attempt by an outstanding scientist at the height of his creative powers, one who had mastered all the sciences of his day, to clarify the ontological status of the objects that the natural sciences study, to outline the aims and tasks of the sciences, and to reflect on "the general nature and advantages of the study of the physical sciences." For Herschel "the more abundant supply of our physical wants, and the increase of our comforts" were neither the only nor the greatest benefits stemming from advances in science. To the question *Cui bono?*, Herschel gave the answer:

> The successful results of our experiments and reasonings in natural philosophy, and the incalculable advantages which experience, systematically consulted and dispassionately reasoned on, has conferred in matters purely physical, tend of necessity to impress something of the well weighed and progressive character of science on the more complicated conduct of our social and moral relations. It is thus that legislation and politics become gradually regarded as experimental sciences; and history, not, as formerly, the mere record of tyrannies and slaughters . . . but as the archive of experiments, successful and unsuccessful, gradually accumulating towards the solution of the grand problem — how the advantages of government are to be secured with the least possible inconvenience to the ground.[94]

Herschel's concerns in 1830 are the same as those that Herbert Spencer[95] and J.S. Mill[96] confronted later in the century: the limits of state action. The dilemma of modernity had not escaped Herschel. Education was a contrivance to mold everyone alike, yet it was necessary to realize one's individuality. Similarly, the help of the state — the greatest force in reducing individuality — was necessary to develop one's individuality to the fullest. Although Herschel could personally afford to resist the state's encroachments in his scientific researches, he was well aware that this was not possible in many scientific enterprises.[97]

Herschel's views on progress reflect the British syncretism of French and German attitudes. He amalgamated a French enthusiasm and confidence in the arts and sciences with a more somber, Germanic view of man's nature — the Christian legacy — and enlisted science for the moral development of man. For Herschel, progress consists in the ethicization of man, but it is propelled by the advancement of knowledge and the accretion of material goods.[98]

The early Victorians tended to reject monistic approaches. Their practical bent had given them a healthy respect for complexity. Although they appreciated the French confidence in mathematics and quantification, they rejected the linear, quasimechanical French view of progress. The introduction of "German thought" into England by Coleridge and, later, by Carlyle, and its dissemination by Wordsworth and others,[99] presented them with an alternative. They were attracted to views of progress and progression as diversity, and they displayed a propensity towards the biological, that is, the genetic, analogy, rather than the mechanical one. I do not mean to imply that Herder's views of progress as diversity were widely known in England during the first third of the nineteenth century.[100] Rather, I would suggest that the British had become acquainted with models of progression and development similar to Herder's in their extensive studies of the origin and diffusion of languages.[101] The biological metaphor implies that change is imperceptible and slow and, in contrast to that of the rectilinear Turgot-Condorcet model, the tempo cannot be altered. Revolutions alter tempo. Evolutionary rather than revolutionary views resonated with the dominant British philosophical and political position regarding continuity.[102] If progress is not linear, its dynamics must be accounted for. The search for causes — *verae causae* — is a British attitude that became identified with the essence of the Newtonian tradition. More than anyone else, Herschel with his *Discourse* was responsible for the legitimatization of this British attitude during the nineteenth century.[103]

There is another aspect of the *Discourse* that should be commented upon: the high praise Herschel confers on Francis Bacon. Throughout the *Discourse*, Herschel constantly refers with approbation to Bacon. The general precepts of scientific methodology that Bacon had formulated but often could not illustrate, in the *Novum Organum* are all presented in the *Discourse*, and their usefulness and fruitfulness demonstrated by Herschel with historical examples that had come to light since Bacon's time. Part of the explanation is of course Herschel's pride in his great predecessors. As Herschel himself indicated

> It is to our immortal countryman, Bacon, that we owe the broad announcement of this grand and fertile principle; and the development of the idea, that the whole of natural philosophy consists entirely of a series of inductive generalizations, commencing with the most circumstantially stated particulars, and carried up to universal laws, or axioms, which comprehend in their statements every subordinate degree of generality, and of a corresponding series of inverted reasoning from generals to particulars, by which these axioms are traced back into their remotest consequences, and all particular propositions deduced from them[104]

Further on in the *Discourse* Herschel also commented that

> By the discoveries of Copernicus, Kepler and Galileo, the errors of the Aristotelian philosophy were effectually overturned on a plain appeal to

the facts of nature; but it remained to show on broad and general principles, how and why Aristotle was in the wrong; . . . This important task was executed by Francis Bacon, Lord Verulam who will, therefore justly be looked upon in all future ages as the great reformer of philosophy . . . [105]

But I suspect that Herschel's great respect for Bacon was due to more than the identity of their nationality and their common interest in philosophy. For Bacon, man's noblest ambition for power found expression in his

> endeavor to establish and extend the power and dominion of the human race itself over the universe . . . Now the empire of man over things depends wholly on the arts and sciences. For we cannot command nature except by obeying her.[106]

In the sciences, "human knowledge and human power meet in one." Bacon thus also represented the tension between the *vita contemplativa* and the *vita activita*, between transcendence and utility, between the sacred and the secular. To find a balance and resolution between these dichotomies was a pressing problem for Herschel throughout his life. Also, if to command Nature men must know her, how is this knowledge to be disseminated?[107] Would Bacon have been as influential had he not been Lord Chancellor? For Herschel, Lyell, Darwin, and others these questions became enmeshed with others: What were their civil obligations? What were their responsibilities for the direction of the Geological Society, the British Association for the Advancement of Science, and for science policy at the Treasury and the Admiralty? What were their responsibilities for public disseminatation of the theological implications of their scientific findings?

Resolving the tension between the requirement of independence as a scholar and the necessity of interdependence as a social being is the perennial moral problem of the intellectual, and was no less so for the early Victorian scientist: perhaps more so, in fact, since at issue once again was man's conception of the Deity. Laplace, believing he had proved the stability of the solar system, when queried by Napoleon where God was in his *Mécanique Céleste*, could answer "I have no need of this hypothesis." What he had meant was that God was unnecessary to account for the operation of inanimate nature since secondary laws could explain everything. The force of the argument was not lost on Herschel.[108] It was John Herschel who in a famous letter to Lyell indicated that he accepted "a natural in contradiction to a miraculous"[109] process of species creation and thus endorsed the search for secondary laws to explain biological phenomena and the origin of species. Herschel was forthright in making public these views. Lyell, on the other hand, had been less than candid in his *Principles of Geology*.[110] Darwin was fully aware in 1838 that the correctness of his "theory" meant that God had been banished from the organic world. This challenge to the orthodoxy of Victorian England proved a formidable psychological barrier for Darwin, and was undoubtedly a factor in his delay in publishing his findings.

The particular stand taken was an individual matter, explainable only in terms of that individual's life. The same is true for the explanation of the degree of his participation in societal affairs. But whatever the attitudes of the early Victorian scientists were in these matters, each of them believed that his intellectual concerns were with important, even ultimate things. The charismatic quality of their work had brought the early Victorian men of science into association with other powerful individuals.[111] These scientists came to see themselves as participating in a social class of economically and politically powerful individuals.[112] And this was true for all the members of the intellectual elite. It was thus inevitable that competition over power and resources would develop among themselves and with the men of power.

Intellectuals create traditions — that is, creations and interpretations of reality — and so do men of power. S. Eisenstadt[113] has suggested that each one of these builders of traditions — intellectuals and men of power — create both symbols and organizations and that there exists a mutual interdependence between intellectuals and political authorities. In Eisenstadt's view, the political authorities of a society seek to be legitimized by intellectuals, religious or secular. Conversely, intellectuals and intellectual organizations generally require the protection of political institutions as a condition of their survival. Eisenstadt contends that the intellectual and political authorities seek to maintain the maximal autonomy for themselves, while at the same time each tries to achieve maximal control over the other. The model is not invalidated by the early Victorian situation.

EPILOGUE

In my exploration of the early Victorian scientific scene, I have tried to locate some of the factors that explain the success of the leaders of the scientific community — Lyell, Herschel, Whewell, Babbage, *et al.* — in their activities to reform and transform the scientific and educational institutions. I have suggested that their successful efforts in revitalizing mathematical research in Britain (by adopting the French research tradition and declaring the Newtonian research program moribund) gave them the necessary confidence and momentum to challenge the social order of the intellectual sphere. They solved the problems attendant upon growing differentiation by importing institutional solutions that had proved successful in France and Germany and by taking control of those institutions in which their scientific prestige and intellectual standing left the opposition disarmed. They also capitalized on the unique features of the British system: its freedom of the press and its freedom of expression. They used the literary reviews to present the implication of the new scientific knowledge and to initiate debates on controversial issues. By conforming to most of the norms of the social order they were able to integrate their intellectual creations as enduring elements within the Victorian construction of reality.

In his great history of European thought, Merz, pondering on the uniqueness of England's scientific spirit during the nineteenth century, noted that

> . . . It is the individualism of the English character, the self-reliant strength of natural genius, which comes out most strongly in its great examples of scientific work . . .; [and] it is interesting to note how even in this sphere, which more than any other seems to bear an international and cosmopolitan character, the genius of the nation strongly asserts itself, . . . England, the country of greatest individual freedom, has been the land most favorable to the growth of genius as well as eccentricity, and has thus produced a disproportionate number of new ideas and departures . . . [114]

His conclusions still stand.

NOTES AND REFERENCES

1. Many such attempts have been made, nevertheless, such as the following: D. C. Somervell, *English Thought in the Nineteenth Century*, 5th ed. (London: Methuen, 1940); G.M. Young, *Victorian England: Portrait of an Age*, 2nd ed. (Oxford: Oxford University Press, 1953); W.E. Houghton, *The Victorian Frame of Mind 1830–1870* (New Haven: Yale University Press, 1957); G.S.R. Kitson Clark, "The Romantic Element, 1830 to 1850," in *Studies in Social History: A Tribute to G.M. Trevelyan*, J.H. Plumb, ed. (London: Longmans, 1955).

2. Arnold Thackray, "Natural Knowledge in Cultural Context: The Manchester Model," *American Historical Review*, vol. 79 (1974), pp. 672–711.

3. W.F. Cannon, "Science as Norm of Truth," *Journal of the History of Ideas*, vol. 25 (1964), pp. 487–502.

4. See for example, T. Kelly, *George Birkbeck: Pioneer of Adult Education*, as well as his

A Select Bibliography of Adult Education in Great Britain including works published to the end of the year 1950 (London: Institute of Adult Education, 1952).

5. In 1818 the Institution of Civil Engineer was established to further "the art of directing the Great Sources of Power in Nature for the use and convenience of mankind." *Annals of Philosophy,* vol. XI (1818), p. 153.

6. See, for example, Henry Roseveare, *The Treasury: The Evolution of a British Institution* (New York: Columbia University Press, 1969). For an account of the emergence of the managerial class in industry, see S. Pollard, *The Genesis of Modern Management: A Study of the Industrial Revolution in Great Britain,* (Cambridge, Mass: Harvard University Press, 1965); A.M. Carr-Saunders and P.A. Wilson, *The Professions* (Oxford: Clarendon Press, 1933); W.J. Reader, *Professional Men: The Rise of The Professional Classes in Nineteenth-Century England* (London: Macmillian, 1966).

7. See Edward Shils, "Intellectuals, Tradition and the Traditions of Intellectuals: Some Preliminary Considerations," *Daedalus,* Spring 1972, pp. 21–35. This entire issue of *Daedalus* is devoted to intellectuals and tradition. The preface by S.R. Graubard states the problem succinctly.

8. N.G. Annan, "The Intellectual Aristocracy," in *Studies in Social History: A Tribute to G.M. Trevelyan,* ed. J.H. Plumb, (London: Longmans, Green, 1955). The intellectual aristocracy of the new intellegentsia was dominated by the descendants of middle-class families, who were predominantly Dissenters. Their religious and/or political opinions in the late eighteenth century placed them at the periphery of established society. The next generation kept their parents' critical perspective on society, even though many of them adopted Anglicanism. Intellectually gifted, they formed the nucleus of the liberally inclined, socially conscious dons, masters, scholars, clergymen, and essayists of the early Victorian period. By the 1860s they formed a major element within the intellegentsia. Annan's essay focuses on the Stephens, Stracheys, Treveleyans and the Darwins.

9. W.F. Cannon, "Scientists and Broad Churchmen: an Early Victorian Intellectual Network," *Journal of British Studies,* vol. 4 (1964), pp. 56–88. See also Cannon's *Science in Culture: The Early Victorian Period* (New York: Science History Publications, 1978), chap. 2, "The Cambridge Network."

10. S. Shapin and A. Thackray, "Prosopography as a Research Tool in History of Science: The British Scientific Community 1700–1900." *History of Science,* vol. 12 (1974), pp.1–28.

11. Annan,[8] "The Intellectual Aristocracy," p. 244.

12. *Life, Letters and Journals of Sir Charles Lyell, Bart,* edited by his sister-in-law, Mrs. Lyell,2 vols. (London: John Murray, 1881), vol. I, p. 356.

13. Shapin and Thackray,[10] These authors also quote L.B. Namier, *England in the Age of the American Revolution* (London: Macmillan, 1961): "For specialisation necessarily entails distortion of mind and loss of balance and the characteristically English attempt to appear unscientific springs from a desire to remain human. . . . What is now valued in England is abstract knowledge as a profession, because the tradition of English civilization is that professions should be practical and culture should be the work of the leisured classes." The tension between tutor and professor at Cambridge in the 1830s was already noted by A. De Morgan in "English Science," *The British and Foreign Review,* vol. I (1835), pp. 134–157. See also S. Rothblatt, *The Revolution of the Dons: Cambridge and Society in Victorian England* (New York: Basic Books, 1968).

14. "All my respect for the Superintendent's talents (and it is not little) does not preclude me from believing, that neither is a judge of the wants of the astronomer, nor of those of the navigator. Place him in an observatory, he would be as much at a loss to direct an instru-

ment to a star, as he would be to navigate a vessel from his own observations." James South, *Refutations of the Numerous Misstatements and Fallacies contained in a paper presented to the Admiralty by Dr. Thomas Young,* (London, 1827), p. 7.

15. Quoted in South,[14] p. 43.

16. In the 1830s Oxford and Cambridge, although they were emerging from their previous lethargy, were still "to be considered, lst, as seminaries of education for the young nobility and gentry of the realm; 2ndly, as nurseries for the Established Church and for the learned professions; and 3rdly, as schools for the advancement and developement of science, and the deeper researches of literature."(In *Letters to the English Public on the Condition, Abuses, and Capabilities of the National Universities No. 1. By a Graduate of Cambridge* (1836), quoted in: Peter Allen, *The Cambridge Apostles: The Early Years* [Cambridge: Cambridge University Press, 1978], pp. 200-201). Corroborating this view is T.J. Hogg's article,"The Universities of Oxford and Cambridge," *Westminster Review,* vol. 15 (1831), pp. 56-69. This is also the contemporary assessment by De Morgan. In the *Memoir,* he recalled,"I first began to know the scientific world in 1828. The forces were then mustering for what may be called the great battle of 1830. The great epidemic which produced the French Revolution, and what is yet the English Reform Bill, showed its effect on the scientific world. . . [11] (In *Memoir of Augustus De Morgan* by his wife Sophia Elizabeth De Morgan [London: Longmans, 1882], p. 4.). The "great battle of 1830" centered around the efforts to reform the Royal Society. For De Morgan's account of that epic see his "Science and Rank," *Dublin Review,* vol. XIII, (1842), pp. 413-448. For the Scottish universities, see G.E. Davie, *The Democratic Intellect,* (Edinburgh: Edinburgh University Press, 1961).

17. Annan[8] in "The Intellectual Aristocracy" states:

> But in the sixties two objectives vital to their class and, as they rightly thought, vital to their country, united them. They worked tirelessly for intellectual freedom within the universities which, they thought, should admit anyone irrespective of his religious beliefs, and for the creation of a public service open to talent. If they can be said to have had a Bill of Rights it was the Trevelyan-Northcote report of 1853 on reform of the civil service and their Glorious Revolution was achieved in 1870, when entry into public service by privilege, purchase of army commissions and the religious tests were finally abolished.

18. See, for example, R.M. Hartwell, "Economic Change in England and Europe, 1780-1830," in *The New Cambridge Modern History, Volume IX — War and Peace in Age of Upheaval, 1793-1830,* ed. C.W. Crawley (Cambridge: Cambridge University Press, 1965).

19. J. Bass Mullinger, *A History of the University of Cambridge* (New York: A.D.F. Randolph & Co., 1889.)

20. J.L. Clive, *Scotch Reviewers: The Edinburgh Review, 1802-1815* (London:Faber and Faber, 1957).

21. See Neil McKendrick, "Josiah Wedgwood and Factory Discipline," *The History Journal,* vol. 4 (1961), p. 3. Also "The Role of Science in the Industrial Revolution: A Study of Josiah Wedgwood as a Scientist and Industrial Chemist," in *Changing Perspectives in the History of Science,* eds. M. Teich and R. Young, (London: Heinemann, 1973). Recall also Engels' later description of London: "We know well enough that this isolation of the individual — this narrowminded self-seeking — is everywhere the fundamental principle of modern society. But nowhere is it is so shamelessly unconcealed, so self-conscious as in the tumultous concourse of the great city. The dissolution of mankind into monads, each of

which has a separate purpose, is carried here to its furthest point. It is the world of atoms."
(Chap. III of Engels' *The Condition of the Working Class*, quoted in Steven Marcus: *Engels,
Manchester and the Working Class* [New York: Random House, 1974], pp. 146–147.)

22.   T.S. Ashton, *The Industrial Revolution: 1760–1830* (London and New York: Oxford
University Press, 1948).

23.   D.C. Somervell, *English Thought in the Nineteenth Century* (New York: Longmans,
Green, 1950).

24.   James Mill, "Education," *Supplement to the 4th, 5th and 6th Editions of the Encyclopedia Britannica*, vol. IV (Edinburgh: Constable, 1824), p. 11.

25.   Elie Halévy, *The Growth of Philosophic Radicalism* (Boston: Beacon, 1955). See particularly chap. III.

26.   For an overview see E. Storella, "O, What a World of Profit and Delight: The Society
for the Diffusion of Useful Knowledge" (Brandeis University, Ph.D. dissertation, 1969).

27.   R.E. Schofield, *The Lunar Society of Birmingham: A Social History of Provincial
Science and Industry in Eighteenth Century England* (Oxford: Clarendon Press, 1963).

28.   F. Manuel, *The Prophets of Paris* (Cambridge: Harvard University Press, 1962). See
chap. II, "Marquis de Condorcet: The Taming of the Future," pp. 53–103. See also K.M.
Baker, *Condorcet: From Natural Philosophy to Social Mathematics* (Chicago: University
of Chicago Press, 1975).

29.   F. Manuel, *Shapes of Philosophical History* (Stanford: Stanford University Press,
1965), pp. 116–118.

30.   Hajo Holborn, "German Idealism in the Light of Social History," in *Germany and
Europe* (New York: Doubleday and Co., 1970); see also his *A History of Modern Germany:
1648–1840* (New York: Knopf, 1967).

31.   F.K. Ringer, *The Decline of the German Mandarins* (Cambridge: Havard University
Press, 1969).

32.   For a masterly overview, see R. Steven Turner, "The Growth of Professorial Research
in Prussia; 1818–1848 — Causes and Context," in *Historical Studies in the Physical Sciences*,
vol. 3 (1971), pp. 137–182.

33.   Henry Guerlac, *Essays and Papers in the History of Modern Science* (Baltimore: John
Hopkins Press, 1977); R. Hahn, *The Anatomy of a Scientific Institution: The Paris
Academy of Sciences 1666–1803* (Berkeley: University of California Press, 1971); K.M.
Baker, *Condorcet: From Natural Philosophy to Social Mathematics* (Chicago: University
of Chicago Press, 1975); M. Crosland, *The Society of Arceuil: A View of French Science at
the Time of Napoleon I* (London: Heinemann, 1967) and "Science in a Natural Context,"
British Journal of the History of Science, vol. 10 (1977), pp. 95–113, "The French Academy
of Science in the Nineteenth Century" *Minerva*, vol. 16 (1978), pp. 73–102. "The Development of a Professional Career in Science in France," in *The Emergence of Science in Western
Europe*, ed. M. Crosland (London: 1975); R. Fox, "The Rise and Fall of Laplacian Physics,"
Historical Studies in the Physical Sciences, vol. 4 (1974), pp. 89–136 and "Scientific En-
treprise and the Patronage of Research in France", in *The Patronage of Science in the Nineteenth Century*" ed. G.L'E. Turner (Leyden: Noordhoff, 1976). See also C.C. Gillispie,
"The Encyclopédie and the Jacobin Philosophy of Science: A Study in Ideas and Conse-
quences" and L. Pearce Williams, "The Politics of Science in the French Revolution," in
Critical Problems in the History of Science, ed. Marshall Claggett (Madison: University of
Wisconsin Press, 1969), and C.C. Gillispie, "Science and Technology," in *The New Cam-
bridge Modern History* (Cambridge: Cambridge University Press, 1965) vol. IX, ed. C.W.
Crawley.

34.   For a recent view see M. Crosland and C. Smith, "The Transmission of Physics from France to Britain, 1800–1840" *Historical Studies in the Physical Sciences*, vol. 9, 1978 (Princeton: Princeton University Press), pp. 1–61.

35.   For example, controversy between Cuvier and Geoffroy St.-Hilaire on evolution in 1830 and that between Pasteur and Pouchet on spontaneous generation in 1863.

36.   This is the "classic" view first presented in J.T. Merz, *A History of European Thought in the Nineteenth Century* (New York: W. Blackwood and Sons, 1912), vol. III, p. 98.

37.   Holborn,[30]*Germany and Europe*, p. 22. See also, George A. Kelly, *Idealism, Politics and History: Sources of Hegelian Thought* (Cambridge: Cambridge University Press, 1969), which is an invaluable source of information and insight for the problems we are discussing.

38.   See, for example, the list of the contributions of John Playfair, John Leslie, and Henry Brougham to the *Edinburgh Review* in W. E. Houghton, ed., *The Wellesley Index to Victorian Periodicals 1824–1900* (Toronto: University of Toronto Press, 1972). The Library of John Playfair was auctioned after his death. A 106-page catalogue was needed to list his books! *Catalogue of the library of the late John Playfair to be sold by auction, without reserve, by Mr. Ballantyre* (Edinburgh: James Ballantyre, 1820).

39.   G.E. Davie, *The Democratic Intellect* (Edinburgh: Edinburgh University Press, 1961); H. Trevor-Roper, "The Scottish Enlightenment," in T. Besterman, ed., *Studies on Voltaire and the Eighteenth Century*, vol. 58 (Geneva: Institut et Musée Voltaire, 1967), pp. 1635–1658; N.T. Phillipson, "Culture and Society in the Eighteenth-Century: The Case of Edinburgh and the Scottish Enlightenment" in L. Stone, ed., *The University in Society* (Princeton: Princeton University Press, 1975), pp. 407–448, and "Towards a definition of the Scottish Enlightenment," in P. Fritz and D. Williams, eds. *City and Society in the Eighteenth Century* (Toronto: Hakkert, 1973); R. Olson, *Scottish Philosophy and British Physics: 1750–1880* (Princeton: Princeton University Press, 1975). See also the essay review on this book by J.R.R. Christie in *Annals of Science*, vol. 33. (1976), pp. 311–318.

40.   M. Mandelbaum, *History, Man and Reason: A Study in Nineteenth-Century Thought* (Baltimore: The Johns Hopkins University Press, 1971).

41.   Thomas Brown, *Observations on the Nature and Tendency of the Doctrine of Mr. Hume Concerning the Relation of Cause and Effect* (Edinburgh: Mundell and Son, 1806). See also Olson[39] for further discussions of these points.

42.   Gladys Bryson, *Man and Society: The Scottish Inquiry of the 18th Century* (Princeton: Princeton University Press, 1945).

43.   See *Science in France in the Revolutionary Era Described by Thomas Bugge*, edited with Introduction and Commentary by Maurice P. Crosland (Cambridge: Massachusetts Institute of Technology Press, 1969).

44.   *Mémoires de l'Institut National des Sciences et Arts Sciences Morales et Politique*, tome I–V. Vols. I and II contain Cabanis's "Rapports du Physique et du Moral de l'homme." Vol. I also contains Destutt-Tracy's "Considération générales sur l'étude de l'homme, et sur les rapports de son organization physique avec ses facultés intellectuels et morales."

45.   See, for example, *Edinburgh Review* (October, 1809) "Review of Discours sur les Progrès des Sciences, Lettres et Arts depuis 1789 jusqu'a ce jour (1808)" and the *Works of John Playfair* for his reviews in the *Edinburgh Review* of Laplace on the system of the world and on probabilities. See *Quarterly Review*, vol. II (1809), pp. 337–349, for a review of Laplace on optics, and *Quarterly Review*, vol. I (1809), pp. 107–112, for Laplace on capillarity. The scientific magazines, of course, regularly reported on French science. See, for example, the *Philosophical Magazine* and the *Annals of Philosophy*.

46. *Edinburgh Review*, vol. II (1809-10), pp. 142-152 and pp. 418-441.

47. J.B. Morrell, "Individualism and the Structure of British Science in 1830" in Russell McCormack, ed. *Historical Studies in the Physical Sciences*, vol. 3 (1971), pp. 183-204; S.F. Cannon, *Science in Culture: The Early Victorian Period* (New York: Science History Publications, 1978), pp. 178-200.

49. A.N. de Condorcet, *Esquisse d'un Tableau Historique des Progrès de l'Esprit Humain, Ouvrage Posthume de Condorcet* (Paris, 1795); *Condorcet: Vie de Turgot* (Paris, 1786). The first English translation of the *Esquisse* appeared in 1795 under the title *Outlines of an Historical View of the Progress of the Human Mind* (translated from the French) (London: J. Johnson, 1795). An English translation by June Barraclough with an introduction by Stuart Hampshire is available under the title: A. Condorcet, *Sketch for a Historical Picture of the Progress of the Human Mind* (New York; Noonday Press, 1955). Here my concern is with the influence of Condorcet's *Esquisse (Outline)* on the shaping of British science before 1830 or so. Of course, the Malthusian debate traces itself back to the Condorcet of the *Esquisse*. Condorcet himself was aware of the Malthusian problem: "Might there not then come a moment when these necessary laws [of progress] begin to work in a contrary direction; when, the number of people in the world finally exceeding the means of subsistence, there will in consequence ensue a continual diminution of happiness and population, a true retrogression, or at best an oscillation between good and bad?" (*Sketch*, p. 188).

But Condorcet was optimistic as to the solution! See K.M. Baker,[33] *Condorcet* and see also his "Scientism, Elitism and Liberalism: the Case of Condorcet, "*Studies on Voltaire and the Eighteenth Century*, ed. T. Besterman (Geneva: Institut et Musée Voltaire 1967), vol. 55 pp. 129-165. See also C.C. Gillispie[11] Probability and Politics: Laplace, Condorcet and Turgot,[11] *Proceedings of the American Philosophical Society*, vol. 116, (1972), pp. 1-20 and R. Rashed, *Mathematique et Société* (Paris: Herman, 1974).

50. No more succinct history of the Analytical Society can be given than the one to be found in De Morgan's review entitled "Cambridge Differential Notation[11] *Cambridge Journal of Education*, vol. VIII (1834), pp. 100-107.

The simultaneous invention of the theory of fluxions by Newton, and of the differential calculus by Leibnitz, was rich in useful consequences, by the various lights in which it caused every question to be viewed. But the difference of notation was an evil which, to this country, more than counterbalanced all the advantages. The whole continent adopted the symbols of Leibnitz; the English retained those of Newton, and gradually lost their mathematical character. The reason is obvious enough: our neighbours, with a more general and powerful notation, could easily translate all that was done in England; while we, on the contrary, could not, without great difficulty, make the language of fluxions tell us all that was discovered abroad. They had also the advantage of numbers and international communication; we could hardly read their writings, and could not, or at least did not, introduce their new and powerful methods of investigation. And to increase the difficulty, any attempt at innovation was considered as a sin against the memory of Newton.

The University of Cambridge first broke through the mist which hid the whole continent from view. In 1803, Mr. Woodhouse published his *Principles of Analytical Calculation*, in which the notation of Leibnitz was explained and dwelt upon. The impulse was thus given, though not with very great force. In 1813, it appeared from the *Memoirs of the Analytical Society*, a body of juniors, among whom were Messrs. Herschel, Peacock and Babbage, that the change had several zealous advocates. In 1816, these gentlemen published a translation of Lacroix's *Differential Calculus*, with a volume of examples, and in 1817, the second named introduced the *Diffential Calculus* formally into the public examinations. Since that time the new system must be considered as established.

The French influence on the *Analytical Society* is attested by the following: In a letter to Babbage dated July 1, 1812, Herschel addresses him as "Citizen" and signs it 'I remain, Citizen; yours truly J. Herschel." (I thank the Librarian of the Royal Society for permission to quote from this letter in the Herschel Archives at the Royal Society.) Bromhead, when writing to Babbage on March 4, 1816, saw Babbage's efforts in the *Analytical Society* as "Exertions in founding a first class in the Institute" (British Museum Add MS 37182f52).

51. Several members of the group together with Sedgwick and Whewell were responsible for the establishment of the Cambridge University Philosophical Society. Herschel in his presidential speech in 1845 to the British Association for the Advancement of Science indicated that "From this society have emanated eight or nine volumes, full of variety and interest, and such as no similar collection, originating as this had done in the bosom, and in great measure, within the walls of an academic institution, can at all compare with; the Memoirs of the Ecole Polytechnique of Paris, alone excepted."

52. Dr. R. Daston has stressed this point in seminars at Harvard University and in an invited paper at the History of Science meeting in Madison, Wisconsin, October 1978.

53. Baker,[33] *Condorcet*, pp. 120–194.

54. J.F.W. Herschel, *Instructions for Making and Registering Meteorological Observations in Southern Africa and other Countries in the South Seas and also at Sea.* Drawn up for circulation by the Meteorological Committee of the South Afrian Literary and Philosophical Institution. Reprinted for private distribution (available at the British Museum).

55. J.F.W. Herschel, *Edinburgh Review*, July 1850, reprinted in *Essays from the Edinburgh and Quarterly Reviews* (London: Longmans, 1851).

56. The essay is reprinted in J.F.W. Herschel: *Essays from the Edinburgh and Quarterly Reviews, with Addresses and other Pieces* (London: Longmans, Green, 1857).

57. W.R. Atbury, "Experiment and Explanation in the Physiology of Bichat and Magendie," in *Studies in the History of Biology*, eds. W. Coleman and C. Limoges (Baltimore: Johns Hopkins University Press, 1977), pp. 47–132. See also, J.M.D. Olmsted, *Francis Magendie* (New York: Schuman's, 1944) and F.L. Holmes, *Claude Bernard and Animal Chemistry* (Cambridge: Harvard University Press, 1974).

58. E.H. Ackerknecht, *Medicine at the Paris Hospital 1794–1848*, (Baltimore: Johns Hopkins University Press, 1967); O. Temkin, *The Double Face of Janus* (Baltimore: Johns Hopkins University Press, 1977), contains reprints of two valuable articles on the ideologues' influence on physiology: "The Philosophic Background of Magendie's Physiology" and "Materialism in French and German Physiology of the Early Nineteenth Century", pp. 317–345. See also M. Foucault, *Les Mots et les Choses* (Paris: Gallimard, 1966); B. Haines, "The Inter-relations between Social, Biological, and Medical Thought, 1750–1850 – Saint Simon and Comte," *British Journal of the History of Science*, vol. II (1978), pp. 19–35. See also Robert M. Young, *Mind, Brain and Adaption in the Nineteenth Century* (Oxford: Clarendon Press, 1970) and Georges Canguilhem's entry for Cabanis in the *Dictionary of Scientific Biography*, ed. C.C. Gillispie, (New York: Scribners 1971). For the ideologues more generally, see: F. Picavet, *Les Idéologues* (Paris, 1891); S. Moravia, *Il Tramonto dell'Illuminismo: Filosofia e Politica nella Societa Francese 1770–1810* (Bari: Laterza, 1968).

59. Darwin in 1838–1839 commented that "thought, however unintelligently it may be

32 ANNALS NEW YORK ACADEMY OF SCIENCES

seems as much [a] function of organ, as bile of liver." Compare this with Cabanis's statement: "In order to form a correct idea of the operations from which thought results, one
must consider the brain as a particular organ, especially destined to produce it; just as the
stomach and intestines are destined to affect digestion, the liver to filter the bile. . . ."
(P.J.G. Cabanis, *Rapports du physique et du moral de l'homme.* Huitième edition
augmentée de notes etc. par L. Peisse. [Paris, 1844], pp. 137–138, quoted in Temkin,[58] *The
Double Face of Janus,* p. 320.) The Darwin comment is to be found in his *Old and Useless
Notes,* dating from the winter 1838 or spring 1839 [OUN 37], and is reprinted in P. Barrett
and H. Gruber, *Darwin on Man* (New York: Dutton, 1974), p. 394

60. J.S. Mill, *Autobiography,* ed. Currin V. Shields (New York: Liberal Arts Press, 1957);
Hill Shine, *Carlyle and the Saint-Simonian: The Concept of Historical Periodicity* (Baltimore:
Johns Hopkins University Press, 1941). See also Robert Southey, "Doctrine of
Saint-Simon", *Quarterly Review,* vol. 45 (1831), pp. 407–450 and T.P. Thompson, "Saint-
Simonianism &c" *Westminster Review,* vol. 16 (1832), pp. 279–321 for contemporary as-
sessments.

61. F. Manuel, *The New World of Henri Saint-Simon,* (Cambridge: Harvard University
Press, 1956), p. 303.

62. I understand that Dr. Anthony Hyman is at work on a comprehensive biography of
Babbage. See also the essay-review by I. Grattan-Guiness of J.M. Dubbey's, *The
Mathematical Works of Charles Babbage* (Cambridge: Cambridge University Press, 1978) in
the *British Journal of the History of Science,* vol. 12 (1979), pp. 82–88.

63. Maboth Moseley, *Irascible Genius* (London: Hutchinson, 1964).

64. Charles Darwin wrote to Babbage in 1839: "My sister is at present staying with us, will
you be so kind as to allow me to bring her to your party on Saturday, that she may see the
World. Ever yours, most truly, Chas. Darwin" (BM 37191f298).

65. C. Babbage: *On the Economy of Machinery and Manufactures* (London: John Murray, 1832).

66. Andrew Ure, *The Philosophy of Manufactures or an Exposition of the Scientific,
Moral and Commercial Economy of the Factory System* (London: C. Knight 1935), pp.
1–2. See also vol. VIII of *Mixed Sciences,* and vol. 6 of the *Encyclopaedia Metropolitana,*
which includes Charles Babbage's "Introductory View of the Principles of Manufactures"
and Peter Barlow's "Manufactures." Barlow's article contains an extensive description of all
the "self-adjusting" and "self-acting" apparatus attached to the "modern steam engine" (pp.
181–182). See also Richard S. Rosenbloom, "Men and Machines: Some 19th Century
Analysis of Mechanization", *Technology and Culture,* vol. 5 (1964), pp. 489–511; and
W.V. Farrar, "Andrew Ure FRS and the Philosophy of Manufactures," *Notes and Records
of the Royal Society of London,* vol. 27 (1972–73), pp. 229–324.

67. Ure,[66] p. 13

68. O. Mayr, *The Origins of Feedback Control* (Cambridge: Massachusetts Institute of
Technology Press, 1970)

69. A.R. Wallace, "On the Tendency of Varieties to Depart Indefinitely from the Original
Type," *Journal of the Proceedings of the Linnaean Society,* August 1858. Charles Lyell in the
eleventh edition of his *Principles of Geology,* which appeared in 1872, commented that
"when first the doctrine of the origin of species by transmutation was proposed, it was objected that such a theory substituted a material self-adjusting machinery for a Supreme
Creative Intelligence."

70. S.F. Cannon,[48] *Science in Culture,* pp. 31–57. See also M.J.S. Rudwick, "Historical
Analogies in the Geological Work of Charles Lyell", *Janus,* vol. 66 (1977), pp. 89–107. For a

more general background see Peter Hans Reill, *The German Enlightenment and the Rise of Historicism* (Berkeley: University of California Press, 1975).

71.   There were of course impressive historical overviews of science before the 1830s. John Playfair's *Dissertation Second: Exhibiting a general view of the progress of mathematical and physical science, since the revival of letters in Europe*, which was printed in the 4th, 5th, and 6th editions of the *Encyclopaedia Britannica* and is reprinted in *The Works of John Playfair*, 4 vols. (Edinburgh: A. Constable, 1822), is a notable example. Herschel and Peacock during the 1820s contributed entries on the history of mathematics to the *Encyclopaedia Britannica* and the *Encyclopaedia Metropolitana*. For a study of bibiliography in the early Victorian period, see A.N.L. Munby, *A History and Bibliography of Science in England — the First Phase 1833–1845* (Berkeley: School of Librarianship, The University of California, 1968).

72.   W. Whewell, *History of the Inductive Sciences, from the earliest to the present times*, 3 vols. (London: J.W. Parker, 1837). See also his *Newton and Flamsteed, Remarks on an article in number CIX of the QR*, 2nd ed., to which are added two letters, occasioned by a note in number CX of the *Review*, Trinity College, February 3 and 6, 1836, (Cambridge: J.J.J. Deighton, 1836).

73.   David Brewster, *The Life of Sir Isaac Newton* (London: J. Murray, 1831); also in Harper's family library, New York, 1838.

74.   D. Brewster, *Memoirs of the life, writings and discoveries of Sir Isaac Newton* (Edinburgh: T. Constable and Co., 1855). The *Memoirs* were reprinted from the Edinburgh editon of 1855 with a new introduction by R.S. Westfall (New York: Johnson Reprint Corp.,(1965).

75.   F. Baily, *An Account of the Rev. John Flamsteed, the first astronomer royal;* compiled from his own manuscripts and other authentic documents, never before published, to which is added to his British catalogue of stars, cor. and enc. by Francis Baily. Printed by order of the Lords Commissioners of the Admiralty 1835. (Rpt., by Dawsons, London, 1966.)

76.   S.P. Rigaud, *Historical Essay on the first publication of Sir Isaac Newton's Principia* (Oxford: The University Press, 1838). See also Stephen Peter Rigaud: *Correspondence of Scientific men of the seventeenth century including letters of Barrow, Flamsteed, Wallis and Newton, printed from the originals in the collection of the R.H., the earl of Macclesfield* (Oxford University Press, 1841), 2nd edition compiled by A. De Morgan (Oxford: Oxford University Press, 1862).

77.   Augustus De Morgan, *Arithmetical books from the invention of printing to the present time; being brief notices of a large number of works drawn up from actual inspection;* (London: Taylor and Watson, 1847). See also De Morgan's *Memoir*,[16] pp. 400–415, for an almost complete bibliography of De Morgan's work, including his bibliographical essays and historical articles.

78.   Kitson Clark,[1] "The Romantic Element 1830 to 1850," p. 220

79.   J.F.W. Herschel, "Address to the Subscribers to the Windsor and Eton Public Library and Reading Room," reprinted in Herschel,[55] pp. 1–20.

80.   *The Athenaeum* (1838), pp. 423–428. This article commented that "The interest excited in the literary and scientific world resembled that which stirred the political on the occasions of the Grey and Peel Festivals — only that this united all interests, and had the good wishes of all parties." An impression of the dinner can also be gleaned from a letter De Morgan wrote to Herschel on the occasion of William Rowan Hamilton's death: "W.R. Hamilton was an intimate friend whom I spoke to once in my life — at Babbage's about

1830; but for thirty years we have corresponded. I *saw* him a second time at the dinner you got at the Freemason's when you came from the Cape, but I could not get near enough to speak." (*Memoir of A. De Morgan,*[16] p. 333.)

81. If nothing else the Tractarian movement is indicative of the fact that the 1830s was a period of dramatic change. Whether the movement was the English equivalent of the French traditionalism of de Maistre, Bonald and Lammenais, or whether it was aimed at internal reform and personal resistance in the face of unsettling influences, the *Tracts* and the Reform Act marked the early Victorian period as one of confrontation with traditional religious and political values. For a more concrete case of interaction between science and theology, see W. F. Cannon, "The Problem of Miracles in the 1830's," *Victorian Studies,* vol. 4 (1960), pp. 5–32.

82. Quoted in W.H. Brock, "The Selection of the Authors of the Bridgewater Treatises," *Notes and Records of the Royal Society,* vol. 21 (1966), pp. 162–174.

83. *Memoir of Augustus De Morgan,*[16] p. 399.

84. The phenomenon of the "life and letters" of the eminent Victorian scientists is one that merits further inquiry. A valuable preliminary assessment has been made by Dr. Janet Browne in "The Charles Darwin-Joseph Hooker Correspondence: an Analysis of Manuscript Resources and Their Use in Biography," *Journal of The Society for Bibliography in Natural History,* vol. 8 (1978), pp. 351–366.

85. See W.E. Houghton, ed., *The Wellesley Index to Victorian Periodicals: 1824–1900,* vols. I, II, and III (Toronto: University of Toronto Press, 1972–1979). See also John L. Clive, *Scotch Reviewers: The Edinburgh Review 1802–1815* (London: Faber and Faber, 1957), W. Graham, *English Literary Periodicals* (New York: Thomas Nelson, 1930); M. Wolff, "Victorian Reviewers and Cultural Responsibility," in *1859: Entering an Age of Crisis,* eds. P. Appleton, W.A. Madden and M. Wolff, (Bloomington: Indiana University Press, 1959), pp. 269–289.

86. See Playfair's review of Laplace's *Mécanique Céleste,* in the *Edinburgh Review,* vol. 11, 1808), pp. 243–284. It is reprinted in Playfair's *Works.*

87. For example, C. Lyell "Scientific Institutions," *Quarterly Review,* vol. 34 (1826), pp. 153–168, and W. Whewell, "Cambridge Transactions. Science of the English Universities," *British Critic* vol. 17 (1831), pp. 71–90. Note that unsigned articles were carried not only in the literary reviews, but also in semiprofessional journals. Thus the first issue of the *Quarterly Journal of Education* carried (unsigned) articles by De Morgan in praise of the curriculum of the *École Polytechnique* (see the *Quarterly Journal of Education,* vol. 1 [1831]). Peacock's suggestions for university reform at Cambridge were first presented to anonymous articles in the *Cambridge Review.*

88. See, for example, "Observations on the Present State of the Mathematical Sciences in Great Britain," *Annals of Philosophy,* vol. VII (1816), pp. 89–98; "Present State of Education," *Westminster Review,* vol. IV (1825), pp. 147–176; "On the Decline of Science in England,"*The Athenaeum,* 1831, pp. 25–27, as well as the numerous other reviews of Babbage's *Reflections on the Decline of Science in England* in the Victorian reviews.

89. For example, W. Whewell, "Lyell's *Principles of Geology* Vol. 1," *British Critic,* vol. 17 (1831), pp. 180–206 and "Lyell's *Principles of Geology,* Vol. 2," *Quarterly Review,* vol. 93 (1832), pp. 103–132.

90. See the list in the bibliography of M. Millhauser's *Just before Darwin: Robert Chambers and Vestiges* (Middletown, Conn.: Wesleyan University Press, 1959), pp. 229–232.

91. Quoted on p. 434 in A. De Morgan, "Science and Rank", *Dublin Review,* vol. XIII (1842), pp. 413–448. A vivid contrast between French practices and tradition and British

ones is obtained from a comparison between the British Association for the Advancement of Science and the *Congrès Scientifique de France* during the period from 1830 to 1850. Both had somewhat similar membership, both were peripatetic, and both represented an attempt to give the provinces some status. In France, however, the elite in Paris provided both the questions *and* the answers to the questions to be discussed at the Congrès!

92.   J.F.W. Herschel, *Preliminary Discourse on the Study of Natural Philosophy* (London: Longmans, 1830). A fascimile of the 1830 edition with a new introduction by Michael Partridge was issued by the Johnson Reprint Corporation in 1966. For an excellent exposition of Herschel's philosophy of science, see R.M. Blake, C.J. Ducasse, and E.H. Madden, *Theories of Scientific Method: The Renaissance through the Nineteenth Century*, ed. E.H. Madden (Seattle: University of Washington Press, 1960). See also L. Laudan, "Theories of Scientific Method from Plato to Mach: A Bibliographical Review," *History of Science*, vol. 7 (1968), pp. 1–63. W.F. Cannon's "John Herschel and the Idea of Science," *Journal of the History of Ideas*, vol. 4 (1964), pp. 56–88, remains the best introduction to this remarkable man.

93.   "During my last year at Cambridge, I read with care and profound interest Humboldt's 'Personal Narrative.' This work and Sir J. Herschel's 'Introduction to the Study of Natural Philosophy' stirred in me a burning zeal to add even the most humble contribution to the noble structure of Natural Science. (Charles Darwin, *Autobiography* in *Life and Letters of Charles Darwin*, vol. 1, p.55.) And, "When your work on the study of Nat. Phil. came out, I read it as all others did with delight. I took it as a school book for philosophers and I feel it has made me a better reasoner & even experimenter and has altogether heightened my character and made me if I may be permitted to say so a better philosopher". (M. Faraday to John Herschel, 10 November 1832, in *The Selected Correspondence of Michael Faraday*, vol. I [1812–1848], L. Pearce Williams [Cambridge: University Press, 1971], p. 235.) For other influences consult, for example, W. Swainson, *A Preliminary Discourse on the Study of Natural History* (London: Longmans, Green, 1834); and H.T. De La Beche, *How to Observe Geology* (London: C. Knight, 1835). By mid century Herschel's *Discourse* had also become a "classic" in the popular anthologies: see, for example, E.P. Hood, *The World of Anecdote* (London: Hodder and Stoughton, 1870).

94.   Herschel,[92]*Preliminary Discourse*, section 65.

95.   H. Spencer's *Social Statics or the Conditions Essential to Human Happiness* was first published in 1850.

96.   J.S. Mill's *On Liberty* first appeared in 1859; republished in *Utilitarianism, On Liberty, Representative Government* (Everyman's Library, London & Toronto: Dent, [n.d]).

97.   Throughout his life Herschel was involved in science policy, giving advice to the Admiralty, the Treasury, the Board of Longitudes and other governmental institutions. For his involvement on science policy by virtue of membership in the Royal Society see, for example, J. C. Weld, *History of the Royal Society*, vol. II, pp. 360–367, 392–396; and Sir Henry Lyons, *The Royal Society 1660–1940* (Cambridge: University Press, 1944), p. 244. John Cawood's, "The Magnetic Crusade: Science and Politics in Early Victorian Britain", *Isis*, vol. 70 (1979), pp. 493–518, details Herschel's involvement in these early efforts at "big science." Herschel's involvement in University reforms is outlined in A. I. Tillyard, *A History of University Reform* (Cambridge: University Press, 1913), pp. 100–132, and D. A. Winstanley, *Early Victorian Cambridge* (Cambridge: University Press, 1940).

98.   Herschel,[55] "An Address,", pp. 3–7.

99.   René Wellek, *Immanuel Kant in England: 1793–1838* (Princeton: Princeton University Press, 1931). It is interesting to note that Kant's "An Idea of a Universal History in a

Cosmopolitan View" was published in English on two occasions before 1800 (pp. 17–18). For a more recent overview of the diffusions of German philosophy in England, see Cannon,[9] *Science in Culture.*

100. Herder's *"Outlines of a Philosophy of the History of Man* as translated by T. Churchill" had two editions. Both the first edition in 1800 and the second in 1803 were printed by J.J. Hansard for L. Hansard of London. Herder's *Oriental Dialogues* selected from the German dialogues and dissertations of the *celebrated* Herder [italics mine] was published by Cadell and Davies, London in 1801. An English translation of Herder's *Treatise Upon the Orgin of Language* was not published until 1827 by Longmans. H. Tronchon has analyzed the diffusion of Herder's ideas in France and England in his "The Influence of Herder's Ideas through the Translation of Quinet and Churchill," Publication de la Faculté des Lettres de Strasbourg, ser. 2, vol. 15, (1937).

101. Hans Aarsleff, *The Study of Language in England, 1780–1860* (Princeton: Princeton University Press, 1967).

102. In a paper attempting to explain absorption lines in spectra, a difficult enterprise since these lines presented no salient regular relationships to one another, Herschel wrote "We seem to lose sight of the great law of continuity and to find ourselves involved among desultory and seemingly capricious relations, quite unlike any which occur in other branches of optical science." J. Herschel, "On the Absorption of Light by Coloured Media, Viewed in Connection with the Undulatory Theory," p. 402 in *Philosophical Magazine*, vol. 3 (1833), pp. 401–412. Enough comments have been made on the dictum *Natura non facit saltum* that I will not add to them here.

103. M. Ruse, "Darwin's debt to Philosophy: An Examination of the Influence of the Philosophical Ideas of John F.W. Herschel and William Whewell on the Development of Charles Darwin," *Studies of the History and Philosophy of Science*, vol. 6 (1975), pp. 159–181. See also M. Ruse: *The Darwinian Revolution: Science Red in Tooth and Claw* (Chicago: University of Chicago Press, 1979).

104. Herschel,[92] sec. 96, pp. 104–105.

105. Herschel,[92] *Preliminary Discourse*, sec. 105, p. 113ff.

106. Francis Bacon, *The New Organon* in *The New Organon and Related Writings*, ed. F.H. Anderson, (Indianapolis: Bobbs-Merrill Press, 1960), p. 29.

107. The choice of words is deliberate. See the articles by E. Fox Keller that stimulated these remarks: E. Fox Keller, "Gender and Science", *Psychoanalysis and Contemporary Thought*, vol. 1, no.3, pp. 409–433 and E. Fox Keller, "Baconian Science: A Hermaphroditic Birth,"*Philosophical Forum*, in press.

108. In his article on physical astronomy in the *Encyclopaedia Metropolitana* as well as in the *Treatise on Astronomy* and in his more popular essays, Herschel stressed the great accomplishments of Lagrange and Laplace in proving the stability of the solar system. See Herschel's Review in the *Quarterly Review* (1833) of Mary Somerville's *Mechanism of the Heavens* and Nathaniel Bowditch's translation of Laplace's *Mécanique Céleste*, which is reprinted in Herschel,[55] pp. 21–62.

109. The letter is reprinted in Walter F. Cannon, "The Impact of Uniformitarianism – Two Letters from John Herschel to Charles Lyell, 1836–1838," *Proceedings of the American Philosophical Society*, vol. 105 (1961), pp. 301–314.

110. C. Lyell, *The Principles of Geology*, 3 vols. 1st ed. (London: John Murray, 1830–33). For Lyell's religious views see, for example, M. Ruse *The Darwinian Revolution*,[103] pp. 80–83.

111. E. Shils, "Intellectuals, Tradition and the Traditions of Intellectuals: Some Preliminary Considerations," *Daedalus*, Spring 1972, pp. 21–34.

112. Note the tone of Lyell's description of the atmosphere of the exclusive Athenaeum Club, of which he was a proud member: "The *Athenaeum* was very entertaining last night, so many members of the honorable House coming there after the new Reform bill was moved, and giving their opinions *pro* and *con*." (*Life and Letters of C. Lyell*, vol. 1, p. 356.) Darwin was "full of admiration at the Athenaeum," of which he became a member upon his return to England in 1837. In 1838 Lyell wrote to him: "I am very glad to hear that you like the *Athenaeum*. I used to make one mistake when first I went there. Where anxious to push on with my book, after a 'two hours' spiel, I went there by way of a lounge, and instead of that, worked my head very hard, being exalted by meeting with clever people, who would often talk to me very much to my profit. . ."

113. S. Eisenstadt: *Intellectuals and Tradition*, Daedalus, Spring 1972, pp. 1–20.

114. J.T. Merz: *A History of European Scientific Thought in the Nineteenth Century*, 2 vols. (London: W. Blackwood and Sons, 1904, rpt. New York: Publications 1965), pp. 277–281.

The Romantic Tide Reaches Trinity:

Notes on the Transmission and Diffusion of New Approaches to Traditional Studies at Cambridge, 1820–1840

ROBERT O. PREYER
Department of English
Brandeis University
Waltham, Massachusetts 02154

> Therefore your Halls, your ancient Colleges,
> Your portals statued with old kings and queens,
> Your gardens, myriad-volumed libraries,
> Wax-lighted chapels, and rich carven screens,
> Your doctors, and your proctors, and your deans,
> Shall not avail you, when the Day-beam sports
> New-risen o'er awakened Albion. No!
> Nor yet your solemn organ-pipes that blow
> Melodious thunders through your vacant courts
> At noon and eve, because your manner sorts
> Not with this age wherefrom ye stand apart,
> Because the lips of little children preach
> Against you, you that do profess to teach
> And teach us nothing, feeding not the heart.
> — TENNYSON, *Lines on Cambridge of 1830*

I<small>T IS BEGINNING</small> to be clear that the intellectual center of early Victorian thought in a variety of fields of learning was located geographically in Cambridge and specifically among a band of friends and colleagues loosely or closely connected with Trinity College in the second and third decades of the nineteenth century. Some of these men were or became the great representative figures in "natural philosophy," as it was then called, and made major contributions to such developing sciences as geology, zoology and paleontology, mineralogy and optics,

39

meteorology, theory of tides, biology and botany. Most of them were considerable mathematicians and were versed in various branches of astronomy and physics, the models for developing new fields in the study of nature. There is a good deal of impressive recent scholarship, much of it in monograph form, that concerns itself with the work of these men — William Whewell, E.C. Clarke, John Herschel, John Henslow, George Peacock, Charles Babbage, George Airy, and Adam Sedgwick, to name some of the main figures.[1] Most of this work is technical "history of science" and is not as well known as it should be to scholars in other fields. There are, to be sure, several useful and important works on the relation of Victorian science to Victorian values — *Apes, Angels and Victorians* by William Irvine, *Genesis and Geology* by C.C. Gillispie, and some other books have introduced nonspecialists to the general area. But very few historians and humanists have the necessary scientific training to understand at first hand what these scientists accomplished. They are at the mercy of "experts," and experts are frequently "specialists." But the language of natural science in these decades is surely matched, if not overmatched, in difficulty of comprehension by the language of Romantic philosophy and philology employed in such complex and awkward works as Coleridge's *Biographia Literaria* (1817), *The Friend* (1818), the lectures on English and German literature of 1818–19, *The Idea of the Constitution of Church and State* (1816), and the *Aids to Reflection* (1825). Many of Coleridge's "sources" are to be found in German texts of notable difficulty: one thinks of his references to works by Kant, Schelling, Fichte, A.W. Schlegel, Tieck and Jean Paul, or of his allusions to technical philological studies by scholars like Wolf, Bopp, Heyne and Niebuhr, some of whom Coleridge either met or studied with while residing in Germany at the turn of the century. The study of this Romantic amalgam in the master himself is a major academic industry[2]; the study of specifically German Romanticism keeps taking on new leases of life as it is "rediscovered" as the impulse behind Marxist critics like the late Georg Lukacs, and, more recently, of the critic Walter Benjamin, who sought to unite in his work the deepest insights of Marx and Freud.[3] There is a continuing debate on the question whether "modernism" and even "postmodernism" are but exfoliations of this amalgam of Romantic thought, a working out of implications of works composed in the decades before and after the year 1800. To enter into these polemical and complex debates is surely daunting, and one can forgive most historians concerned with the ideas, values and institutions of the Victorian period for attempting to skirt this endless bog of speculation.

But one cannot ignore the fact that the humanist mind of Cambridge, for better or worse, was certainly altered by the marsh lights and vapors that ascended from the Romantic swamplands; and in what follows I will suggest also that these emanations affected the mentality of the distinguished practitioners of natural philosophy who were also at Cambridge when the fog rolled in. My argument has three parts. First, I will bring evidence to show that the classical tutor Julius Hare (1795–1855)[4] and his colleague Connop Thirlwall (1797–1875)[5] played a crucial role in opening up the Cambridge mind to the implications of Romanticism for a variety of studies, among them natural philosophy, history, theology, philosophy and the whole field of language and literature. Second, I will suggest that the existence of a set of institutional arrangements peculiar to Trinity College at the time aided and abetted them; and finally, that the existence of bonds of personal friendship among leading practitioners of the new science of that day — Whewell, Sedgwick and Peacock for example, and humanists like Hare and Thirlwall — had important consequences for the direction taken by scientific studies as well as the liberal arts curriculum at Cambridge.

We can begin with an account of the institutional arrangements that prevailed at Trinity College, relying on the information pieced together by Robert Robson.[6] The Master of Trinity was appointed by the Crown. Bishop William Mansel, a protégé of the Duke of Gloucester, and Spencer Perceval, both of whom had been his pupils at Trinity, was Master from 1798 to 1820, when he was succeeded by Christopher Wordsworth who, with his elder brother the poet, attended Hawkshead Academy before coming up to Trinity (1792) as an undergraduate. His election followed a pattern similar to Mansel's: his candidacy was put through largely by the efforts of a former pupil who had become Speaker of the House (and whose father was Archbishop of Canterbury), Charles Manners Sutton. The appointment was for life and Wordsworth held on to the job until a Tory ministry under Peel came to replace that of the Whig Lord Melbourne in 1841, at which juncture William Whewell began his lengthy tenure (1841–66). The Fellowships, however, were not influenced by national political considerations. Robson relates how two large gifts to Trinity in 1810 and 1811 were employed to augment the income of a number of the College livings, thus leading to a rapid and easy circulation out of their Cambridge quarters of numerous Fellows who might otherwise have stayed on indefinitely occupying positions which were still regarded as "freeholds existing for the benefit of their owners rather than as the means whereby necessary functions were performed for society."[7] These 60 Fellowships, tenable for life, now began to rotate rapidly, and empty places were filled by men who had passed a stiff written examination rather than the old fashioned *viva voce* over sherry in the snug quarters of some Senior Fellow. Although a Fellowship was still not looked upon as the first step in an academic career — there really wasn't such a thing — a remarkable body of Fellows were in and out of Trinity in the 1820s and 1830s, many of them destined to be prominent in the chief learned societies of the era — The Royal Society, the Geological Society, the British Association for the Advancement of Science, The Philological Society of London, and so on. Many of the papers presented in London were worked out during Fellowship years. The rigid barriers that separated Fellows from undergraduates were beginning to break down and the introduction in 1790 of examinations not only for first-and second-year men, but also, in 1818, for third year men, had "wonderful effects . . . in exciting industry and emulation among the young men, and exalting the character of the college," according to J.H. Monk, one of the tutors responsible for this reform.[8] All this tended to stimulate a flow of really bright undergraduates to Trinity, and from these men future Fellows of Trinity were selected. Scientists like John Herschel (Senior Wrangler and First Smith's Prizeman, 1813) and Charles Babbage, his intimate friend, wandered off to London or Slough in pursuit of their avocations. Julius Hare, who went down from Cambridge as M.A. and Fellow of Trinity in 1818, spent a few years in London before Christopher Wordsworth appointed him an assistant tutor on the classical side in 1822. But other Fellows, like their classmate George Peacock (second Wrangler after Herschel) never left the College at all. William Whewell, second Wrangler in 1816, and George Airy, Senior Wrangler in 1823, followed Peacock's path and took up resident Fellowships upon matriculation. By 1830 all these men held appointments as tutors. Hare was soon joined on the classical side by another classmate whom he had known since their school days together at Charterhouse, the erstwhile child prodigy Connop Thirlwall, later to be known as a bishop whose Episcopal charges were referred to as "the chief oracles of the English Church" by Dean Stanley,[9] and whose *History of Greece* remains a monument of Victorian scholarship.

Upon graduating, Hare and Thirlwall had both gone to London to prepare for careers at the bar; Whewell, already Head Tutor, persuaded Hare to return to Trinity as assistant tutor and Hare in turn brought back Thirlwall a few years later.[10] Together with their slightly older friend, Professor Adam Sedgwick, these "young Turks" formed a determined group who proceeded to put their special mark on the younger generations not only at Cambridge as tutors and professors but elsewhere in England through a bewildering variety of publications—textbooks, introductory works in mechanics, tracts on optics, treatises on astronomy, and numerous translations from the work of French physicists and German historians, philologists, theologians, and poets. They were adept at gaining access to the great quarterlies of the day and even managed to publish frequently in a rash of miscellaneous journals controlled by their former students or acquaintances outside the University.[11] This band of friends was blissfully unaware that it contained the chief early Victorian representatives of C.P. Snow's "two cultures"—indeed they seem to have been quite fascinated with one another's activities and often indulged themselves in similar avocations. Whewell wrote hexameters, translated German lyric poetry, edited Grotius, translated the dialogues of Plato, and even published a small book on Gothic churches of Germany. He was also able to impress an undergraduate who devoted most of his time to sport, Charles Darwin, and the future Lord Kelvin.[12] Another pupil, James Clerk Maxwell (1831–79), was almost as fascinated by the thinking of Hare's pupil F.D. Maurice (1805–72) as he was by physics, and joined the undergraduate Apostles club founded by Maurice and John Sterling as a forum in which to discuss the implications of what they had been learning from their classical tutors Hare and Thirlwall. (Alfred Tennyson, on the other hand, developed his scientific interests largely through exposure to his most respected tutor, Whewell)

At one time or another, each of this band of brothers seems to have acted as teacher to the others: Julius Hare had been brought up on the continent and spoke and wrote German with the same fluency as his native tongue. While an undergraduate he tutored Whewell, Thirlwall and Hugh James Rose[13] in German. (Rose soon joined them as tutor and professor of Hebrew.) Interestingly enough, the textbook used was Niebuhr's recent *Römanische Geschichte*, which provided a revolutionary new way of studying the literary remains of antiquity, and which was a book about which their Cambridge instructors had no knowledge at all. These were the figures who taught a generation of students that included literary men like Macaulay, Thackeray, Tennyson, Bulwer-Lytton, W. M. Praed, John Moultrie, and Derwent Coleridge (son of the poet), philologists like John Kemble and Benjamin Thorpe, the first significant Anglo-Saxonists, and a host of clergymen subsequently identified as leaders of the "Broad Church" party in the Anglican communion (as opposed to Low or High Church parties), F.D. Maurice, Richard

Chenevix Trench, John Sterling, and many others who went on to careers in government and the services. All these men admired and were impressed by the formidable and stiff Head Tutor "Billy Whistle" (Whewell); they left affectionate accounts of "Gentleman" Peacock,[14] and especially of the amiable and eccentric Julius Hare, who seems to have been adored by undergraduates as unlike as Monckton Milnes, James Spedding, FitzGerald, Kinglake, Buller and Hallam, to mention a few. But above all, as two early novels by Sterling and Maurice[15] attest (as do many other sources), these were the teachers who persuaded the undergraduates that they need not choose between a heartless High Tory "reactionary" point of view and the "radical" view of the Benthamites or philosophical radicals. (The aggressive pietism of Charles Simeon and his wealthy supporters outside the university was held in contempt by Trinity men: it was not intellectually serious.)

These two positions appeared to be the only alternatives in the years immediately following the Napoleonic wars and even such a traditional Tory as Macaulay was very nearly swept into the Benthamite ranks by the arguments of his fellow student Charles Austin. Early in their Cambridge days both John Sterling and John Kemble adopted the radical position and were prepared to consider a plunge into the youthful atheism of Shelley (who had been expelled from Oxford in 1811). The *via media* between these two unfortunate alternatives was provided largely by Hare and Thirlwall and comes down to us as variously, "Liberal Anglicanism" or "Broad Church theology" — about which more in a moment. I think it is worthy of note that Whewell, Peacock, Sedgwick, and Herschel were all strong opponents of Benthamism and of certain forms of High Tory bigotry; they were happy to support — sometimes in a puzzled fashion — any group of thinkers who could put together a more intelligible alternative. The fact that one of the presumed "discoverers" of such an alternative was brother to the eminently respectable Tory Master of Trinity made it more palatable to some who felt considerable uneasiness about his problematic and possibly disreputable accomplice, Samuel Taylor Coleridge (1771–1834).[16] It helped, of course, that William Wordsworth and Coleridge were both Cambridge men (Trinity and Jesus) and that they had apparently gone through their radical phases at the time of the French Revolution and strongly embraced the Anglican faith from the other side of doubt: perhaps that explained their curious appeal to undergraduates who visited them and were visited in turn by the venerable poets. Hare and Thirlwall used every bit of their prestige and position to reassure doubters, and a specially moving example of how they went about this occurs in the dedication of Hare's major theological work, *The Mission of the Comforter* (1852)[17]:

> To the honoured memory of Samuel Taylor Coleridge . . . who, through dark and winding paths of Speculation, was led to the light, In order that others, by his guidance, might reach that light without passing through the darkness.

Richard Chenevix Trench (1807–86), afterwards Dean of Westminster, Archbishop of Dublin and a great popularizer of the scientific study of language, described his fellow Apostles of 1830 as "that gallant band of Platonico-Wordsworthian-Coleridgeian-anti-Utilitarians,"[18] which nicely conveys the flavor of these Cambridge years. F. D. Maurice, the modest instigator of the Apostles Society, referred to it as "a small society" which "defended Coleridge's metaphysics and Wordsworth's poetry against the utilitarian teaching."[19] Obviously, such a society had the approval of Professor Adam Sedgwick, who thundered against the Benthamites in a commemoration sermon in College Chapel in December 1832: "Utilitarian philosophy, in destroying the domination of the moral feelings, offends both against the law of honour and the law of God. It rises not for an instant above the world; allows not the expansion of a single lofty sentiment; and its natural tendency is to harden the hearts and debase the moral practice of mankind."[20] Frequent attacks on Bentham can be found in Whewell's works and, High Tory though he remained in many matters, it must be remembered that he had been President of the Cambridge Union (Thirlwall was secretary) and occupied the chair when its proceedings were dissolved by order of the high-handed reactionary university authorities of the day. He could understand the angry contempt among undergraduates for such repressive tactics and the consequent vogue, in the 1820s, of philosophical radicalism among gifted undergraduates.[21] As we shall see, Whewell had another reason for supporting the Idealist reaction against Benthamite radicalism: he was one of the very few Englishmen who had read and understood the writings of Kant, Fichte, Schelling and Hegel — in German, a legacy of his undergraduate instruction with his classmate Hare.

Another Apostle, Charles Merivale (1808–93), later to win prominence as the author of two important works on Roman history, has left us with perhaps the most interesting account available of the attitudes and influences operative among undergraduates of the Tennyson years. "We began to think that we had a mission to enlighten the world upon things intellectual and spiritual. We held established principles, especially in poetry and metaphysics, and set up certain idols for our worship. Coleridge and Wordsworth were our principal divinities, and Hare and Thirlwall were regarded as their prophets; or rather in this celestial hierarchy I should have put Shakespere on top of all, and I should have found a lofty pedestal for Kant and Goethe. It was with a vague idea that it should be our function to interpret the oracles of transcendental wisdom to the world of Philistines, or Stumpfs . . . that we picqued ourselves on the name of the 'Apostles.' "[22]

It was a heady combination, to be sure, and before going on to describe the main ingredients in the mixture, I want to call attention to the astonishing effect that it had on young John Stuart Mill when he met it first through the agency of Connop Thirlwall, who in 1825 was still in London and not yet decided on a return to Trinity to take orders and confirm his Fellowship. Mill had been at a loss to locate conservative speakers worth debating with — he had despaired of finding anyone until there appeared "a Chancery barrister, unknown except by a high reputation for eloquence acquired at the Cambridge Union before the era of Austin and Macaulay. His speech was in answer to one of mine. Before he had uttered ten sentences, I set him down as the best speaker I had ever heard, and I have never since heard anyone whom I placed above him." This was Mill's introduction to the Cambridge "Germano-Coleridgians" (his epithet), who in 1828 and 1829 "in the persons of Maurice and Sterling, made their appearance in the [London Debating] Society as a second Liberal and even Radical party, on totally different grounds from Benthamism and vehemently opposed to it; bringing into these discussions the general doctrines and modes of thought of the European reaction against the philosophy of the eighteenth century; and adding a third and very important belligerent party to our contests. . . . "[23]

[II]

Julius Hare has left us many accounts of the debt he felt he owed to Coleridge, and of the debt owed to Niebuhr.[24] Let us attempt in a summary fashion to see what the debts were and how they became interrelated in his mind and in that of the group of thinkers who were attentive to his thought. Specifically, let us ask what the connection is between the Christian philosopher, Coleridge, and Niebuhr, the German classical philologist whose "critical method" appeared to be an eighteenth-century exercise in debunking the received accounts of the origin and early years of the Roman people and their institutions. Niebuhr was frequently attacked as irreligious, and not just impious, by other clergymen of the day; his "critical method" of analyzing the written accounts that survived from the past could easily be turned into a "deconstruction" of even sacred narratives; in short, he appeared to be a standard representative figure of Carlyle's "age of unbelief." A clue is to be found in Niebuhr's curious confidence in his powers of divination, of *intuiting a total pattern in the light of which the individual items and data, the "evidences" from the past, could be assembled and provided with significance.* In 1816 Niebuhr wrote to his intimate friend Madame Hensler,

> I am as certain of the correctness of my views as I am of my own existence. . . . It is not the love of conjecture that has impelled me, but the necessity of understanding, and the faculty of guessing and divining. . . . Further it is not to be expected that everyone, or even that many, should have that faculty of immediate intuition which would enable them to partake in my immoveable [sic] conviction."[25]

Here "intuition," "guessing," and "divination" are part of a new, imaginative scientific procedure for philological investigations. Niebuhr sets forth what Whewell might call a "philosophy of discovery;" a romantic epistemology in which an "idea" or "guess" proceeding from the subjective consciousness penetrates the surface phenomena (in this case Livy's *Annals*) and is able to unlock an inner structure of meanings whose presence had not previously been suspected. Niebuhr's seemingly impossible task was to investigate the prehistory of Rome by a new scrutiny of the fanciful and unreliable sources that Gibbon and Voltaire had cast aside as useless. His "guess" was that Herder and Wolf had been correct in their hypothesis that the earliest literatures in any society would be oral folk-ballads: these "lost lays" were presumed to have provided the "matter" of Norse saga and Homeric epic. (Macaulay, another Trinity Fellow, was so struck with this notion that he hastened to provide *his* version of the *Lays of Ancient Rome* in 1842.) Niebuhr's problem was to disentangle this posited ballad literature from all the later accretions that had grown up around it. His problem was, in this respect, much like that faced by the Higher Critics, who were trying to devise stylistic, historical, and even geographical and imagistic tests that would aid them in sorting out various strata of authorships in both Old and New Testament writings. One procedure he employed was taken over from the new field of comparative mythology. If Livy or some other chronicler reported a story that had analogues in the folklore of other nations, Niebuhr concluded that the material was mythic; if no analogues existed, the anecdote could be taken as local in origin and therefore related, however distantly, to an historical reality. It would be a folk-creation of the Roman people. Thus, "The preservation of Romulus and Remus is a fable, and may pass from the heroic poetry of one people into that of another, or may arise in many places, as it was told of Cyrus in the East, of Hebis in the West; but the rape of the Sabines relates to traditions of another kind." It had a local reference and Niebuhr went on to "divine" that Rome was originally two cities inhabited by the Sabine and Roman peoples. The story of the rape reflected the historical fact that "no right of intermarriage yet subsisted between two cities." (We are further told that when a union between the two cities was effected, its symbol was the double Janus. The word Janus originally meant "door"; the double Janus was a reference to the gate that joined together the two settlements.[26]) In this fashion Niebuhr extrapolated out "internal evidence" from "useless" old chronicles and was able to recover many lost "hypothetical facts" concerning early social and institutional arrangements. By the use of *comparative studies* of the early literatures and institutions of other peoples — a study rapidly advancing with the advent of Germanic philology, the revived interest in Icelandic and Norse sagas and in the surviving fragments in Anglo-Saxon, Old German, and Old French, not to mention Sanskrit, Aramaic and Basque — new data were pouring in from all parts of the globe. Gaps in one history might be filled by analogy to a more documented record of that stage of development that had survived from another land or from a different language.

Niebuhr's first two volumes of the projected three-volume history appeared in Germany in 1816 and it was this text that was read by Thomas Arnold at Oxford and Hugh James Rose, William Whewell, and Connop Thirlwall at Cambridge. Augustus William Hare, then a Fellow and Tutor at New College, Oxford, and the close friend of Arnold of Oriel had begun the "cult of Niebuhr" at the sister university. In 1825 the two Hare brothers decided on a joint undertaking, the publication of a collection of fragments, notes, aphorisms, and essays based on their reading and tutoring activities. Their publisher, John Taylor, had printed a similar collection, Coleridge's *Aid to Reflection* in the late Spring of 1825, and Augustus suggested that several proofsheets might go to "old Coleridge," remarking to Taylor that "their fragmentary character would not displease him; if they were sent to him interleaved with blank pages, he might be tempted occasionally to add a note, a limitation, a deduction or a guess which would double the value of the book to thinkers. . . ."[27]

This work, *Guesses At Truth By Two Brothers*, was delayed while Julius arranged for publication of Landor's *Imaginary Conversations* and busied himself with the translation of the *Römanische Geschichte*. *Guesses* came out in two volumes in 1827 and had a striking success. After the early death of Augustus in 1834 it was revised and expanded in 1838 and again in 1847–48, going through a number of printings and editions well after Julius' death in 1855. (*The History of Rome,* Volume 1, translated by Julius Charles Hare and Connop Thirlwall, appeared in 1828 and again in 1829; the second volume was published in 1832 and Hare's *A Vindication of Niebuhr's History of Rome from the Charges of the Quarterly Review*, in 1829.[28]) *Guesses At Truth* contains repeated references to Coleridge and provides a contemporaneous running account of how his meaning came to be related to that of the many German writers quoted or explicated in *Guesses*. Hare refers frequently to correspondence and conversations (sometimes quoted from memory) with Coleridge, Wordsworth, Landor, DeQuincey, Schleiermacher, Niebuhr, and other German philologists and Biblical scholars; he also writes interestingly on Kant ("On the Sublime"), Wolf, Goethe, Schlegel, Schiller, Herder, Tieck, Savigny, and Hegel (as well as deMaistre and Bonald). A typical notation is the following acute distinction between the Neoclassical notion of Ideality and that of the Romantics:

> The common notion of the Ideal, as exemplified more especially in the Painting of the last century, degrades it into a mere abstraction. It is assumed that to raise an object into an ideal, you must get rid of everything individual about it. Whereas the true ideal is the individual, purified and potentiated, the individual freed from everything that is not individual in it, with all its parts pervaded and animated and harmonized by the spirit of life which flows from the center.[29]

There follows a criticism of John Locke's *Essay on Human Understanding* for maintaining that all ideas are abstractions based on sense impressions—precisely the doctrine Romanticism sets out to refute and which Kant was thought to have definitely overthrown.

This passage finds a curious echo in William Whewell's *On the Philosophy of Discovery*,[30] where we read that Locke's *successors* were the villians who rejected ideas and settled for sensations as the sole content of consciousness. Kant, wrote Whewell, "exposed the untenable nature" of this, and Fichte, overreacting, would have one believe that *all* knowledge is ideal rather than sensational, a view to which Whewell rejoins, in true schoolmaster fashion,

> But when the ideal element of our knowledge was thus exclusively dwelt upon it was soon seen that this ideal system no more gave a complete ex-

planation of the real nature of knowledge than the old sensationalist doctrine had done. Both elements, Ideas and Sensations, must be taken into account.[30]

Whewell went on to indicate that Schelling and "then his follower Hegel" concocted a specious argument to show how the race advances from perception to the idea, from fact to theory. But his entire enterprise of writing the *History of the Inductive Sciences From the Earliest to the Present Time*[31] invalidated such a claim: in historical fact the developments in the various sciences did not unfold in that neat ahistorical dialectical progression. At any given moment, "all Truths include an Idea and a Fact. The Idea is derived from the mind within, the Fact from the world without."[32] This proposition is illustrated elsewhere by reference to a set of "fundamental Rights of Men" that are universal and "flow naturally from the Moral Nature of Man" — these natural rights are "Ideas" (or conceptions) "derived from the mind within." The "Facts" (that is, interpretations or definitions) of these natural rights, since they are the product of the varied circumstances and histories of particular societies, will vary considerably. "The Fact [the implementation of the idea of Natural Rights] is supplied by the Law of the Society in which we live and the trains of events which have made that law what it is. The Moral Nature of Man is moulded into shape by the History of each Nation; and thus, though we have, in different places, different Laws, we have everywhere the same Morality."[33] Whewell, it can be seen, is attempting to incorporate the new Romantic conceptions of the mode of existence of ideas in the old vessels of Neoclassicism (including natural theology, toward which he did not object quite so strongly as did Julius Hare, to whom he dedicated a volume of his sermons). As one might expect, Whewell attacked Comte on the grounds that he was "unhistorical" and that "his pretensions to discoveries are, as Sir John Herschel has shown, absurdly fallacious," whereas the doctrine of the Saint-Simonians of alternating *critical* and *organic* periods of development can be given more credence by historians than can the "unhistorical" Comtean dialectic of change as an invariable and universal progress through three stages, a notion that is easily exploded by reference to historical data since ideas are consonant with actualities.[34]

Whewell's "Fundamental Ideas" are what the *activity of mind* contributes to knowing; and it is of course this *activity of mind* (we "half-create what we half-perceive," noted Coleridge) that is the mark of the Romantic revolution in sensibility. Indeed Coleridge and his followers defined imagination as *the intuitive prehension of a unified theme in the light of which isolated actions and persons could be understood* — and this is a plausible definition of a scientific theory.

Hare quotes the following passage from Wordsworth as an example of how a Romantic *idea* or "divination of the whole" alters the quality of consciousness of an imaginative thinker: "When the mind is fully possest with the idea of a work, it will carry out that idea in all its details, preserving a unity of tone and character throughout. . . . "[36] It is contended that the product of such a full or imaginative consciousness has a *different kind of significance and a different kind of structure* than what we obtain by the normal mode of ordering and arranging a set of particulars into a "prosaic" statement. Prosaic statements are the product of the understanding, and of course have their place in the economy of reason — they provide an orderly storehouse for what we already know and a rational and rhetorical method of establishing relations between a variety of previously identified phenomena.[37] What Wordsworth and Coleridge insisted upon was the *logical priority* of an imaginative or "full" mode of perception in the *process of bringing something hitherto unknown into the realm of consciousness*, at which point it could be discussed and reasoned about, having now an objective existence, a form or shape, however "darkly visible." There are many who can then undertake the further processes of understanding and placing, but the men of imagination are rare creatures, frequently isolated from the concerns of "practical" men by an intuitive recognition that they seek to explore more profound questions than others are aware of. Such "explorers" tend to become aware of one another, whatever their calling. And nowhere in literature is there a more intimate and compelling recognition scene than that which took place in nocturnal intimacy between Wordsworth and his fellow voyager in conceptual space, Sir Isaac Newton:

> And from my pillow, looking forth by light
> Of moon or favoring stars, I could behold
> The antechapel where the statue stood
> Of Newton with his prism and silent face,
> The Marble index of a mind for ever
> Voyaging through strange seas of Thought, alone.
>
> Prelude, III, ll. 58–63 (1850 edition)

The more scornful theologians and philosophers of Oxford felt that the results of natural science were trivial and did not require this sort of imaginative reason among its practitioners — that natural science was relatively unimportant ("With matter it began, with matter it will end; it will never trespass into the province of mind," wrote Newman in *The Idea of a University*). This was not a belief shared by Trinity College, Cambridge, where Newton could be depicted as another Ancient Mariner whose questing mind resembled that of Coleridge.

It is true, of course, that Wordsworth continued to hold strong prejudices aganist "the wandering Herbalist" and the chap "who with pocket-hammer smites the edge / Of luckless rock." Walter Cannon notes with some amusement that, so far as Wordsworth was concerned, botany and mineralogy were nothing more than inert "sciences of classification," which "gave worthless, superficial knowledge; whereas true sciences should show wider and deeper interrelationships, should lead to an understanding of the system of the world and therefore should lead eventually to God."[38] This anti-eighteenth-century prejudice aside, it is evident that Romantic poets fully accepted astronomy, mathematics, and geology as capable of arousing (and demanding) an imaginative response from practitioners and onlookers as well. They belonged in the same company as those who put forth the new philology, with its insistence on the exercise of a full imaginative consciousness in the investigation of the life of the past. The new idea in philology was to bring to life the consciousness of unknown generations who had employed a particular language within which was imprinted a storehouse of clues concerning human experiences, political and social arrangements, and communal attributes. Wilhelm von Humboldt bluntly defined philology as "the knowledge of human nature as exhibited in antiquity." Niebuhr, more precisely, described philology as "the *introduction* to all other studies [preserving] an unbroken identity through thousands of years with the noblest and greatest nations of the ancient world, with the work of their minds, and the cause of their destinies, as if there were no gulph dividing us from them."[39] A vivid and living reconstruction of the Idea — the total *Gestalt* — of a nation was at last possible, for as Vico was the first to have insisted, languages survive all other testimonies and speak with more authority than the pyramids of the real life of the past. In the study of language it seemed possible to establish a meeting place between the subjective energies of consciousness and the still living consciousness of the *Volk* embedded in the deep structures of a language. Hare and Coleridge clearly believed that attributes of consciousness, whether subjective or externalized in language, were part of a single Reality, which they named God or Spirit. It would be hard to find a natural philosopher of their era who would not give fervent assent to that proposition.

Metaphysical visions such as these energized the fields of history and language study and, I suggest, spilled over into a variety of scientific studies. The "new" philology and historiography are related in explicit and implicit ways to the new specializations and subfields that were developing in geology, fossil paleontology, physics, biology, and astronomy, to mention a few fields where renewed activity was visible. These Trinity tutors had been educated along Neoclassical lines. This signifies, broadly speaking, that they shared common assumptions such as the belief that a keen investigator, probing beneath surface appearances, would sooner or later come upon evidence of an underlying rational structure of the sort Newton posited. The difference that Romanticism (as a specific historical movement) stood for—one that bound men together in generational bonds whatever their "discipline"—was a perception that superficial appearances, wherever studied, concealed quite a different kind of order than had previously been suspected. Behind the veil lurked (for Keats) the totally mysterious face of Moneta, impenetrable depths of mystery, explicable, if at all, by a mind and sensibility of a different order than that which could casually congratulate the God who said, "let Newton be, and all was light." As Baudelaire maddeningly—and accurately—noted, "Romanticism is precisely situated neither in choice of subject nor in exact truth, but in a way of feeling." Hugh Honour quotes these words in his new work, *Romanticism* (New York, 1979), adding that this fundamental change in attitude toward life was a response to the disruptions occasioned by the French Revolution and the subsequent diffusion of Kant's philosophy, which was "perhaps the most important intellectual event since the Protestant Reformation"(p.11). It is this conjuction of influences that arrived at Trinity and was being articulated by the Hares and Whewells who were undergraduates together in the first decade of the century. We can recognize, once again, how Romanticism juxtaposes its "higher" or "deeper" awareness with the preexisting mode of consciousness.

[III]

Hare followed Coleridge in asking how and in what sense poetic meanings differed from prosaic ones. By way of answer, both men pointed to the greater complexity and concentration of poetic (imaginative) apprehensions, which resulted from a *simultaneous experience* of multiple layers of structural ordering—an order of sound, of figure, of syntax, of narration, of logic. Something like this simultaneous juxtaposition of several layers of consciousness is present in the consciousness of any person who makes a "breakthrough" discovery.

Coleridge was the first English literary critic to enunciate *qualitative* and *quantitative* norms for estimating the completeness and value of written utterances so structured.[40] For our present purposes, however, it is only necessary to insist on his view that whatever knowledge mankind may be said to possess of Ideas and Forms is accounted for by the Imagination, which seeks, in its apprehension of the world, to pass beyond the simpler structures, which satisfy the Understanding, and struggles, as best it can, to body forth in symbolic forms what it dimly apprehends or intuits as the essential life or Idea informing objects, persons, and events. In a famous definition, Coleridge spoke of life itself "as the power which discloses itself from within as a principle of *unity* in the many. . . . a power that unites a given all into a whole that is presupposed in all its parts."[41] An example would be that sense of our self as an identity that persists from earliest childhood and maintains itself through all the changes, developments, and vicissitudes of adolescence, manhood, maturity, and old age. What persists is the Idea of self, our "character," despite variations in "personality" (the varying images of self we choose to exhibit to others). The Idea, then, is "that conception of a thing which is not abstracted from a particular state, form, or mode, in which the thing may happen to exist at this or that time; nor yet generalized from any number or succession of such forms or modes; but which is given by the knowledge of its ultimate aim."[42] "Ideas" so defined resemble very closely that total constellation of significances which the *Oxford English Dictionary* attempted to provide for each English word, and of course this heroic work of scholarship was an offspring of the philology taught to Trinity men by Hare and Thirlwall. Coleridge and Hare insist that Ideas are not a German metaphysical concept but rather belong to the structure of human minds, participate in and explain the actions and thinking we engage in. As an example of the mode of existence of the "Idea of moral freedom," Coleridge offers this homely illustration:

> Speak to a young Liberal, fresh from Edinburgh or Hackney or the hospitals, of free-will as implied in free agency, he will perhaps confess with a smile that he is a necessitarian, — proceed to assure his hearers that the liberty of the will is an impossible conception, a contradiction in terms . . . Converse on the same subject with a plain, simpleminded, yet reflecting, neighbour, and he may say . . . "I know it well enough if you do not ask me." But alike with both supposed parties . . . if we attend to their actions,

their feelings, and even to their words, we shall be in ill luck, if ten minutes
pass without having full and satisfactory proof that the idea of men's moral
freedom possesses, and modifies their whole practical being, in all they say,
in all they feel, in all they do and are done to . . . We speak, and have a
right to speak, of the *idea itself as actually existing in the only way a principle can exist, — in the minds and consciousness of the persons.* . . . "[43] [my
italics]

Ideas are constitutive elements of individual minds and govern instinctive behavior. It is in this sense that they are "the most real of all
realities"; "it is the privilege of the few to possess an idea: of the generali-
ty of men, it might be more truly affirmed that they are possessed by
it."[44] It followed then that "true" history, since it records the thoughts and
actions of individuals in the past, is governed not so much by "abstract
laws and psychological determinism," but rather by a consciousness of
laws, that is, by Ideas.

Many of the passages referred to earlier are discussed, paraphrased,
made the subject of short essays ("the language of exact science and the
language of poetry" being one such) or reduced to gnomic or aphoristic
form in *Guesses At Truth.* It is a fascinating compendium of the shards of
thought of the early nineteenth century and a main source for the intellectual history I have been attempting to trace. Like Coleridge, Hare
was "a master of the fragment, the isolated thought, the marginalium,
and though it has been argued that one must be a peculiarly systematic
thinker in order to write a really fine fragment, still it is true that system
in the obvious sense is the product of the Understanding and that Reason
clothes itself most naturally in the aphorism, the lyrical paragraph, the
aperçu. This is partly because it is only by moments that one can glimpse
transcendental truth and partly because the need is not for the dead level
of comprehensiveness but for the depth, the intensity, the *O altitudo* of a
living faith."[45]

Guesses At Truth is dedicated to Coleridge and Wordsworth and is
preceded by two mottoes. The first reads in its entirety "The best divine is
he who well divines"; the second is a quotation from Bacon: "As young
men, when they knit and shape perfectly, do seldom grow to a further
stature, so knowledge, while it is in aphorisms and observations, it is in
growth; but when it once is comprehended in exact methods, it may per-
chance be further polished and illustrated, and accommodated for use
and practise; but it increaseth no more in bulk and substance."

Hare's method of juxtaposition and analogy and cross-reference from
one level of activity to another could be illuminating as well as ob-

fuscating. He conveys the excitement of one who knows himself to be part of a worldwide movement, the borders and edges of which might have to be located in a variety of disciplines spread across the *speculum mentis*. Dimly visible in one place, sharply clear in another, a vast overall pattern seemed to be taking shape. *Guesses At Truth By Two Brothers* was a work in progress.

[IV]

I want to close with a few more indications of how these speculations affected the scientists of the day. We have already spoken of the relation of Ideas to knowledge in the thought of Niebuhr, Coleridge, Hare, and Whewell, and little more need be added on that subject. The great figures of Victorian science, men like Faraday, John Herschel, W.R. Hamilton, and James Clerk Maxwell, refused to accept "common sense" notions of matter and continued to entertain Romantic perceptions of a pattern "far more deeply interfused," whose design and significance could not be ascertained by the mere application of some mechanical system. They had intimate and imaginative knowledge of the language of mathematics, a marvel and a continuing mystery; "doing science" did not imply a contempt for epistemology or metaphysics. They also had before them the mystery of the spoken languages now being investigated by comparative philologists as a sort of treasure trove containing evidence of all that mankind had known and experienced in response to a variety of circumstances and events, many of which were "extinct" and *unavailable as experience to the modern consciousness*, but extant as a sort of "genetic bank" of human possibility to be drawn upon by great scholars and poets, geniuses with the power to bring these "lost" modes of consciousness into imaginative or hypothetical forms of existence and so make them available to the operations of intellect. Hare's formulation of this imaginative possibility of human consciousness can scarcely be improved on:

> Now a language will often be wiser, not merely than the vulgar, but even than the wisest of those who speak it. Being like amber in its efficacy to circulate the electric spirit of truth, it is also like amber in embalming and perserving the relics of ancient wisdom, although one is not seldom puzzled to decipher its contents. Sometimes it locks up truths, which were once well known, but which in the course of ages have passed out of sight and been forgotten. In other cases it holds the germs of truths, of which, though they

were never plainly discerned, the genius of the framers caught a glimpse in a happy moment of divination. . . . often it would seem as though rays of truth, which were still below the intellectual horizon, had dawned upon the Imagination as it was looking up to heaven.[46]

Hare's uncle, it is worth recalling, was that Sir William Jones who in 1786 was the first to remark that there was an apparent connection between Sanskrit, Latin, and Greek — "they point to the same source." This oft-quoted observation is usually taken as the starting point for the modern study of historical and comparative linguistics. In a similar moment of "divination" five years later, William Smith, a practical land surveyor, linked together his intimate familiarity with rock strata exposed in cuttings and the presence in them of animal remains, thus providing the starting point for a developmental science of fossil paleontology. By 1800, Cuvier had borrowed the old Linnaean system from the science of botany in order to classify the orders of fossils that Smith had begun to arrange along a sequential geological time scale.

Explanations seemed to start up everywhere if one took an historical and comparative approach similar to that of students of language. In 1803, Friedrich Schlegel had discovered Sanskrit manuscripts in a library in Paris and five years later published his momentous *Über die Sprache und Weisheit der Indien*. Rasmas Rask, the great Danish philologist, managed to produce the first usable grammars of Old English (1817) and Old Norse (1811); Franz Bopp in 1816 began his epochal work in the comparative linguistic study of Persian, Sanskrit, Greek and German. In 1821, Champollion solved the riddle of the Rosetta Stone by way of Coptic, thus opening up hieroglyphic and "demotic" Egyptian for future study. From 1819 and continuing through 1837, editions of Jacob Grimm's *Deutsche Grammatik* appeared, laying the basis for the comparative study of the totality of Germanic languages. Shortly before Darwin's *Origins*, a modest German school teacher, Johann Zeuss, laid the foundations of Celtic linguistics by the discovery of interlinear commentaries on ancient Latin bibles — commentaries made by Irish priests, although the books in question were found in Würzburg, St. Gall, and Milan. [47] It would seem that the greatest discoveries of the nineteenth century had to do with what Whewell called *The Philosophy of Discovery*; and that "philosophy," "expressed in a compact manner, and detached from the reasonings on which they rest," is found in the "Aphorisms Concerning Ideas" and "Aphorisms Concerning Science," of which the following are a sample:

I

Man is the interpreter of Nature, Science the right interpretation.

II

The *Senses* place before us the *Characters* of the Book of Nature; but these convey no knowledge to us, till we have discovered the Alphabet by which they are to be read.

III

The *Alphabet*, by means of which we interpret Phenomena, consists of the Ideas existing in our own minds; for these give to the phenomena that coherence and significance which is not the object of sense.

VIII

The Conceptions by which Facts are bound together, are suggested by the sagacity of discoverers. This sagacity cannot be taught. It commonly succeeds by guessing; and this success seems to consist in framing several *tentative hypotheses* and selecting the right one. But a supply of appropriate hypotheses cannot be constructed by rule, nor without inventive talent.

XI

Hypotheses may be useful, though involving much that is superfluous, and even erroneous; for they may supply the true bond of connexion of the facts; and the superfluity and errour may afterward be pared away.

XII

It is a test of true theories not only to account for, but to predict phenomena. [48]

Whewell dedicated works to Julius Hare, William Wordsworth, and other friends of Romantic persuasion. He would not have objected to the Platonism of Coleridge since he was himself a close student and translator of that philosopher; in addition, he was an announced Kantian, a careful student of the *Critique of Pure Reason*, which he read in its original language. His "Fundamental Ideas" are what the activity of mind contributes to knowing and he likens many of them to Kant's forms of intuition, which make meaningful perception possible. But he does not believe that all such Ideas are simply stored up in the mind *a priori*: "we cannot term them *innate* ideas," he writes. "It is not the *first*, but the most complete and developed condition of our conceptions which enables us to see what are axiomatic truths in each province of human speculation."[49] He seems to be arguing that Ideas only emerge as a product of a long *history* of trial-and-error experimentation in a given field of study.

The *development* of the clear and distinct notions we need are traced
in the great two-volume *History of the Inductive Sciences;* the successor
volumes are entitled *The Philosophy of the Inductive Sciences Founded
Upon Their History.* Here then is an astonishing "consilience of inductions" (Whewell's coinage) between humanist and natural scientist at
Trinity College. For Whewell's science is developmental: *"There are
scientific truths which are seen by intuition, but this intuition is progressive."*[50] A recent commentator puts it this way: "Ideas that in fact
organize and systematize whole bodies of general propositions are
precisely what, for Whewell, are to be counted as sciences."[51] The
History shows how very late in the history of a science come these progressive revelations. Kepler's discovery of the laws of motion of the
planets, and Newton's theory of gravitation are such discoveries.
Whewell's friend Lyell's *Principles of Geology* published some seven
years before the *History* appeared may or may not have contained such
an Idea; Darwin's *Origin of Species* was published after the third and final
edition of the *History* appeared in 1857 and Whewell apparently opposed
those views, although we think of them as providing just such an Idea as
Whewell was trying to describe as marking epochs in scientific history.
Surely it is evident that the "idea" in Coleridge, in Kant, in Hare and in
Whewell is the same sort of structure all Romantics were attempting to
build at the turn of the century: a bridge over the widening abyss that
separated subject from object, self from society, politics from
psychology, and inner from outer experience. By sensibility a man of the
High and Dry "right wing" persuasion, Whewell had become intellectually persuaded that rationalist versions of nature as an ordered, objective
system hidden just beneath a variety of surface accidents were inadequate, and he contended with great force that in actual fact men do not
think rationally, that "induction is not a rational process, the reverse of
deduction; but that it consists of an imaginative or intuitive *guess* by the
scientist which turns out to be verifiable."[52] This view horrified John
Stuart Mill, and when Whewell's young friend Augustus de Morgan also
objected, he received this reply:

> My object [in the *History of the Inductive Sciences*] was to analyze as far as
> I could, the method by which scientific discoveries have really been made;
> and I called this method Induction, because all the world seem to have
> agreed to call it so. . . . But I do not wonder at your denying these devices a
> place in Logic: and you will think me heretical and profane, if I say, *so
> much the worse for Logic.*[53]

This tough-minded historical approach is well exemplified in Whewell's famous account of how Kepler *imposed* the idea of an ellipse on the data of planetary observations and found that it solved a number of problems. To use the title of his friend's book, it was a "guess at truth" — a guess that worked. From time to time scientists guess right; and very rarely that guess provides us with knowledge of *necessary truths, axioms of science,* or what he usually referred to (using Kantian terminology) as the *Fundamental Idea* necessary to each field. Unlike Kant, who believed he could set out for posterity the limited number of necessary categories that we require in thinking, Whewell seemed to assert that necessary truths have to be discovered, and once found, are seen as inevitable (and necessary), but that probably there were many more such to be disclosed by future generations of workers. As it turned out, the leading scientists of the next generations had equally strong metaphysical powers and valued themselves, as Faraday did, not as "physicists" or "scientists" — he deplored these terms that his friend Whewell had concocted — but as philosophers. Perhaps Faraday's lifelong membership in a pietistic Protestant sect had something to do with his boldest idea, which was to imagine a cosmos in which there were vibrations without vibrating matter, that is, a motion and a property without a phenomenal container. This "immaterialist" notion deeply shocked those contemporaries who had hailed him as the practical scientist who discovered the electric motor (1821) and a decade later, the electric dynamo. As Joseph Agassi observed, "Faraday's theory of the world as comprised of fields of force in empty space was so revolutionary that even his disciples rejected it — at least in the first instance. Einstein made it respectable. . . . it does not assume that space houses matter, and thus it looks idealistic and idealism is traditionally hostile to science."[54]

It looks as though we may have to revise that last phrase when we consider how very many Kantian scientists like Whewell and Oersted were lurking about the premises. And there was always the delightful James Clerk Maxwell, in religion and social concerns a serious follower of F.D. Maurice, whose thought seemed to John Stuart Mill all a muddle of idealistic moonshine. Yet in 1856, Maxwell had this to say: "I find I get fonder of metaphysics and less of calculation continually, and that my metaphysics are fast settling into the rigid, high style, that is about ten times as far above Whewell as Mill is below him."[55] The expert on early Victorian science, Walter F. Cannon, reported that "far from being a period of 'Baconian' influence, the second quarter of the nineteenth century was the period when German Idealism had its strongest impact on

British science."[56] I think that influence retained its strength long after the men who gave it such a strong impulse at Trinity College were gone and forgotten.

Is there a moral to our early Victorian tale? I believe that there is and that it is worth attending to. The thinkers we have encountered were interesting precisely because they "saw" more in what interested them than others had noted. But when we look for the source of this power of "vision" we note that it depends on the richness of the resources from which they draw — including a stock of ideas lying well beyond the boundaries circumscribed by formal definitions of given fields or disciplines. Moreover, they seem to draw inspiration from areas that reputable experts shy away from as productive of biases — metaphysics, religion, historical awareness, and sensibility to the arts. But surely by now we are all aware that there are built-in biases in some of our own "value-free" methods and procedures? We all know that the late Victorian notion that facts are just out there ready to be picked up, classified and matched with observation-statements derived from them is pure hokum. What we see and how we see are matters governed by concepts, and concepts have theories lurking behind them. Facts, as has been remarked, are "value-laden." The bias of many of our "objective" modern methods and modes of comprehension (mechanical systems analysis, game theory, communication modeling, and so on) is that they may be *inherently incapable of utilizing the wisdom to be derived from historical, metaphysical, and artistic knowledge.* The range of vision of the early Victorian thinkers was extensive precisely because they knew how to use the "tacit knowledge" which came to them from these fields that no longer nourish many modern sensibilities. We need to learn from what is strange, other, and mysterious — but which was taken for granted by our Victorian forebears. If their problem was "superstition," then surely William James was correct: our problem is "dessication." Where there is no vision, the life of the mind becomes trivial.

NOTES AND REFERENCES

1. The most useful studies are by W. F. Cannon: "John Herschel and the Idea of Science," *Journal of the History of Ideas,* vol. 22 (1961), pp.215–239; "Normative Role of Science in Early Victorian Thought," *Journal of the History of Ideas,* vol. 25 (1964) pp. 487–502; "History in Depth: The early Victorian Period," *History of Science,* vol. 3 (1964), pp. 20–38. See also R. Olson, *Scottish Philosophy and British Physics 1750–1880* (Princeton: Princeton University Press, 1975) and M. Reich and R. Young, *Changing Perspectives in the History of Science* (Dordrecht: D. Reidel, 1973).

2. The best introduction to the Coleridge problem is to be found in G.N.G. Orsini, *Coleridge and German Idealism* (Carbondale: Southern Illinois University Press, 1969). Two major projects are: E.L. Griggs, *Collected Letters of Samuel Taylor Coleridge* (Oxford, 1965 et seq.) and Kathleen Coburn, ed. *The Notebooks* (London, 1957 et. seq.). Hans Aarsleff, in *The Study of Language in England 1780–1860* (Princeton: Princeton University Press, 1967), provides a detailed confirmation of the thesis that Trinity College provided the impetus for the creation of the Philological Society of London (in 1842, Connop Thirlwall presiding) and its subsequent great project, the O.E.D. (see esp. pp. 211–263).

3. See the continuing set of essays written for the *New York Review of Books* by Charles Rosen; Harold Bloom's collection, *Romanticism and Consciousness* (New York, 1970), reprints essays by such important participants in this debate on modernism and romance as Bate, Hartmann, Abrams, Frye and Wimsatt, among others. David Thorburn's, *Conrad and Romanticism* (New Haven: Yale University Press, 1976) offers a brisk account of the state of the question at Yale.

4. Julius C. Hare (1795–1855) was one of the four sons of Georgiana Hare-Naylor, an intellectual lady of high rank and aristocratic connections who took her family to live in Weimar in 1804 and lived abroad frequently thereafter. For the early life and the complicated intermarriages between members of the Hare family, Stanleys, Maurices, Arnolds, see A.J.C. Hare's *Memorials of a Quiet Life*, 4th ed. (London: Strahan, 1873). Julius Hare attended Charterhouse with Grote and Thirlwall, the future historians of Greece, entered Cambridge in 1812, was elected Fellow of Trinity College in 1818 and took up residence as Assistant Tutor in 1822. Adam Sedgwick, William Whewell, Kenelm Digby, Connop Thirlwall, and Hugh James Rose were his especial undergraduate companions. John Sterling, his student and later curate, married into his family: Hare edited his *Remains*, which were sneered at by Carlyle in his version of *The Life of John Sterling*, a moving and misleading work. In 1832 Hare resigned from Trinity to marry and accept the family living at Herstmonceux, Sussex. After more than a year abroad with his friends Walter Savage Landor and Christian Bunsen, then secretary to Niebuhr at the Prussian Legation in Rome (and between 1842 and 1854 Prussian Ambassador to Court of St. James), he returned to his library of more than 14,000 volumes and continued to write *The Victory of Faith* (1840), *The Mission of the Comforter* (1846), *Pamphlets on Church Questions* (1855) and a number of "Vindications" of favorite authors and friends who were under attack by various parties: Niebuhr (1829); Coleridge (1835); Martin Luther (1844); and Christian Bunsen (1840). See also: C.B. Sanders, *Coleridge and the Broad Church Movement* (Durham, N.C., 1942); Robert Preyer, "Julius Hare and Coleridgian Criticism," *Journal of Aesthetics and Art Criticism*, vol. XXI (1957), pp. 449–460; Robert Preyer, "Victorian Wisdom Literature: Fragments and Maxims," *Victorian Studies* (March, 1963), pp. 245–262; G. F. McFarland, "Julius Charles Hare: Coleridge, DeQuincey, and German Literature", *Bulletin of John Rylands Library*, vol. 47 (1964) pp. 165–197. My personal copy of Niebuhr's *History of Rome* translated by Hare and Thirlwall in three volumes ("New edition-1851") contains the bookplate of Charles Dickens.

5. See J.J.S. Perowne, *Remains, Literary and Theological, of Connop Thirlwall*, 2 vols. (London, 1877) and *Letters*, 2 vols. (London, 1881); J.C. Thirlwall, Jr., *Connop Thirlwall: Historian and Theologian* (New York, 1936); Robert Preyer, "The Histories of Grote and Thirlwall" in *Bentham, Coleridge, and the Science of History* (Bochum, Germany, 1958). Thirlwall's *History of Greece* appeared in the years 1833–47. He was made Bishop by a Whig administration on the recommendation of Brougham by way of J.S. Mill, probably to spite the Tory High Church party. He is buried in the same grave as his schoolfellow, the

notorious Benthamite, George Grote. Each lost track of the other after Charterhouse and had no idea they were almost simultaneously to publish massive histories on the same general subject. Thirlwall's translation of Schleiermacher's "Essay on the Gospel of St. Luke" contained a full account of the problem of the origins and transmission of the sacred texts; it began the persecution that led him to resign as Tutor from Trinity in 1834.

6. Robert Robson, *Ideas and Institutions of Victorian Britain* (London, 1967).

7. Robson,[6] pp. 318–319.

8. Robson,[6] p. 324. But see Tennyson's poem above.

9. Arthur P. Stanley, *Review Article*, posthumous publications of J.C. Hare, *Quarterly Review*, vol. 95 (1855), 1–28.

10. Stanley,[9] p. 5.

11. W. F. Cannon,[1] (p. 216) provides a list of textbook titles. George Airy did a series on optics, mathematics and astronomy; Whewell on mechanics and dynamics, Robert Willis, Whewell's friend, made professor in 1837, did two more textbooks on mechanics in the 1840s. The intention of the group of scientific friends (Airy was Plumian Professor of Astronomy, Babbage Lucasian Professor of Mathematics, Whewell Professor of Mineralogy, Sedgwick Professor of Geology) was to open up Trinity science to new continental developments, notably French calculus and analysis, of the very recent years. The university appeared to have dozed since Newton's *Principia*. For journalism: *The Quarterly Review* was excellent on scientific affairs, and it also printed several articles on Niebuhr (see especially Tom Arnold's "plant" in July 1822, written before he could read German, incidentally!). J.S. Mill opened the pages of the *Westminster Review* to Sterling, Maurice and other "Germano-Coleridgians" he met through them; Julius Hare had access to *Ollier's Literary Miscellanies* as early as 1820 (containing four essays on German literature). He was a friend of John Taylor, publisher of Keats, Coleridge, DeQuincey and Landor, and of Hare's books. Taylor opened the pages of his *London Magazine* to Hare as early as 1824. Students or acquaintances of Hare were the principal contributors to *Knight's Quarterly Magazine*, among them Macaulay, Praed, Derwent Coleridge, W.S. Walker, and Moultrie. F.D. Maurice and John Sterling were editors of *The Metropolitan*, which merged with the *Athenaeum and Literary Chronical* in 1828, and friendly reviews of Tennyson's *Poems*, *Guesses At Truth* and *Children of Light* soon appeared. Sterling took over the editorship from Maurice (1825–29) and printed a series ("Museum of Thoughts") by Hare. Thirlwall and Hare had their own technical journal as well: *The Philological Museum* (2 vols., 1832–33). Hugh James Rose, another college friend, edited *The British Magazine* in 1835 and printed the important essay by Hare "Samuel Taylor Coleridge and the English Opium Eater," vol. VII (January 1835).

12. William Whewell (1794–1866), Master of Trinity (1841–66), twice Vice Chancellor of Cambridge (1842, 1855), twice widowed, and twice a Professor (Mineralogy and Moral Philosophy), FRS and Fellow of Trinity (1820), member of British Association, the Royal Society, The Geological Society and the Philological Society of London, winner of Chancellor's Prize for poetry (1814), etc. was the son of a master carpenter and had come up on scholarship to Trinity in 1812. He contributed the terms "ion," "anode" and "cathode" to Michael Faraday; named the "Eocene," "Miocene," and "Pliocene" ages for his friend Adam Sedgwick, and introduced the terms "scientist" and "physicist." His fame rests on the *History of the Inductive Sciences*, 3 vols. (1837) and *The Philosophy of the Inductive Sciences, Founded upon Their History*, 2 vols. (1840), still impressive monuments of Victorian learning. His nonscientific publications include: *The Elements of Morality, Including Polity*, 2 vols. (1845); *Platonic Dialogues for English Readers* (1859–61); translations of

Grotius, and with J.F.W. Herschel, Julius Hare and E.C. Hawtrey, *English Hexameter Translations from Schiller, Goethe, Homer, Callinus, and Meleager* (1847) and the early *Architectural Notes on German Churches, With Remarks on the Origin of Gothic Architecture* (1830). Isaac Todhunter has written an invaluable assessment of his scientific and literary endeavors; Mrs. Stair Douglas had edited *The Life and Selections from the Correspondence* (1881). A joint essay by Robson and Cannon in *Notes and Records, Royal Society of London*, vol. 19 (December, 1964), pp. 168–191 is a model of its kind and provides a contemporary version of Whewell's achievements.

13. Hugh James Rose became Professor of Hebrew after some years of study at German universities. To the dismay of many old friends, he suffered a violent change of heart and joined the Oxford Malignants (as Thomas Arnold called them) and in *Protestant Theology in Germany* directed four polemical sermons against the Broad Church.

14. A grateful pupil wrote, "Never was there a Tutor of Trinity . . . more affectionately remembered by his pupils." Quoted in Robson,[6] p. 323. Other well known philologists in the Hare-Thirlwall connection are William Smith, editor of the *Dictionary of Greek and Roman Antiquities* (1842); J.W. Donaldson, author of *The New Cratylus* (1839), which introduced English scholars to the grammatical work of Franz Bopp, George Long, R.C. Trench, John Kemble, passim. See on this subject Aarsleff.[2] Trench's *English Past and Present* (1845) and *On the Study of Words* (1851) were just as influential in America as in England.

15. John Sterling, *Arthur Coningsby* (London, 1833). F. D. Maurice, *Eustace Conway* (London, 1834). See also the college poems of Alfred, Lord Tennyson – "A character," "The Poet," sonnets to John Kemble and Blakesley. There was of course a third, but socially unacceptable alternative: Charles Simeon, the evangelical preacher, was rapidly acquiring outside funds to buy up livings for presentation to Low Church pietists like himself.

16. Coleridge and Wordsworth first met in 1795; their joint *Lyrical Ballads* date from 1798; both were in Germany in 1798–99. At Göttingen, Coleridge, with his usual flair and perversity, studied natural history and physiology under the great Blumenbach, the New Testament with Eichorn (perhaps the leading Higher critic), and Gothic with Professor Tychsen, one of the new Germanic philologists. His serious study of Kant probably dates from his return. A famous passage in *Biographia Literaria*, ("The writings of the illustrious sage of Konigsberg . . . took possession as with a giant's hand. After fifteen years' familiarity with them, I still read these and all his other productions with undiminished delight.") chapt. IX shows how sharp and permanent was his response to "Kant's philosophy, perhaps the most important intellectual event since the Protestant Reformation," according to Hugh Honour's *Romanticism* (New York, 1979), p. 11.

17. Hare also wrote a very fine *Vindication of Coleridge* (1835) against the charges of DeQuincey, a *Vindication of Luther* against Newman and W. G. Ward; a *Vindication of Bunsen* (also against the Oxford Movement attacks, notably of Pusey). The translation of Niebuhr was dedicated to Bunsen. Few recall that Hare's final tribute to his old friend Thomas Arnold was to supply all the footnotes and references for Arnold's posthumous volume on Roman history.

18. R.C. Trench, *Letters and Memorials*, vol. 1 (London, 1888), p. 10.

19. Frederick Maurice, ed., *The Life of Frederick Denison Maurice, Chiefly Told in His Own Letters*, 3rd ed. vol. 1 (London, 1884) p. 176.

20. Robson,[6] p. 321.

21. See especially chap. 3 in *John Stuart Mill's Autobiography* (New York: Columbia University Press, 1924), where Mill speaks of Charles Austin at Cambridge. "The effect he

produced on his Cambridge contemporaries deserves to be accounted an historical event; for to it may in part be traced the tendency toward liberalism in general, and the Benthamic and politico-economic form of it in particular, which showed itself . . . from this time to 1830. . . . Through him I became acquainted with Macaulay, Hyde and Charles Villiers, Strutt (now Lord Belper), Romilly . . . and various others." Eustace Conway, in Maurice's novel of that name, arrives home on vacation a confirmed Benthamite; Sterling's tiresome hero has also been surrounded by Dostoevskian atheist-radicals capable of intellectual murder.

22. Judith Merivale, ed., *Autobiography and Letters of Charles Merivale, Dean of Ely* (London, 1899), pp. 98–99.

23. *Autobiography of John Stuart Mill* with Preface by J.J. Coss (New York, 1924), pp. 87–88, 90–91.

24. F.M. Brookfield, *The Cambridge "Apostles"* (New York, 1906). E.H. Plumtre, "Memoir" in *Guesses At Truth* (1871 ed.); Augustus J.C. Hare, *Memorials of a Quiet Life* (London, 1872); Ronald E. Prothero, *The Life and Correspondence of Arthur Penrhyn Stanley* (New York, 1894); Charles R. Sanders, *Coleridge and the Broad Church Movement* (Durham, N.C., 1942); Duncan Forbes, *The Liberal Anglican Idea of History* (Cambridge, 1952). Two articles by G.F. McFarland, "The Early Literary Career of Julius C. Hare" and "Julius Hare: Coleridge, DeQuincey, and German Literature," *Bulletin of John Rylands Library*, vol. XLVI (1963–64) and vol. XLVII (1964); Robert Preyer, *Bentham, Coleridge and the Science of History* (Bochun, Germany, 1958); "The Dream of A Spiritualized Learning" in *Geschichte und Gesellschaft in der amerikanischen Literature* (Heidelberg, 1975), pp. 62–85; "Victorian Wisdom Literature: Fragments and Maxims," *Victorian Studies* (March, 1963), pp. 245–262.

25. McFarland,[24] p. 175.

26. B.G. Niebuhr, *The History of Rome*, translated by Hare and Thirlwall (Cambridge, 1828) vol. 1, p. 250.

27. McFarland,[24] p. 169. From an undated letter fragment in Rylands Library (Eng. Ms. 1238) found by Professor McFarland.

28. *The Vindication of Niebuhr* was a 60-page pamphlet published by John Taylor, who also published Volume 1 of the *Roman History* translation in 1828. A revised German language edition then appeared and it was so different from the first edition that Hare and Thirlwall started all over with a new Volume 1 (1829) and a Volume 2 in 1832.

29. Julius and Augustus Hare, *Guesses At Truth By Two Brothers* (Boston: Ticknor & Fields, 1865) p. 435 There are many editions; footnotes are to this easily available Ticknor and Fields edition. (Contributions by Augustus are infrequent and distinguished by a different initial; he died soon after the first edition appeared.)

30. William Whewell, *On the Philosophy of Discovery: Including the Completion of the Third Edition of the Philosophy of the Inductive Sciences* (New York, 1971), p. 308. This is a reprint of the 1860 edition, which includes additional chapters attacking Comte (chap. XXI) and "Mr. Mill's Logic" (chap. XXII), and a curious Appendix on Hegel's attack on Newton's *Principia*, among other matter.

31. In *Guesses At Truth* (p. 222), Hare refers to "the great *History of the Inductive Sciences*, in which one of Bacon's worthiest and most enlightened disciples has lately been tracing the progress of scientific discovery throughout the whole world of nature." This is preceded (p. 219) by an interesting explanation of why the terminology of Science "is almost wholly Greek. . . . The plastic nature of the language — its words really coalesce and are not merely tacked together — fits it for expressing the innumerable combinations, which

it is the business of science to detect. And as Science is altogether a cosmopolite, less connected than any other mode of intellectual action with the peculiarities of national character, . . . it is well that the vocabulary of Science should be common to all the nations that come and worship at its shrine." He also adds: "Of all words however the least vivacious are those coined by Science . . . " Since Hare greatly admired his friend's efforts at constructing a scientific nomenclature, we can presume that he understood the necessity for working out a deliberately "dead" or unmetaphorical diction. Both would agree that "It is Poetry, the Imagination, in one or other of its forms, that produces what has life in it." Emerson and Hare both jotted down Coleridge's dictum, "I would endeavour to destroy the old antithesis of Words and Things; elevating as it were Words into Things and living Things too." Scientific language is deliberately *iconophobic*, it has no intention of bridging a gap between self and nature by a symbolic prehension of a unifying force or spirit that is present, so to speak, on both sides of the equation. Living language is symbolic for Coleridge, not allegorical. And he insists that "a symbol . . . is characterized by translucence of the Special in the Individual or of the General in the Especial. . . . above all by the translucence of the Eternal, through and in the Temporal. It always partakes of the Reality which it renders intelligible; and while it enunciates the whole, abides itself as a living part in that Unity, of which it is the representative." (*Statesman's Manual*, [Pickering, 1816], p. 37) It is a famous and difficult definition of the "unifying spirit" that the imagination bodies forth in such romantic, symbolic masterpieces as Caspar David Friedrich's *The Cross in the Mountains* (1807–08) or Turner's first great symbolic masterpiece, *Snowstorm: Hannibal and His Army Crossing the Alps* (1812). It is what the Romantic poets sought after in their verse and proclaimed in their devotions – an enhanced image, a unified, simultaneous experience of many levels of consciousness and of the reconciliation of opposites.

32. William Whewell, *The Elements of Morality, Including Polity*, vol. 1 (New York, 1845), p. 77.

33. Whewell,[32] p. 78.

34. Isaac Todhunter, *William Whewell*, vol. 1 (London, 1876), p.240

35. Robert E. Butts, *William Whewell's Theory of Scientific Method* (Pittsburgh, 1968), p.5

36. Julius and Augustus Hare,[29] p. 377.

37. See Robert Preyer, "Victorian Wisdom Literature: Fragments and Maxims," *Victorian Studies* (March 1963), for further details on the short forms that are most appropriate for rendering instantaneous insights into very deep matters. Coleridge was, like Hare, fascinated with the visionary act of mind as a necessary precondition for the commencement of the operations of discursive reflection. The most truculent aphorisms on this subject are to be found, however, not in Coleridge or Hare, but in William Blake's little known Manifestoes of 1788: "There Is No Natural Religion" and "All Religions Are One": "Man by his reasoning power can only compare and judge of what he has already perceived." "If it were not for the Poetic or Prophetic character the Philosophic and Experimental would soon be at the ratio of all things, and stand still, unable to do other than repeat the same dull round again." And, "As none by travelling over known lands can find out the unknown, So from already acquired knowledge Man could not acquire more: therefore an universal Poetic Genius exists." G. Keynes, ed. *Poetry and Prose of William Blake* (London, 1957), pp. 147, 149.

38. Many passages in *Guesses At Truth*[29] are concerned with precision of thought and words that are mischievous because they have no precise meaning and "poorly fulfill their office of being a sign and guide of thought . . . "(p. 231). "In proportion as every word is the

distinct, determinate sign of the conception it stands for, does that conception form part and parcel of the nation's knowledge" (p. 234). Julius Hare discusses the question of new coinages: "When any new conception stands out so broadly and simply as to give it a claim for having a special sign to denote it — if no word for the purpose can be found in the extant vocabulary of the Language . . . " He praises Coleridge and DeQuincey for reviving old terms and for specific new coinages, especially in metaphysics (p. 235). Whewell, of course, was frequently consulted by notable scientists — among them Bell, Faraday, Lubbock, Lyell, Murchison, and Owen — for an appropriate nomenclature for new phenomena whose nature ought to be "fixed" of fossilized in an invariant form, sterilized so far as possible from metaphysical or other sorts of "suggestivity." Whewell's "Aphorisms Respecting The Language of Science" (in *The Philosophy of the Inductive Sciences)* have been studied in an essay by James Paradis, "William Whewell's New Scientific Lexicology."

Whewell believed that the classical languages of Greek and Latin (with their flexible prefixes and suffixes) provided a variety of modifications for conceptual declensions not susceptible to further shifts in signification occasioned by historical developments in syntax and meanings ascribed to particular terms; he was trying to construct a *deliberately antipoetic use of language.* This "dead" language of science had been employed in the eighteenth century poetic diction to which Wordsworth and Coleridge took violent exception. Periphrastic terms like "the loquacious race" [frogs] and "bearded product" [corn] were formed on precisely the same principle as the classifications of Linnaeus, that is, the name of the genus together with the distinguishing characteristic. Many of the early "antiscientific" rumblings of Wordsworth have to do with a false application of scientific diction systems to literary works, which require a full use of all the life and suggestivity and implication that a "living," developing language can bear. Horne Tooke's *Diversions of Purley* was the linguistic handbook of the Utilitarians, approved by Erasmus Darwin, Bentham, James Mill, Macintosh, Brougham, Hazlitt, and John Stuart Mill. They believed that Tooke's far-fetched etymologies "proved" that every word could be traced back to a sensation and consequently that there was no need for "innate ideas" or "spirit". The attack from German philology noted that many words have 50 meanings, some of which are clearly a product of the history of usage and none of which is "correct" and "rational", except in relation to a particular context. Syntax and historical contexts were the key to understanding languages as they developed over time. It took awhile for partisans of the two polar views of language to realize that each had its place in the economy of reason, a lesson more quickly assimilated at Cambridge than elsewhere.

39. B.G. Niebuhr, *History of Rome,* p. ix.

40. Samuel Taylor Coleridge, *Biographia Literaria,* chap. XIV. A new edition will be out very shortly.

41. Quoted in Robert O. Preyer, "Julius Hare and Coleridgian Criticism," *Journal of Aesthetics and Art Criticism,* vol. 15 (June 1957), p. 452.

42. Samuel Taylor Coleridge, *On the Constitution of the Church and State According to the Idea of Each, Lay Sermons,* the Pickering edition (London, 1839), p. xi.

43. Coleridge[42] pp. 17, 19.

44. Coleridge[42] p. 12.

45. Dwight C. Culler, *The Poetry of Tennyson* (New Haven,1977), p. 157.

46. Julius and Augustus Hare,[29] pp. 234–235.

47. Alert Victorian minds were quick to pick up the connections between philology and the natural sciences. G.H. Lewes made one such connection shortly after the appearance of Darwin's *Origin of Species:*

The development of numerous specific forms, widely distinguished from each other, out of one common stock, is not a whit more improbable than the development of numerous distinct languages out of a common parent language, which modern philologists have proved to be indubitably the case. Indeed there is a very remarkable analogy between philology and zoology in this respect: just as the comparative anatomist traces the existence of similar organs, and similar connections of the organs, throughout the various animals classed under one type, so does the comparative philologist detect the family likeness in the various languages scattered from China to the Basque Provinces, and from Cape Comorin across the Caucasus to Lapland—a likeness that assures him that the Teutonic, Celtic, Wendic, Italic, Hellenic, Iranic, and Indic languages are of common origin, and separated from the Arabian, Aramean, and Hebrew languages, which have another origin. Let us bring together a Frenchman, a Spaniard, an Italian, a Portugese, a Wallachian, and a Rhaetian, and we shall hear six very different languages . . . yet we know most positively that all these languages, are offshoots from the Latin which was once a living language, but which is now, so to speak, a fossil. (G.H. Lewes, *Studies in Animal Life* [New York, 1860] pp. 102–103.)

48.   William Whewell, *The Philosophy of the Inductive Sciences Based on Their History* pt. II (London, 1967), pp. 443, 468–469. (Vol. VI in "The Historical and Philosophical Works of William Whewell," ed. G. Buchdahl and L.L. Laudon.)

49.   R.E. Butts,[35] pp. 6, 7.

50.   Butts, [35] p. 16.

51.   Butts, p. 16.

52.   Walter F. Cannon, "William Whewell, FRS (1794–1866). II. Contributions to Science and Learning," *Notes and Records: Royal Society of London* vol. 19 (December 1964) p. 186.

53.   Cannon,[52] p. 187

54.   Joseph Agassi, *Faraday As A Natural Philosopher* (Chicago, 1971), p. 8.

55.   W.F. Cannon, "History in Depth,"[1] pp. 23–24.

56.   Cannon, [55] p. 23.

Filaments, Females, Families and Social Fabric: Carlyle's Extension of a Biological Analogy

JEFFREY L. SPEAR
Department of English
Princeton University
Princeton, New Jersey 08540

I. AGGREGATION

THOMAS CARLYLE is one of those Victorian sages whose figures of speech are themselves part of their arguments. He is careful, in John Holloway's words, to "cash"[1] many otherwise obscure expressions; that is, to turn them by explanatory restatement into an idiosyncratic technical vocabulary. Yet there are allusions, metaphors, and symbols that he leaves undefined except by context and the reader's own awareness of the source, or at least the general realm of discourse, from which Carlyle derived them. *Sartor Resartus* (The Tailor Retailored, 1833–34) displays Carlyle's special rhetoric at its most extreme. In the guise of recording the attempts of an editor to prepare for an English public the fragmentary "philosophy of clothes" of the very German Professor Diogenes Teufelsdröckh, Carlyle incorporates contemporary ideas and vocabularies including scientific ones into a new version of the universe of analogies in which microcosm and macrocosm reflect one another.[2] In the central metaphor of *Sartor Resartus* flesh becomes clothing to the soul as cloth to the body, name to the self; and nature herself becomes "what the Earth-Spirit in *Faust* names it, *the living visible garment of God.*"[3]

My starting point is literally a thread in Carlyle's argument, the "organic filament," one of the symbols unexplained by Carlyle and unexplicated by his commentators, although it gives its name to a chapter of *Sartor* that is regularly quoted in discussions of Carlyle's "organicism."

69

The "organic filament" is a term for that elemental fiber from which the embryo develops according to various (conflicting) precellular theories of reproduction. The implicit and explicit analogies to organic filaments in Carlyle's work provide a key to his concept of a natural social order and the role of the hero in history as expressed in *Sartor Resartus, The French Revolution* (1837), *Past and Present* (1843), and his essays, especially *The Latter-Day Pamphlets* (1850). By making individual and social development analogous, Carlyle was able to equate biblical ethics with natural law without invoking the authority of a specific religion.

In the chapter on the "genesis" of Teufelsdröckh in the first book of *Sartor Resartus*, Carlyle compares his hero's mental development from sensation through thought, fantasy, and force to the action of "organic elements and fibers shoot[ing] through the watery albumen" of an egg (*SR*, p. 88); but it is in "Organic Filaments," the parallel chapter on social "genesis" in the third book, that the elemental fiber becomes the basis for an extended analogy. Carlyle shows us the filaments in a "Fire-whirlwind," linking the first vision of Ezekiel with that favorite symbol for birth of a new order out of the ashes of the old, the Phoenix, the embodiment of the dual nature of fire: destruction and regeneration, whether of body or spirit. Unlike the traditional thread of fate, the organic filament has the power to spin, to form, itself in this vision of historical ontogeny.

> 'Little knowest thou of the burning of a World-Phoenix, who fanciest that she must first burn-out, and lie as a dead cinereous heap; and therefrom the young one start-up by miracle, and fly heavenward. Far otherwise! In that Fire-whirlwind, Creation and Destruction proceed together; ever as the ashes of the Old are blown about, do organic filaments of the New mysteriously spin themselves. . . .' Let us actually look, then: to poor individuals, who cannot expect to live two centuries, those same organic filaments, mysteriously spinning themselves, will be the best part of the spectacle. (*SR*, pp. 244–245)

Carlyle may well have derived the term "organic filament" from the chapter on Generation in the *Zoonomia* (1794) of Erasmus Darwin. Darwin, rejecting the contention of the Comte de Buffon that new life was formed from organic particles gathered into the reproductive organs of both sexes from all parts of the body and then combined in the womb, argued instead that the fetus develops from "living filaments," that is, from single strands of muscle or nerve tissue secreted from the blood and possessing "irritability," the capacity to be excited into motion or development.

The living filament is part of the father, and has therefore certain propensities or appetencies, which belong to him; which may have been gradually acquired during a million generations, even from the infancy of the inhabitable earth; and which possess such properties, as would render, by the apposition of nutritious particles, the new fetus exactly similar to the father. . . . But as the first nutriment is supplied by the mother, and therefore resembles such nutritive particles as have been used for her own nutriment or growth, the progeny takes in part the likeness of the mother.[4]

Development of male or female form is a propensity of the filament that Darwin, in a Shandean extension of Aristotelian biology, with its dichotomy of male form and female matter, attributes to the imagination of the father at the moment of impregnation. Such influence of an idea upon organic form well suits Carlyle's Germanic idealism; indeed Darwin's theory was known to Schelling and Novalis, but the specific source of Carlyle's vocabulary is less important than the fact that his audience would recognize "organic filament" as a term for the basic constituent of life.[5]

As nature endows the merest thread of living matter with the power to weave itself into the astounding complexity of a human being, so, Carlyle implies, may individuals acting on their own volition make up an organic whole — an "aggregate" in both the zoological sense (distinct animals united into a common organism) and the legal one (any civil association of individuals). Social life is not merely the total of separate lives in a given location, but the "aggregate of all the individual men's lives who constitute society; History is the essence of innumerable biographies" ("On History," vol. XXVII, p. 86). (By "essence" Carlyle means specifically "the living principle round which all detached facts and phenomena . . . would fashion themselves into a coherent whole, if they are by any means to cohere," that is, the philosophical equivalent of an organic filament ["Historic Survey of German Poetry," vol. XXVII, p. 342]). Carlyle's primary interest in the organic filament is not, of course, in reproduction itself, but in the possible analogy between the constitution of individuals and the reconstitution of society. With pre-Darwinian evolutionary biology in mind, Carlyle's remarks on tradition in "Organic Filaments" gain new force as the cultural equivalent of man's physical inheritance.

Hast thou ever meditated on that word, Tradition: how we inherit not Life only, but all the garniture and form of Life; and work, and speak, and even think and feel, as our Fathers, and primeval grandfathers, from the beginning, have given it us? — Who printed thee, for example, this unpretending

Volume on the Philosophy of Clothes? Not the Herren Stillschweigen and Company; but Cadmus of Thebes, Faust of Mentz, and innumerable others whom thou knowest not. (*SR*, p. 246)

The analogy between the development of an individual from an organic filament and the issue of a new body politic from the filaments spun by many lives implies that society, like the individual, must have a proper human form if it is to live, and that its form will be the incarnation of an idea. "Such is SOCIETY the vital articulation of many individuals into a new collective individual . . . a second all-embracing life wherein our first individual life becomes doubly and trebly alive . . ." ("Characteristics," vol. XXVIII, p. 12). Both the individual and society are physical manifestations of the laws of nature. The essence, or living principle, of society is its collective belief, its religion. ("Religion originates by Society. Society becomes possible by Religion" [*SR*, p. 215]). Loss of faith is the social equivalent of degenerative disease, and Carlyle turns to the language of physiology and experimental science to describe it:

'For the last three centuries, above all for the last three quarters of a century, that same Pericardial Nervous Tissue (as we named it) of Religion, where lies the Life-essence of Society, has been smote-at and perforated, needfully and needlessly; till now it is quite rent into shreds; and Society, long pining, diabetic, consumptive, can be regarded as defunct; for those spasmodic, galvanic sprawlings are not life; neither indeed will they endure, galvanise as you may, beyond two days.' (*SR*, p. 232)

The reformation of the social world depends upon the renewed health of the internal or spiritual one. However much Teufelsdröckh may believe that nature is the "Living Garment of God," that inner conviction is no longer supported by a "mythus," a communal belief, a sacred story and its catechism. In their absence it is duty that must transform conviction into action. When doubt is overcome by action, the need to decide made subservient to duty, then internal chaos becomes cosmos. By becoming yourself a small world in action, Carlyle implies, you contribute to the development of new order in the larger world. The same principle of order works on all levels, although the historical progress whereby society evolves is too broad for one person to comprehend wholly and too lengthy for one person to experience.

To discern and nurture the organic filaments of a new order means to develop what is natural in human relations. The most basic of ties, essential to the continuation of any society, are those between men and

women, adults and their children. The family with its traditional structure of affections and obligations is an implicit prime model for Carlyle's
concept of social order. Carlyle thinks of marriage as a perennial institution, as he makes clear in his letters: ". . . I would stand by my argument
that the Covenant of Marriage m[ust] be perennial; nay that in a better
state of society there will be other pere[nnial] Covenants between man
and man, and the home-feeling of man in this world of his be all the
kindlier for it. . . ."[6] Carlyle uses the relative health of marriage as a
measure of social vitality. The marriage contract becomes a pattern for
those social relationships that require, in the metaphor of *Sartor Resartus*, new clothes.

Beyond the family, the organic filament of political order must embody man's natural reverence for a superior, the "Hero-Worship" that
connects each person to the next in a "heroarchy" of mutual obligation
and dependence. Any society that attempts to substitute a form of
government or a set of abstract rules for this organic truth denies nature,
whose

> small still voice, speaking from the inmost heart of us, shall not, under ter-
> rible penalties, be disregarded. No one man can depart from the truth with-
> out damage to himself; no one million of men; no Twenty-seven Millions
> of men. . . . *Un*nature, what we call Chaos, holds nothing in it but vacu-
> ities, devouring gulfs. (*P & P*, vol. X, pp. 142–143)

A false gospel, like the "Mammonism" of Carlyle's England, which would
make the pursuit of economic self-interest a social good and substitute a
"cash nexus" for the obligations of mutually recognized interdependence,
ignores the laws of nature that give society its human form. The mill
owner, faced with starving workers yet arguing that he has paid what the
contract calls for, echoes Cain's unholy question. Denied or rejected,
these laws of nature assert themselves in terrible ways.

> The forlorn Irish Widow applies to her fellow-creatures, as if saying,
> 'Behold I am sinking, bare of help: ye must help me!' . . . They answer,
> 'No, impossible; thou art no sister of ours.' But she proves her sisterhood;
> her typhus-fever kills *them*: they actually were her brothers, though denying
> it! Had human creature ever to go lower for a proof? (*P & P*, vol. X, p. 149)

Just as on the level of interpersonal relationships the eternal truth of the
social bond will assert itself in terrible ways if denied, so too must essential ties to the past be maintained if a new order is to emerge without a
period of social chaos, of "anarchy." It is futile to sever all ties to the past

because some are oppressive and to attempt to create a new social order by legislative fiat. To do so destroys the natural growth of organic filaments "cutting asunder ancient intolerable bonds; and, for new ones, assiduously spinning ropes of sand" (FR 1, vol. II, p. 220).

II. THE EXAMPLE OF FRANCE

The French Revolution is a gargantuan object lesson, detailing what follows from the refusal of necessary reform and exemplifying the process of historical change given in *Sartor Resartus* and such early essays as "Signs of the Times," "Characteristics," and "On History."[7] Sansculottism is the nemesis roused by the systematic violation of what Carlyle insists are the natural laws of social organization. Revolution is the fate of a society in which the hierarchy of classes has ceased to embody mutual dependence and responsibility, becoming instead a mechanism of exploitation. On the general level, Carlyle's history records the failure of the "realized ideals" of Catholicism and Kingship, which have become shams, their "principle of Life" flowing into the anarchic upheaval that now destroys them. With the decline and fall of the old regime come the successive efforts of Mirabeau, Danton, and Napoleon to restore viable order to shape what Carlyle describes as the natural tendency toward organization, the organic filaments of a political and social structure. But in its specifics, Carlyle's history is that collection of biographies, that amalgam of individual lives and motivations promised in "On History" (1830). Historical figures replace the fictional characters of *Sartor* in exemplifying moral states and ideas. As Carlyle weaves the separate strands of biography together to make his tapestry of revolution, he relies upon the "natural" values assigned to women and the family to draw scenes designed to reveal the moral character of historical figures and to suggest judgments upon them. Indeed, Carlyle's neofeudal conception of a mutually responsible hierarchy of classes assumes an analogy to the family, although Carlyle never invokes the traditional patriarchal justifications of monarchy.

In the case of Mirabeau, there are conventional reasons for sketching his family background. The scandals of his youth affected his relationship with the court, and quarrels with his father in part account for the fact that he was elected to the Assembly from the Third Estate. But the anecdote from the life of Robespierre's deputy, Couthon, that Carlyle chooses to tell is not in any conventional sense related to his public life. Imbedded in "The Book of the Law" (bk. V, chap. 2) — a title suggesting an ironic contrast between the Legislative Assembly charged with drafting a French Constitution and the framers of the Pentateuch — the anecdote implies that the Mountain faction in general, and Couthon in particular, are hardly people to be regulating the conduct of others.

> There too is Couthon, little dreaming *what* he is; — whom a sad chance has paralysed in the lower extremities. For, it seems, he sat once a whole night, not warm in his true-love's bower (who indeed was by law another's), but sunken to the middle in a cold peat-bog, being hunted out from her; quaking for his life, in the cold quaking morass; and goes now on crutches to the end. (*FR* 2, vol. III, p. 207)

Carlyle's treatment of Couthon smacks of the British Protestant's suspicion of things Catholic in general, and French in particular, as does his attempt to epitomize the decadence of the court of the dying Louis XV in the figure of DuBarry, "a wonderfully dizened Scarlet-woman" waited upon by the royal valet she has named La France. But Carlyle's use in *The French Revolution* of the traditional values of the family, and particularly of the nurturing, life-giving role of woman, goes well beyond the conventional stuff of satiric comedy suggested by these first examples.

The captive state of the French monarchy in Paris, the prelude to the abortive flight of the King and Queen to the frontier, and their subsequent execution, grew from what conventional historians call the *journées* of October 1789; but to Carlyle it is the "Insurrection of Women," the inevitable response of the nurturers of mankind to a combination of insult, starvation, and the breakdown of authority — political and domestic.

> Men know not what the pantry is, when it grows empty; only house-mothers know. O women, wives of men that will only calculate and not act! Patrollotism is strong; but Death, by starvation and military onfall, is stronger. . . . Will Guards named National thrust their bayonets into the bosoms of women? (*FR* 1, vol. II, p. 250)

What follows is a vision of the world turned upside down: of women gathering, impressing others into their ranks, storming buildings, seizing arms, and like an irresistible force of nature marching to Versailles, demanding: "Du pain, et parler au Roi." "It is the cause of all Eve's Daughters, mothers that are, or that ought to be" (*FR* 1, vol. II, p. 256). The women occupy the Assembly, conduct a parody of legislative debate, and can only be removed to the galleries by the distribution of food. The "Strong Dame of the Market" is forced from the President's chair so that debate can continue upon a new penal code. France herself is in labor.

> To such length have we got in regenerating France. Methinks the travail-throes are of the sharpest! — Menadism will not be restrained from occasional remarks; asks, 'What is the use of Penal Code? The thing we want is Bread.' (FR 1, vol. II, p. 273)

A picture of mothers baking bread for their families with children around them in the folds of an Alpine valley was central to Teufelsdröckh's vision of a maternal Nature restored to him in "The Everlasting Yea" of *Sartor Resartus*. The Insurrection of Women is the antithesis of that vision, an inevitable response of the nurturers of mankind to the combination of deprivation and the breakdown of legitimate authority capable of responding to their needs. Their cry for bread is a motif that runs through the rest of Carlyle's history. The actions of the women are at once admirable and terrible, courageous and insane. They are a mixture of Judiths, taking morally justifiable action when authority is paralyzed and the men overawed, and Menads, drawing their energy from the most instinctual level of the psyche. Their will is to live and to foster life, but the constant parallel Carlyle draws between these women and the Menads emphasizes the potential for violence that finally breaks out when the guards fire on the crowd, which then storms the palace to the very door of the Queen's chamber, tearing guards apart almost as the Menads of old.

The personified form of these women, Judith and Menad, is the Girondist Demoiselle Théroigne. A Pallas Athena (but no virgin) who disarms the guards at Versailles with "soft arms," she is also a cannoneer who goes armed herself, leads troops into battle, and even handles diplomatic negotiations. But after being caught, stripped and beaten by the Jacobin women in the Tuileries, she goes mad and spends her last 33 years in asylums. Carlyle introduces her and suggests her fate by imagining himself looking in vain for her "brown eloquent beauty" in the procession of May 4, 1789:

> . . . pike and helm lie provided for thee in due season; and, alas, also straight-waistcoat and long lodging in the Salpêtrière! Better hadst thou stayed in native Luxemburg, and been the mother of some brave man's children: but it was not thy task, it was not thy lot. (FR 1, vol. II, p. 135)

Carlyle's final image of Demoiselle Théroigne is an inversion of the myth of Actaeon, suggesting that she has violated her own sexual nature: "Brown-locked Diana (were that possible) attacked by her own dogs, or she-dogs" (FR 3, vol. IV, p. 154).

Théroigne is no sooner removed in straight-waistcoat from the stage of Carlyle's history than Charlotte Corday makes her entrance, writing falsely to her father that she has fled to London, and then making her way to Paris. Carlyle treats the encounter between "hapless, beautiful Charlotte and hapless, squalid Marat" as a central emblem of revolutionary anarchy and its perversion of the natural urge for connection. It is his prelude to the Terror. Marat to Carlyle is the Simon Stylites of revolution, the fanatic, demonic parody of a hero. It takes the request of Marat's brother after the assassination for his musket to remind Carlyle "that Marat too had a brother and natural feelings." Charlotte Corday is Marat's fanatical contrary, convinced that in slaying Marat she saves a hundred thousand. Carlyle draws them together, she from the west of France, he from the east, as by a perverse elective affinity into an antithesis of a sexual embrace. ". . . Charlotte has drawn her knife from the sheath; plunges it, with one sure stroke into the writer's heart. 'À moi, chère amie, Help, dear!' no more could the Death-choked say or shriek." (FR 3, vol. IV, p. 169). The act being perverse, the results are the reverse of the "angelic-demonic" Charlotte's intent.

> 'Day of the Preparation of Peace?' Alas, how were peace possible or preparable, while, for example, the hearts of lovely Maidens, in their convent-stillness, are dreaming not of Love-paradises and the light of Life, but of Codrus' sacrifices and Death well-earned? That Twenty-five million hearts have got to such temper, this is the Anarchy; the soul of it lies in this: whereof not peace can be the embodiment! The death of Marat, whetting old animosities tenfold, will be worse than any life. O ye hapless Two, mutually extinctive, the Beautiful and the Squalid, sleep ye well, —in the Mother's bosom that bore you both! (FR 3, vol. IV, p. 172)

Young Adam Lux, presumably a proper "light of Life" for the maiden Charlotte's convent dreams in normal days, inspired by the sight of virgin beauty going peacefully to her death, declares "that it were beautiful to die with her" and is, in fact, guillotined for defending her in pamphlets. Their story is a perverted comedy, substituting death and social disintegration for marriage and social revival.

Clearly one of the tragedies of revolution for Carlyle is that it diverts women from their nurturing, domestic roles into assertive public ones that provoke nemesis. Carlyle's allusions to biology and mother nature, his depiction of women forced to seek their family's bread, and of Demoiselle Théroigne and Charlotte Corday forced or lured into fatal perversity—all assume that males are naturally active and females passive. Nevertheless, although Carlyle's female martyrs of the Terror exhibit the noble passivity of so many heroines of Victorian fiction, his model of womanhood, Madame Roland, was notably forceful. She fed the mind as well as the body and engaged in active partnership with her husband. A woman of the middle class, at once intellectual and domestic, she maintained a radical salon in the early days of the Revolution and was actually the author of many writings that bore her husband's name. In prison she charmed the guards with conversation as "frank and courageous as that of a great man." At the foot of the scaffold she asked "for pen and paper, 'to write the strange thoughts that were rising in her.' " Her last wish, as Carlyle tells it, is to die first, to ease the fear of the man with whom she was to be executed. She is Carlyle's model of woman the nurturer.

> Honour to great Nature who, in Paris City, in the Era of Noble-Sentiment and Pompadourism, can make a Jeanne Phlipon [Roland], and nourish her to clear perennial Womanhood, though but on Logics, *Encyclopédies*, and the Gospel according to Jean-Jacques! . . . She left long written counsels to her little Girl; she said her Husband would not survive her. (*FR* 3, vol. IV, p. 211)

Indeed, upon hearing of his wife's execution, Roland left his sanctuary and killed himself. "I wished not to remain longer on an Earth polluted with crimes."

These individual lives, history incarnate, are also emblems "of that grand Miraculous tissue, and Living Tapestry named *French Revolution*" (*FR* 2, vol. III, p. 185). Turning from the fate of individuals to that of the larger body of society, we find Carlyle again building upon biological analogy to explain how the Girondist devotion to abstract principle inevitably lost out to the organization of the Jacobin Mountain.

> . . . all stirs that has what the Physiologists call *irritability* in it: how much more all wherein irritability has perfected itself into vitality, into actual vision, and force that can will! All stirs; and if not in Paris, flocks thither. (*FR* 2, vol. III, p. 19)

There is in nature an "Aggregative Principle" that expresses itself, in a time of weak central authority in France, through the formation of clubs, of which the Jacobin Club grows preeminent. Its "Mother Society" in Paris "breeds and brings forth Three Hundred Daughter Societies . . . Jacobinism shoots forth organic filaments to the utmost corners of confused dissolved France; organising it anew: — this properly is the grand fact of the Time" (*FR* 2, vol. III, p. 110).[8]

> Fatal-looking! Are not such Societies an incipient New Order of Society itself? The Aggregative Principle anew at work in a Society grown ob-

solete, cracked asunder, dissolving into rubbish and primary atoms? (*FR* 2, vol. III, p. 34)

Jacobin filiation forms a new constitution, not made by august Senators, "but by nature itself out of the want and efforts of Twenty-five Millions of men." The organic filament, the most elemental form of life, transforms its own irritability into organized growth. An analogous social impulse moves individuals to form the natural bonds of the family and to create larger organizations that serve a common purpose. Nations are themselves "collective individuals." A France invaded and seemingly helpless in her "dead cerements of a Constitution" rends them, "and she fronts you in that terrible strength of Nature, which no man has measured, which goes down to Madness and Tophet" (*FR* 3, vol. IV, p. 1). But this rising of France in response to invasion, like the insurrection of women provoked by hunger and insult, is an instinctual response to the threat of death and has no permanent form.

Twice in the course of the Revolution it seemed to Carlyle that men who understood the organic movement of the history in which they were engaged and who had the power of mind and will to help shape that movement to a stable end, a proper human form, nearly took command, but proved heroes out of phase with history. One was Mirabeau, who died worn out by his labors before achieving the power that might have enabled him to forge a constitutional monarchy; the other was Danton, who, "like Mirabeau, has a natural *eye*, and begins to see whither Constitutionalism is tending, though with a wish in it different from Mirabeau's" (*FR* 2, vol. III, p. 19). Carlyle presents a Danton who saw the necessity of extreme measures during a period of foreign invasions, but who resisted those who, in effect, substituted Terror for government. He was, in other words, a man endowed with the capacity to govern. The rejection of Danton became the rejection of government that led the Revolution to devour itself.

Carlyle's doctrine of hero worship is analogous in one respect to the traditional mystery of the king's two bodies: that is, in addition to possessing a private being, the hero has a collective aspect or body politic. The realized hero is the product of a double election: heavenly election — for the hero is one of Carlyle's "intrinsic symbols," worthy to be followed because he manifests an aspect of the eternal "through the Time-Figure (*Zeitbild*)!" (*SR*, p. 223)—and mundane election, the worship of his followers. The hero is to his followers as the organic filament is to the nutritive material in the womb: he gives them shape; they given him political substance. Confident that government is impossible without him, Danton refuses to plot against those who oppose his more moderate course of action. He returns from the spring greenness of the "everlasting Mother" and his "household loves" in Acris and is arrested. Tried, he pits his being against the power of his accusers. Were Danton to be acquitted, says Carlyle, were the "amorphous Titan" to take definite political shape, the whole history of France would be altered.

> For in France there is this Danton only that could still try to govern France. He only, the wild amorphous Titan; — and perhaps that other olive-complexioned individual, the Artillary-Officer at Toulon, whom we left pushing his fortune in the South? (*FR* 3, vol. IV, p. 258)

Thus Carlyle looks past Robespierre to Napoleon, reminding us that all through *The French Revolution* he has quietly, with brief references, presented the gestation of the hero to whom France will soon give birth. Danton's execution leaves Robespierre alone on center stage, but Carlyle thinks him, like Marat, the antithesis of a hero, a demonic force that is in essence a negation. He is a man consumed with "feminine hatred," the very model of a Jesuit or a Methodist parson, a man whose very purity is evidence of his narrowness. "Green," or "seagreen" Carlyle constantly calls him, seizing upon accounts of his complexion to associate him with envy and a kind of effeminacy, as if he were afflicted with the "green sickness" of adolescent girls. (While Carlyle can accept a good measure of "male" intellect in women (Madame Roland, for example), his ascription of female attributes to a man is consistently derogatory.) Having loosed anarchy once more, Robespierre and his associates are devoured by it, executed by a Committee fearful of its own safety. The Terror is ended; the Jacobins suppressed.

Carlyle's Napoleon is a less admirable man than either Mirabeau, who exhausted himself working against the tide of history, or Danton, whose public body, in effect, deserted him at his trial, leaving him to fall in near tragic isolation. But Napoleon is heroic because he drills sansculottism into order. On the "anniversary of that Menad-march, six years ago" he gives the necessary "Whiff of Grapeshot" (*FR* 3, vol. IV, p. 320). He tames the French Revolution "that it may become *organic*, and be able to live among other organisms and *formed* things . . ." (*Heroes and Hero Worship*, vol. V, p. 240).

III. REVERSION

One of Carlyle's purposes in using scientific vocabularies in his writing is to suggest that moral and physical laws are analogous, having the same divine source. But he compares what inevitably happens in the natural world not only with what happens in the social world, but also with what ought to happen. The organization of life in the womb suggests an analogy to the organization of the family, which in turn provides a model for the proper organization of industry and the state. By drawing metaphors suggesting Erasmus Darwin's optimistic theory of development and the more systematic speculations of the German *Naturphilosophen*,[9] Carlyle could make what he wanted to see happen sound inevitable. The social "chaos of being," like the nutritive or organic particles in the womb must be organized by an incarnate essence — an organic filament, a Captain of Industry, a Hero. But on the social level it is possible that the wrong choices will be made: the Captains of Industry may deny their duty, the people may call forth a sham rather than a Hero. If they do, then England, like France, will have to undergo a period of chaos, a social death, before it can be reborn.

Carlyle posits the choice between social life and social death in *Past and Present* (1843), which sets an organic medieval order with a place for everyone and most everyone in his place against a modern world on the verge of chaos because it is replacing essential social ties with the "unnatural" principles of competition, supply and demand, cash nexus. "Men cannot live isolated: we *are* all bound together, for mutual good or else mutual misery, as living nerves in the same body. No highest man can disunite himself from any lowest" (*P & P*, vol. X, p. 286).

Just as the marriage vow formalizes a natural tie and reinforces it with social sanctions, both positive and negative, so, Carlyle hoped, might the emerging industrial order reject individual contracts for labor and return to the model of the household. Such relationships would be permanent, or nearly so, and the employer would thus be hiring a worker and not just so many hours of work. Such a "fabric of law" would make the duties of government what they should be, those of a father, not those of a constable enforcing such limits on everyone's pursuit of self-interest as might be necessary to keep the machinery of society in gear.

But what Carlyle saw as the greatest act of statemanship of his day, the repeal of the Corn Laws in 1846, led not to the strengthening of Sir Robert Peel's power, but to his downfall. The revolutions of 1848 arrived and passed, leaving neither a stable new order nor a revivified kingship in place of the forces Carlyle associated with chaos and anarchy. At mid-century Carlyle saw, instead of social rebirth on the familial and neo-feudal model of hierarchical interdependence, instead of a new social marriage, an apocalyptic divorce:

> From the 'Sacrament of Marriage' downwards, human beings used to be manifoldly related, one to another, and each to all. . . . But henceforth, be it known, we have changed all that, by favour of Heaven: 'the voluntary principle' has come-up, which will itself do the business for us; and now let a new Sacrament, that of *Divorce*, which we call emancipation, and spout-of on our platform, be universally the order of the day! ("The Present Time," *Latter Day Pamphlets*, vol. XX, p. 25)

Always convinced that mere ballot-box elections could not express the real will of a people, Carlyle found to his horror that the people had in truth elected not the industrial duke, fit hero of his latter-day epic "Tools and the Man," but the speculator, Hudson the Railway King. The "Apotheosis of Hudson" to his satanic "bad eminence" is, to Carlyle's dismay, the product of actual hero worship.

> Hudson solicited no vote; his votes were silent voluntary ones, not liable to be false: he *did* a thing which men found, in their inarticulate hearts, to be worthy of paying money for; and they paid it. . . . Without gratitude to Hudson, or even without thought of him, they raised Hudson to his bad eminence, not by their voice given once at some hustings under the influence of balderdash and beer, but by the thought of their heart, by the inarticulate, indisputable dictate of their whole being. ("Hudson's Statue," vol. XX, pp. 264–265)

In the 1830s Carlyle could write with equanimity of the two centuries that would be required for the establishment of the new order, and have Teufelsdröckh point to organic filaments "mysteriously spinning themselves" as the best feature of the evolving historical spectacle. But the spectacle proved unedifying. In place of the idyllic vision in *Sartor Resartus* of Swiss housewives sustaining and nurturing their families, he sees the nightmare reality of a London with 30,000 distressed Needlewomen who never learned to sew; "three-million paupers rotting in forced idleness, *helping* said Needlewomen to die . . ." ("The Present Time," vol. XX, p. 27). The concept of development carries the seeds of its opposite, reversion, and in his later work Carlyle reverses the direction of his biological analogies and stresses the fall from order into chaos and elemental particles.

Carlyle saw that he was destined to die in Moab with England yet "ruleless, given up to the rule of Chaos, in the primordial fibres of its being" ("Hudson," vol. XX, p. 258). His vision of the future through the mouthpiece of Gathercoal is of apes, marching into Chaos, "that land of which Bedlam is the Mount Zion" ("Jesuitism," vol. XX, p. 331), for like the kingdom of Satan, that of Chaos can only parody divine order. In the pessimism of his late years, Carlyle's social extension of the organic filament becomes a thing of the past, the last such growth coming from the "Martyr Heroism" of Jesus ("Jesuitism," vol. XX, p. 332). The "firm regimented mass" Carlyle called upon his Captains of Industry to organize in *Past and Present* was to be an organic growth based upon a mutual recognition between men who know in their hearts that they must be led and those whose conscience prompts them to leadership. Certainly the world of Carlyle's late essays may reflect his own disappointments more than he realized. Nevertheless the inorganic, atomized, disgusting world of filth, decay, disease and death of *The Latter-Day Pamphlets*, wherein men are as animals, was Carlyle's portrait of what he believed the collective social and religious imagination of his countrymen had created through the worship of Mammon. It is not a "chaos of being," from which order will naturally emerge, but simply chaos. The biological analogies in *The Latter-Day Pamphlets* form images not only of divorce, but of decadence — a return to primary atoms in the fibers of being, men reduced to apes. If mankind can evolve, so can it degenerate.[11] Carlyle's social extensions of the biological terms he learned early in the century anticipate both possibilities.

NOTES AND REFERENCES

1. John Holloway, *The Victorian Sage* (New York: Norton, 1965), p. 37. Holloway's is the standard analysis of Carlyle's rhetoric.

2. Although an idealist and outspoken opponent of what he saw as the mechanistic reductivism of Victorian science, Carlyle was trained in mathematics and Newtonian physics, reviewed and translated scientific works, and through both denunciation and metaphor displayed in his work familiarity with a wide range of scientific ideas. He seems, however, to have made little attempt to keep his knowledge current after creating his unique version of *Naturphilosophie* in the 1830s. For the influence of this early training on his work see Carlyle Moore's "Carlyle: Mathematics and 'Mathesis' " in *Carlyle Past and Present*, eds. K. J. Fielding and Roger L. Tarr (London: Vision Press, 1976), p. 61ff and "Carlyle and Goethe as Scientist" in *Carlyle and His Contemporaries*, ed. John Clubbe (Durham, N.C.: Duke University Press, 1976), p. 21ff. G.B. Tennyson published a calendar of Carlyle's reviews, translations and encyclopedia articles in *The Sartor Called Resartus* (Princeton: Princeton University Press, 1965), p. 331ff — the best analysis of the structure of *Sartor Resartus*. Frank M. Turner records Carlyle's friendships with Victorian scientists and his influence upon them in "Victorian Scientific Naturalism and Thomas Carlyle," *Victorian Studies*, vol. XVIII (1975), pp. 325–343. These studies do not treat Carlyle's biological imagery, perhaps because he made no special study of the subject.

3. *Sartor Resartus*, ed. Charles F. Harrold (New York: Odyssey Press, 1937), p. 55. All references to *Sartor Resartus* will be from this edition. Quotations from his other works will be from the Centenary Edition of *The Works of Thomas Carlyle*, ed. H.D. Traill (New York: Scribner, 1899) and will be cited by title, volume and page number in the text. The

metaphor of flesh as cloth is a venerable one, going back at least as far as Porphyry's commentary *On the Homeric Caves of the Nymphs*, whose weaving of purple robes he interpreted as the formation of flesh on the bones. Later it became a figure for the incarnation (see George Herbert's poem "The Bag.") Carlyle's play upon the word "tissue" as both cloth and flesh revives its root in the French and ultimately Latin verbs "to weave," *tisser, texere*, this last being the root of "text," the clothing of Carlyle's ideas and the material upon which Teufelsdröckh's Editor demonstrates his sartorial skill.

4. Erasmus Darwin, *Zoonomia* (Dublin: P. Byrne, 1800), vol. I, p. 596. In later years Carlyle remarked: "What they call Evolution is no new doctrine. I can remember when Erasmus Darwin's Zoonomia was still supplying subjects for discussion, and there was a debate among the students whether men were descended from an oyster or a cabbage." Carlyle's conversation was recorded by Moncure Conway in *Thomas Carlyle* (New York: Harper, 1881), p. 84.

5. Although Darwin's vocabulary is the closest I have found to Carlyle's, it is not identical, and there is a wide range of possible sources for one or another of his terms. Carlyle wrote on Diderot and may very well have known his comprehensive article on "fiber" in the *Encyclopédie*. For a survey of fiber theories see Thomas Hall's *Ideas of Life and Matter*, vol. 2 (Chicago: University of Chicago Press, 1969), pp. 5–107.

6. Carlyle to Leigh Hunt, 18 July 1833, *The Collected Letters of Thomas and Jane Welsh Carlyle*, vol. 6, ed. Charles R. Sanders *et al.* (Durham, N.C.: Duke University Press, 1977), p. 418.

7. For a study of the immediate social and political context of *The French Revolution*, see Patrick Brantlinger's chapter on "The Lessons of Revolution" in *The Spirit of Reform: British Literature and Politics, 1832–1867* (Cambridge: Harvard University Press, 1977).

8. Philip Rosenberg connects the organic filaments of *The French Revolution* with those of *Sartor Resartus* in a different context in *The Seventh Hero* (Cambridge: Harvard University Press, 1974), p. 126.

9. For a brief account of the developmental hypotheses of the *Naturphilosophen* with whom Carlyle shared roots in Goethe, Herder and Schelling, see Stephen Jay Gould, *Ontogeny and Phylogeny* (Cambridge: Harvard University Press, 1977).

10. Jules P. Siegal discusses animal imagery in his *"Latter-Day Pamphlets*: The Near Failure of Form and Vision" in Fielding and Tarr,[2] pp. 155–176.

11. For a Darwinian analysis of reversion see Edwin Lankester, *Degeneration, a Chapter in Darwinism* (London: Macmillan, 1880).

Darwin and Landscape

JAMES PARADIS

Department of Humanities
Massachusetts Institute of Technology
Cambridge, Massachusetts 02139

In the ancient world, nations, and the distinctions of their civilization, formed the principal figures on the canvass; in the new, man and his productions almost disappear amid the stupendous display of wild and gigantic nature.

— ALEXANDER VON HUMBOLDT, *Personal Narrative*

[I]

I N THE *Journal of Researches* (1839), Charles Darwin's two versions of South American landscape reflect the aesthetic idealism of Romantic art and the system-building traditions of the geological and natural sciences.[1] On the one hand, Darwin carries a substantial personal narrative throughout the text, describing his own emotional impressions of the rich physical detail of South America. He refers to these emotions, which he feels vividly, as "relic[s] of an instinctive passion," identifying them with the organic pleasures of "the savage returning to his wild and native habits."[2] Darwin, in this personal perspective, unifies the diverse elements of the landscape through the power of emotion, creating a sense of the harmony and integrity of nature that is essential to landscape art, whether literary or visual. On the other hand, Darwin systematically organizes the same physical detail of the terrain into classes of biological and geological fact, modeled upon the descriptive patterns of earlier theoretical works of such authors as Alexander Humboldt and Charles Lyell. In his technical descriptions, Darwin dismantles his initial phenomenological experience in nature and sorts the elements of the landscape into categories of fact. In the text of the *Journal*, Darwin consciously shifts from one descriptive mode to another, weaving his less formal aesthetic impressions with the more structured enumerations of specific landforms, fossils, and species. Two idealizations of the perceived landscape thus emerge in Darwin's voyage account, each a distillation of the common subject matter of visible nature, manifested in the separate but not entirely distinct vocabularies of personal meditation and systemic science.

The effect of Darwin's double representation of his voyage experience is to link the visual splendor of natural landscape with the schematic organization of its many specific forms. From the visual unity of the observer's momentary perception, we move to the conceptual diversity of his accumulating record of fact, unified through the force of generalization. From the impressionistic result of a landscape of selected forms in approximate spatial relationships with one another, we move to a more elaborate and structured result of nature as the geographical and temporal distribution of thousands of specific forms. This narrative movement from the aesthetic impulses of Darwin's "instinctive passion" to the analytical responses of Darwin's directed intellect is a descriptive pattern repeated throughout the *Journal*.

As a term, "landscape" historically has taken a number of sometimes conflicting meanings, and I should like to make some preliminary distinctions before I proceed. Landscape is primarily an aesthetic term, traditionally referring both to the artistic genre of landscape painting and to the more general literary notion of an expanse of scenery taken in at a glance. In this essay, I intend to use the term in its more general sense, the way Samuel Taylor Coleridge used it in his poem, "Lines Composed While Climbing Brockley Coomb" (1795): "Ah, what a luxury of landscape meets / My gaze!"[3] Darwin used the term similarly in the *Journal of Researches* to refer to sudden, striking sense impressions of scenes taken in at a glance.[4] For both Coleridge and Darwin, as for most early nineteenth-century Romantics, landscape was a term of human perception, most often the complex sense impression of some striking moment of experience in physical nature. It was an aesthetic category for the physical nature mediated through sight and feeling: the landscape of the emotions. Both Coleridge and Darwin, moreover, assumed that the physical origins of this emotional response could be recreated through memory in poetic and narrative description. Hence, the elaborate attempts of both men at physical description of the picturesque.

Darwin's idea of landscape, however, took on some strikingly new dimensions in the late 1830s as his technical knowledge of physical nature began to expand. In a notebook passage written the month he discovered natural selection (October 1838), Darwin extended the concept of landscape; he reflected on William Wordsworth's Preface to the *Lyrical Ballads* (1800) and acknowledged the poetic dimension of his own developing vision of geological process: "I a geologist, have ill-defined notion of land covered with ocean, former animals, slow force cracking surface etc truly poetical. (V. Wordsworth about the sciences being sufficiently habitual to become poetical.)"[5] This suggests not only that Darwin's early sense of natural beauty was shaped by the Romantic tradition of Wordsworth and Coleridge, but also that Romantic aestheticism continued to color Darwin's developing physical theories of nature.

Throughout the period when he was writing the Transmutation Notebooks and the *Journal of Researches*, Darwin was fascinated with the aesthetic category of the "sublime"; that is, the emotional sensation of vastness one felt upon viewing an immense landscape from a great distance. The term is used repeatedly by Darwin in the notebooks and *Journal*, and he returns to this vision of "grandeur" in the famous "entangled bank" passage at the end of his *Origin of Species*. If, as has been noted by Walter Cannon, Darwin's argument in the *Origin* was preponderantly a visual, demonstrative exposition, I would add that it was conceived as a species of landscape, with emotional overtones of the sublime, the chord on which it ends. To follow Darwin, Cannon argues, one must "see, above all, . . . the whole surface of the earth at all times."[6]

This large view, however, had swelled through the power of Darwin's abstractions to a scale that eclipsed all earlier Romantic notions of landscape. Its unity was no longer the result of an immediate act of perception in physical nature, but, rather, the result of Darwin's aesthetic response to a mental landscape founded upon generalization. It was a landscape of metaphors such as "species," of a multitude of forms ineluctably flowing and diversifying over the face of the earth through the immensity of time. "The map of the world ceases to be a blank," Darwin cor.cluded at the end of the *Journal*; "it becomes a picture full of the most varied and animated figures." It was thus a landscape that combined the precision of the cartographer with the aesthetic sense of the landscape artist, thereby establishing an unprecedented concept of natural space and organic position. The relationships among the forms of such landscape composites as those of Darwin's Galapagos Archipelago were specific and precise in systematic ways that contrasted markedly with the spontaneous harmony that seemed always to obtain among forms in Romantic landscape descriptions, landscapes which, as Wordsworth observed, revealed "the passions of men . . . incorporated with the beautiful and permanent forms of nature."[7]

Although the two perceptions of landscape were by no means unconnected in Darwin's imagination, the new global perspective created a formidable discontinuity between human knowledge and sensation for those who did not comprehend Darwin's complex conceptual structure. The difficulty of Darwin's mature vision led in the Victorian age to a feeling that the representations of physical nature in literature conflicted with those in the natural sciences.[8] Already by 1800, in his Preface to the *Lyrical Ballads*, Wordsworth had drawn a contrast between the immediate, poetic sensation of nature and the more removed reconstructions of the systematic observer. Wordsworth's sense of the discontinuity between natural knowledge and poetic sensation was reiterated in a variety of Victorian works, notably in the essays and *Autobiography* of John Stuart Mill and the essays and poems of Matthew Arnold. Both Mill and Arnold believed that there was a primary division between the physical universe of the sciences and the emotional world of art.

I would like to examine more closely the manner in which Darwin's early Romantic vision of landscape blended with his developing views of the physical organization and history of organic life. The traditions both of Romanticism and the natural sciences had inspired Darwin as a young man;[9] until at least the age of thirty, the age at which he published his *Journal of Researches*, Darwin, who had grown up in the midst of the Romantic era, took great "pleasure" in his reading of the works of "Milton, Gray, Byron, Wordsworth, Coleridge, and Shelley."[10] Nevertheless, in the last years of his *Beagle* voyage, through the period he worked on his Transmutation Notebooks and his *Journal of Researches* —that is, from the mid-1830s on—Darwin, as he approached thirty, found his views of nature's physical order dramatically changing. He recalled in his *Autobiography* that his "love for science gradually preponderated over every other taste" during this period. The theoretical background of Darwin's changing aesthetic views, his new concepts of natural space, and the manner in which his theoretical views departed from certain nineteenth-century Romantic assumptions about physical landscape, will be the primary topics of my essay.

[II]

Behind Darwin's landscapes in the *Journal of Researches* were the twin forces of nineteenth-century Romanticism and natural science. Behind these, in turn, lay the natural philosophies of Samuel Taylor Coleridge and John Frederick Herschel, whose discussions of human perception help us to understand the order the Romantic poet and the naturalist created from their respective experiences in physical nature. Coleridge, whose study of human perception and imagination in *The Friend* (1809–10) and *Biographica Literaria* (1817) revealed the architecture of English Romanticism, was obliged to examine the new descriptive thrust of the natural sciences, the materials of which partially belonged to the repertory of Romantic nature. An enthusiast of contemporary science, yet its severe critic as well, Coleridge sought to balance the physical world of the natural philosopher with the spiritual world of the artist and thus to provide a unified theory of human knowledge and perception. Herschel, by contrast, believed that ontological questions were unanswerable and raised grave doubts about the accuracy and indeed value of unaided human perception itself. A vigorous intellectual descendant of Bacon and of eighteenth-century Rationalism, Herschel raised the inquiries of the intellect to the scale of global nature, made Newton's *vera causa* the basis of natural explanation, and installed technical language systems or nomenclatures at the operational centers of various systems of natural knowledge. Herschel's Baconian distrust of sense experience and emotion as avenues to physical truth and causality was characteristic of an attitude that in the nineteenth century led to a simple but profound divergence of viewpoint between poet and natural philosopher.

Coleridge assumed that nature was a reality divided by the act of human perception into the *forma formans* or "inward principle of whatever is requisite for the reality of a thing," and the *forma formata*, or "material nature," the "sum total of all things, as far as they are objects of our senses."[11] These twin facets of nature were disclosed to one's "inner" (spiritual) sense and one's "outer" (material) sense, both of which could operate simultaneously or selectively, according to the preference or prejudice of the perceiver. The natural philosopher, Coleridge argued, made it his business to study material, phenomenological nature, "resist[ing] the substitution of [final for efficient causes] as premature, presumptious, and preclusive of all science."[12] Hence, the natural philosopher accepted the *forma formans*, the world of the inner sense, as a category beyond man's capacity to experience, a category that could be summed up as the Deity discovered in the design.

In Coleridge's metaphysical scheme, both inner and outer senses could operate at once, and it was possible to discover the simultaneity and continuity of spiritual and material phenomena and thus to apprehend the reality of nature. This cardinal Romantic notion was based on the assumption that the mind itself was a part of the *forma formans*, and that the reality behind physical nature was continuous with the human intellect. The "ideas of the mind" and the "laws of nature," Coleridge argued, were the "subjective and objective poles of the same magnet, that is, of the same living and energizing reason."[13] The seeker of knowledge, whether poet or natural philosopher, was driven by a primary desire to discover the unity and harmony of all existence. The chemist, for example, must ultimately confront the indecomposable substances of material nature in his effort to extend the boundaries of his discipline. These same substances, Coleridge held, were the code of a higher reality — they were "the symbols of elementary powers, . . . the exponents of a law, which, as the root of all these powers, the chemical philosopher, whatever his theory may be, is instinctively laboring to extract."[14] Hence, in spite of his disregard or disdain for metaphysical causes, the natural philosopher labored instinctively to discover them.

The "instinct" Coleridge spoke of was the source of the human drive to transcend distinctions between subject and object, an instinct "by which, in every act of conscious perception, we at once identify our being with that of the world without us, and yet place ourselves in contradistinction to that world." This instinct could not exist without "evolving a belief that the productive power, which in nature acts as nature, is essentially one (that is of one kind) with the intelligence, which is in the human mind above nature. . . ."[15] Such radical union of mind with the powers of nature joined the perceiver with the perceived, intellect and emotion becoming elementary forces of nature.

Given Coleridge's Romantic vision of mind and nature, one is not surprised at his criticisms of certain trends in chemistry and the natural sciences, where descriptive systems that had begun to bypass metaphysical considerations had also begun to analyze the familiar organic nature that had been the physical domain of the poet. In his "Hints towards a More Comprehensive Theory of Life," a *tour de force* of Romantic metaphysics, Coleridge attacked the new "French theory of chemistry," which had brought clarity and order to chemical science largely through the establishment of a new nomenclature. Suspicious of the detour around metaphysical questions taken by Lavoisier and his colleagues at the French Academy in 1787, Coleridge objected to the growing practice among natural philosophers in which nomenclatures were "substituted . . . even in common conversation," for the far more "philosophic" language which the human race had "abstracted from the laboratory of nature."[16] This brilliant linguistic reductionism, Coleridge warned, had now "become the common road to all departments of knowledge," and, under the delirious excitement of intellectual revolution, the chemist now sought through experiment "to penetrate . . . the secret recesses, the sacred adyta of organic life."[17] In their use of fixed and precise language, the new chemists structured their experience and knowledge in a verbal system that rigidly excluded the ontological language of the Romantic internal landscape. Language was now the great barrier between poet and natural philosopher.

In *The Friend*, Coleridge rejected Linnaean taxonomy as an activity unworthy the name of science, for while Linnaeus had brought, through language, order to the immense botanical world, he fell short of true philosophical achievement: "The master light is missing."

> All that can be done by the most patient and active industry, by the widest and most continuous researches; all that the amplest survey of the vegetable realm, brought under immediate contemplation by the most stupendous collections of species and varieties, can suggest; all that the minutest dissection and exactest chemical analysis, can unfold, all that varied experiment and position of plants [can lead to, is . . .] little more than an enormous nomenclature; a huge catalogue, well arranged, and yearly and monthly augmented, each with its own scheme of technical memory and its own conveniences of reference.[18]

The activities of artificial taxonomists like Linnaeus were part of a blind mnemonic impulse that Coleridge believed would destroy language itself. The specialist operated in a highly disciplined system of verbal reference that excluded the emotional, sensuous language of the artist. The linguistic revolution in the sciences of chemistry, geology, biology, not to mention a dozen emerging disciplines, ignored ontology. Yet, all these surveying, collecting, dissecting, and augmenting activities, Coleridge believed, contributed only to the superficial systematizing of nature; they gave little insight into the deeper structure and reality of things. The chemical philosopher who "reduces the chemical process to the positions of atoms," Coleridge observed, "would doubtless thereby render chemistry calculable, but . . . he commences by destroying the chemical process itself, and substitutes for it a *mote dance* of abstractions."[19] Linnaeus, whose artificial system of classification was based on sexual characteristics of plants, failed to discover, Coleridge noted, the "central idea of vegetation itself," by means of which he might have discovered the "constitutive nature and inner necessity of sex itself." Coleridge concluded that "the full applicability of an abstract science ceases, the moment reality begins."[20]

The most cogent answer to Coleridge's critique of the limits of natural philosophy is found in John Herschel's seminal book on the method and philosophy of science, *Preliminary Discourse on the Study of Natural Philosophy* (1830), which Darwin claimed had profoundly influenced his own early thought. Where Coleridge sought to validate ontology as the proper object of human perception, Herschel rejected the idea that reality was the proper, or indeed possible, subject of the natural philosopher. Science, Herschel held, "is the knowledge of many, orderly and methodically digested and arranged, so as to become attainable by one."[21] In order to accumulate and then to arrange such an extensive fund of knowledge, one had first to agree on terms and names. The study of nature, Herschel held, was one which began with the inventory of nature's content, the first step toward gaining control:

> Before we can enter into any thing which deserves to be called a general and systematic view of nature, it is necessary that we should possess an enumeration, if not complete, at least of considerable extent, of her materials and combinations; and that those which appear in any degree important should be distinguished by names which may not only tend to fix them in our recollection, but may constitute, as it were, nuclei or centres, about which information may collect into masses. The imposition of a name on any subject of contemplation, be it a material object, a phenomenon of nature, or a group of facts and relations, looked upon in a peculiar point of view, is an epoch in its history of great importance.[22]

The construction of a nomenclature, particularly one based on natural classification, enabled the investigator to position things in relationships with one another, fixing categories for what Coleridge had called the *forma formata*, the material nature of things perceived by the senses.

Herschel departed from metaphysics in his theory of causality, which eliminated considerations of ontology. Using Newton's term, *verae causae*, for "causes recognized as having a real existence in nature, and not being mere hypotheses or figments of mind," Herschel argued that events in nature could only be explained by *verae causae*.[23] Hence, the phenomenon of shells found at great heights above the sea should not be traced to such "figments of fancy" as the "plastic virtue" of the soil, but should find explanations in common processes already observed to have taken place — in this instance, geological elevation. Natural conditions were to be explained in terms of physical processes actually observed in nature. This restriction on causality eliminated Coleridge's *forma formans*, divorcing mind and spirit from physical nature. Herschelian standards of causality eliminated human intuition as the sufficient basis for determining truth, locating the study of physical processes in an empirical method that did not acknowledge Coleridge's inner sense.

As Bacon had, Herschel argued that the senses were most imperfect instruments of observation, not only because they could not provide precise, calculable results, but also because they were not adapted to the physical scale of natural process:

> The mechanism of nature is for the most part either on too large or too small a scale to be immediately cognizable to our senses; and her agents in like manner elude direct observation, and become known to us only by their effects. It is in vain therefore that we desire to become witnesses to the processes carried on with such means, and to be admitted into the secret recesses and laboratories where they are effected.[24]

What was for the Romantic artist the cardinal link between truth and sensation, became for Herschel a link of doubtful dependability. While Herschel's argument that the primary processes of physical nature were invisible was not in itself in conflict with Romanticist metaphysics, his argument was based on physical scale rather than on possible spiritual origins. The processes of physical nature were hidden from the senses because they occurred on magnitudes of space and time that human physiology was not equipped to apprehend. Hence, Herschel argued that the data of human perception could be improved through repetitious observation, the average being more precise than any single instance could be.[25] Periodic observation established a statistical average, which Herschel believed improved the accuracy of any single observation, a method that was completely opposed to the Romantic belief in the uniqueness of each individual experience and moment in nature.

Coleridge's and Herschel's conflicting theories of perception explain how the physical basis of Romantic landscape art was unintentionally subverted by a growing body of physical theory that rigorously analyzed, defined, and, thus, laid claim to the same geological and biological spaces of nature. The Romantic landscapes transmitted through the poetry of Coleridge and Wordsworth were the reconstructions of intense moments in physical nature, in which sense perceptions, colored by human emotion and the inner, intuitive senses, revealed the unity of all existence. These landscapes were characterized in part by their (1) human scale and orientation (even in the instance of the sublime),[26] (2) qualities of permanence and harmony, and (3) emotional and spiritual content. They were landscapes of sensation and emotion, created through direct sense impressions that led to higher realms of the ideal, to vision itself, and, hence, they embodied the spirituality of man. The Romantic terrain was one with the observer, joined in an essential unity of subject and object.[27] It was the landscape of intuition, the landscape Wordsworth spoke of in his 1800 Preface to the *Lyrical Ballads*, in which he held that poetry reflected the "passions of men . . . incorporated with the beautiful and permanent forms of nature."[28] For Wordsworth, as for Coleridge, physical nature was the emblem of permanence and harmony, joined with the mind in the act of perception in which "almost suspended, we are laid asleep / In body, and become a living soul."[29]

In contrast to the landscapes of the Romantic artist, the physical nature of the nineteenth-century naturalist, as transmitted through the views of Herschel and, particularly, Darwin, was characterized by its (1) global scale, (2) impermanence and, for Darwin, disharmony, and (3) calculated physical organization in space and time. No less the product of the imagination than its counterpart, the naturalist's landscape was conceptual and architectonic, extending beyond moments of perception to the continental scope of land formations and species distribution, to the temporal dimension of the prehistoric past. The new natural schemes in the astronomy of the Herschels, the geology of Hutton and Lyell, the natural history and paleontology of Cuvier, Lamarck, and Darwin all operated in a physical nature in which the human figure did not constitute the center or the standard of measurement. Beside such systems, the human moment shrank to the least significant of factors.

The universe that was envisioned as a consequence differed from that palpable, sensuous texture of objects and emotions created by the Romantics. Herschel was opposed to Romantic notions; in fact, he attributed to them an inaccuracy of sense impression and identified them with a certain self-indulgence. The vast order of the universe invited one, stoically, "to merge individual feeling," and thus to achieve the tranquility of mind that would make one's thoughts "less accessible to repining, selfish, and turbulent emotions"—which was Herschel's view of Romanticism. Herschel found Deity in the great "order and design" argument of the eighteenth-century Deists.[30] In what he called the "emergency" of the intellect abroad in the material world, Herschel summoned nothing like Coleridge's "inner sense," but rather elicited the corrective capabilities of "instrumentation": "In this emergency we are obliged to have recourse to instrumental aids, that is, to contrivances which shall substitute for the vague impressions of sense the precise one of number, and reduce all measurement to counting."[31] But just as Herschel's universe was to be constructed with minimal assistance from the senses, it was destined thereafter to remain inaccessible to the senses, an abstraction removed from direct experience. Even as he labored under the conviction that he confronted the material, concrete world, the natural philosopher was preoccupied with converting it to numbers, symbols, and terms — abstractions that could not be experienced in a unique, individual way. His computed result remained outside the realm of the senses and beyond the grasp of art.

Like Herschel, Darwin, particularly in the early voyage years, felt a sense of the overall harmony of nature. But unlike Herschel, Darwin discovered in nature a sensation of pleasure and a source of vivid emotion rather than quietude. If ultimately Darwin was to abandon even his early idea of the harmony of physical nature, finding disequilibrium where he once found balance and permanence, it was not until after he had viewed the wilderness of South America with the enthusiasm of a Romantic.

[III]

To complement his account of the natural history and geology of South America and the South Seas in the *Journal of Researches*, Darwin created a wilderness of great splendor and beauty. His landscape descriptions are not the casual traveler's observations of attractive scenery: they are more an artist's account of his remarkable experience with sensation itself. Darwin beheld the physical features of the land with the eye of an aesthete, becoming with his predecessor Alexander von Humboldt one of the first European naturalists to view the foreign wilderness from a Romantic perspective.[32] The land, which was the source for Darwin of a bewildering riot of sensations, became the natural focus of his work — that is, the land both as some mysterious entity and presence and as the result of ages of geological and organic process. Darwin's powerful emotional bond with the land moved him to recreate moments and scenes that had left deep emotional impressions on him. He delighted in the novel and unfamiliar strangeness of foreign settings, in sensation itself rather than in some classical or formal aesthetic of beauty, in the radical contrast between the foreign landscape and the familiar English countryside. "The island [Porto Praya] would generally be considered as very uninteresting," he observed at the outset of his voyage, "but to any one accustomed only to an English landscape, the novel aspect of an utterly sterile land possesses a grandeur which more vegetation might spoil."[33] In giving priority to sensation, Darwin aligned himself with the tradition of Coleridge, Wordsworth, and the Romantics, for whom, as Wordsworth had pointed out, "the feeling therein developed gives importance to the action and situation and not the action and situation to the feeling."[34]

Of his intellectual and spiritual models, Darwin considered Humboldt, along with John Herschel, one of the two most important of his life, being especially attracted to Humboldt's bold imagination, spirit of exploration, and humanity. Friend and collaborator of Goethe, naturalist and Romantic, Humboldt prepared a brilliant account of his travels to the Americas, *Personal Narrative of Travels to the Equinoctial Regions of the New Continent During the Years 1799–1804*, which inspired Darwin's hope of a Romantic voyage of discovery undertaken to advance the noble cause of science. By the time Darwin arrived off the coast of Brazil in December 1831, Humboldt had seized his imagination:

> From what I have seen, Humboldt's glorious descriptions are & will for ever be unparalleled: but even [Humboldt] with his dark blue skies & rare union of poetry with science which he so strongly displays when writing on tropical scenery, with all this falls far short of the truth. The delight one experiences at such times bewilders the mind; if the eye attempts to follow the flight of a gaudy butter-fly, it is arrested by some strange tree or fruit; in watching an insect one forgets it in the stranger flower it is crawling over; if turning to admire the splendour of scenery, the individual character of the foreground fixes the attention. The mind is a chaos of delight, out of which a world of future and more quiet pleasure will arise. I am at present fit only to read Humboldt; he like another sun illumines everything I behold.[35]

To the end of his voyage, Darwin saw the physical landscape in the illumination of Humboldt's sun, and was most anxious to do justice to his predecessor in his own descriptions. The voyage was the only extended "chaos of delight" of Darwin's life, and it became his single richest source of ideas and sensuous recollections. Returning to England in 1835 and then marrying Emma Wedgwood in 1839 after a not-so-romantic courtship, Darwin settled into his now-legendary sedentary life of research and writing at Down by 1842.

The aesthetic responses of Darwin to the South American landscape are the deeply felt manifestations of a sensuous bond between perceiver and the perceived. Often, as in the famous instance of Darwin's sudden vision in the temple of the Amazon forest, the sensation of the moment leads to some more profound reality and truth, some sense of natural supernaturalism determined through intuition:

> Among the scenes which are deeply impressed on my mind, none exceed in sublimity the primeval forests undefaced by the hand of man; whether those of Brazil, where the powers of Life are predominant, or those of Tierra del Fuego, where Death and Decay prevail. Both are temples filled with the varied productions of the God of Nature: no one can stand in

these solitudes unmoved, and not feel that there is more in man than the mere breath of his body.[36]

Although Darwin repudiated this specific passage forty years later in his *Autobiography*, it was consistent with the frequent imaginative flights of the *"Beagle* Diary," where Darwin recorded unforgettable scenes often reminiscent of Percy Shelley's titanic landscapes.[37] High in the Cordilleras, near the continental divide, Darwin looked back in March of 1835 over his incredible path of ascent, feeling a flood of powerful sensation. He later recorded his experience, noting "the wild broken forms, the heaps of ruins piled up during the lapse of ages, . . . the inanimate mass." Pleased with his solitude in the moment of infinity, Darwin recalled feeling a sensation similar to that experienced in "watching a thunderstorm, or hearing in the full orchestra a chorus of the *Messiah.*"[38]

In the *Journal of Researches*, most of the landscape description is drawn from Darwin's *"Beagle* Diary." Throughout both the diary and the *Journal*, Darwin borrowed copiously from the Romantic lexicon, making frequent use of such words as "savage," "gloom," "grandeur," "sublime," "wildness," "vivid," "solitude," and so on. Although such words were common enough in the eighteenth-century aesthetic vocabulary, Darwin's emphasis on sensation, pleasure, and vividness gave them the Romantic associationist emphasis. Filled with a sense of mystery on the plain of Patagonia, Darwin reflected on time, another Romantic preoccupation, and quoted from Percy Shelley's "Mount Blanc":

> All was stillness and desolation. Yet, in passing over these scenes without one bright object near, an ill-defined but strong sense of pleasure is vividly excited. One asked how many ages the plain had thus lasted, and how many more it was doomed thus to continue.
>
> > *None can reply — all seems eternal now.*
> > *The wilderness has a mysterious tongue,*
> > *Which teaches awful doubt.*[39]

Darwin quoted Shelley only in the 1845 edition of the *Journal*, well after he had worked out his idea of natural selection, and, indeed, this fact probably explains why he did not complete Shelley's line, which emphasized how "faith" was taught as well as doubt when the mind confronted the infinities of existence. In this passage, however, Darwin's sensations of space and time are distorted by intuition. A landscape of mystery is found to exist beyond the physical surface of reality. With his "strong sense of pleasure," a phenomenological experience of the specific moment in time when time itself seems near-annihilated, Darwin feels the mystery and permanence of the land itself, the sense that its vast secrets are part of some sphere of existence utterly foreign and beyond reach of the intellect. Such states of intense sensation and emotion were the sources of Wordsworth's notion of the "spots of time," the "state of vivid sensation" of the poetic impulse.[40] These states were located by both Darwin and Wordsworth in the human instincts, or what Wordsworth called the "primary laws of our nature." Such sensations, Wordsworth held, unified human passion with the "beautiful and permanent forms of nature."[41]

The dualism of physical nature as both moment of experience and conceptualized structure was well known to Darwin. It was largely this awareness that enabled him to write descriptively of both his intellectual and emotional life in the wilderness. Initially, Darwin felt little, perhaps not any, conflict over turning back and forth between his professional accounts of an area's natural history and his evocations of the same landscape's fleeting moments of beauty. The simpler landscape elements were capable, he observed as his voyage was coming to an end, of inspiring exquisite sense impressions of natural beauty. The evanescence of such powerful sources of pleasure, the passing of the *Beagle* voyage itself, indeed, his own sensation of aging, all impressed Darwin with the great paradox of man's—and the naturalist's—inability to seize a given moment or to distill it in human knowledge. One is reminded of Coleridge's observation that "the full applicability of an abstract science ceases the moment reality begins," as Darwin realizes, with quiet regret, that his great voyage is coming to an end:

> Learned naturalists describe these scenes of the Tropics by naming a multitude of objects & mentioning some characteristic feature of each. To a learned traveller, this possibly may communicate some definite ideas: but who else from seeing a plant in an herbarium can imagine its appearance when growing in its native soil? Who, from seeing choice plants in a hot house, can multiply some into the dimensions of forest trees, or crowd others into an entangled mass? Who, when examining in a cabinet the gay butterflies, or singular Cicadas, will associate with these objects the ceaseless harsh music of the latter, or the lazy flight of the former—the sure accompaniments of the still glowing noon day of the Tropics. . . . In the last walk I took, I stopped again & again to gaze on such beauties, & tried to fix for ever in my mind, an impression which at the time I knew must sooner or later fade away. The forms of the Orange tree, the Cocoa nut, the Palms, the Mango, the Banana, will remain clear and separate, but the thousand beauties, which unite them all into one perfect scene, must perish: yet they will leave, like a tale heard in childhood, a picture full of indistinct, but most beautiful figures.[42]

Darwin's "entangled mass," a recurrent image that appeared at the end of the *Origin* some twenty years later, was initially a Romantic landscape image of a complex moment in life; two decades later, in the *Origin*, Darwin would perceive its unity in light of the laws of organic nature. Darwin felt a great sense of loss as he turned homeward in 1836, a regret not unlike that of Wordsworth in the Immortality Ode, over the passing of a compelling and vital way of life and the impending loss of a sense of the bright "radiance" and "splendor" of things. This sensation of time was a Romantic preoccupation, leading to the mixing of memory and desire.

We find in Darwin's passage the temporal emphasis of the Romantic artist, who discovers time in the self, as opposed to the emphasis of the natural historian, who seeks time in external nature. For the Romantic observer, time was the sensation of one's own aging amid the permanent forms of nature, which altered on a cyclical, seasonal basis. For the naturalist, and preeminently the evolutionist, time was measured in terms of the progressions of inorganic and organic nature. The perceived landscape was for both Romantic artist and naturalist an emblem of time, but the artist emphasized the complexities of the moment amid the tangled flux of nature, as Darwin had the day he prepared for his voyage home. This sensation of time was identified with personal loss, as in Wordsworth's "Tintern Abbey" and Immortality Ode. Time in the naturalist's nature was beyond the possibilities of human sensation, the geological and evolutionary record dwarfing the human moment. The great temporal progressions of evolution radically differed from poetic or natural time; in evolutionary theory a concept of earth history replaced what in the traditionally agrarian society of England had been the essentially atemporal cycle of death and renewal — the pastoral pattern that gave permanence to the landscape. Evolutionary geology and evolution reversed the Romantic emphasis on the permanence and harmony of landscape.

The other Darwin of the *Journal of Researches*, the geologist-naturalist, seeks a more profound understanding of the structure of natural phenomena in the Herschelian sense, with emphasis on the *verae causae* of events. In this approach he seeks to fix natural phenomena in a vast physical and chronological system that cannot be experienced in any direct and immediate sense. In his *Autobiography* Darwin recalled how during his voyage he first surveyed the chaos of rock formations and fossils at given points, "always reasoning and predicting what will be found elsewhere," until light began to dawn on the district, and "the structure of the whole [became] more or less intelligible."[43] This method

of abstraction or idealization enabled Darwin to discover physical organization and unity in both inorganic and organic nature.

Darwin traced the historical path from poetic to scientific nature in his
M Transmutation Notebook in 1838, while he was completing the first
edition of his *Journal of Researches*. Tying Auguste Comte's positive
stage of human intellectual development to the *verae causae* of John
Herschel, Darwin associated Herschel's illustration of fanciful causation
and Comte's theological and metaphysical stages of thinking with the
poetic state of mind. Herschel's *verae causae* raised the human intellect to
the positive stage by disciplining human perception and imagination
with reason. The implication in the passage was that such discipline was
accomplished at the expense of the artistic imagination:

> [Physical phenomena are interpreted by primitive man] as first caused by
> will of Gods. or God[.] secondly that these are replaced by metaphysical
> abstractions, such as plastic virtue, etc. (very true, no doubt savage at-
> tribute thunder & lightning to Gods anger. — (: more poetry in that state of
> mind: The Chileno says the mountains are as God made them, — next step
> plastic natures accounting for fossils). & lastly the tracing facts to laws
> without any attempt to know their nature.[44]

The "plastic nature" was a reference to Herschel's discussion of fanciful
causality in the *Preliminary Discourse*, while the remarks on the various
stages of human explanation were clear references to Comte's Law of
Three Stages. Herschel had cited as an example of a fanciful cause the
metaphysical, transcendentalist explanation that "plastic nature" accounted for the ocean shells on mountain peaks. He then compared this
explanation to what he believed was the *vera causa* — geological uplift-
ing. Darwin's similar tracing of facts to laws without attempts to know
their nature led out of the reference of human time, directly to the global
continuum of nonontological nature, a nature no longer given meaning
by the human presence and spiritual vision.

In addition to emphasizing the temporal and physical aspects of
organic phenomena, the naturalist also verbally reconstructed his experience without using a traditional narrative structure. He abandoned personal chronology. Humboldt, in the Preface to his *Personal Narrative*,
noted the difficulties of the field naturalist as narrator:

> I had left Europe with the firm intention of not writing what is usually called
> the historical narrative of a journey, but to publish the fruit of my inquiries
> in works merely descriptive; and I arranged the facts, not in the order in
> which they successively presented themselves, but according to the rela-
> tions they bore to each other.[45]

This no-longer striking notion was felt by Humboldt to be an innovation, as indeed it was; it identifies one of the significant formal or structural differences in the landscape literature of art and science. What Humboldt meant by "merely descriptive" was that he intended to represent his experience not in the more personal, human context of its temporal sequence relative to him, but rather in its more abstract connection with other similiar facts. Significance of event was not measured in human terms.

Modern scientific knowledge, Humboldt argued, had developed to where it "is scarcely possible to connect so many different materials with the narration of events; and that part which we may call dramatic gives way to dissertations merely descriptive."[46] The abandonment of the "dramatic" element not only denied the centrality of man as the measure of significant experience, a centrality Coleridge had argued to be the heart of all deep knowledge, but it also eliminated the human scale of time. The same descriptions, however, possessed a world systemic status, enabling the naturalist to distill and accommodate through abstract association vast collections of facts. Although the naturalist ignored the personal element of his experience — in some sense denied even its reality — he was thereby able to partition his experience into facts and then to link them with the similar experiential groupings of other researchers. The naturalist ignored the drama of the living moments in nature, seeking instead the abstract temporal and physical unity of the system.

Combining materials from the personal, chronological "*Beagle* Diary" with other materials from field notebooks, Darwin's *Journal of Researches* was a composite that integrated the narrative of his own personal progress with the nondramatic account of local natural histories. This integration of materials gathered in different frames of mind accounts for the two visions of landscape in the *Journal*. The most fitting metaphor for Humboldt's and Herschel's theories of natural description is that of mapping — the positioning of natural entities in relation to one another and not in relation to the observer himself. This method removed the observer from the center of landscape. Darwin, also the great mapper, closed his account of the *Beagle* voyage with the metaphor of a world coming into existence: "the map of the world ceases to be a blank; it becomes a picture full of the most varied and animated figures. Each part assumes its proper dimensions . . ."[47] What Darwin describes here is both a concrete landscape, full of distinct, palpabl organic forms, and an abstract physical scheme, formally organiz d into so many lines, districts, and positions. Darwin is alternately at the center of the whole and at its distant periphery.

Darwin's natural history map in the *Journal* organized the contents of landscape into geographical relationships that from the perspective of the artist seem highly abstract and precise. Bahia Blanca with its fossils, the Galapagos Islands with their aboriginal forms, and Keeling Island with its coral structures, all emerge as aesthetic landscapes, with tonalities of sensation derived from Darwin's view or feeling of the whole, and as natural histories in which they are analyzed and reconstructed into local "oeconomies" and considered in relation to larger geographical patterns of species distribution.

The Galapagos Archipelago, perhaps the most famous of Darwin's episodes in the second edition of the *Journal*, was Darwin's "world within the world," with a great mystery brooding over its strange, exotic landscape of primitive forms. Taking into consideration the possible history of shifting land masses, Darwin visually imagined a geological theater in which species origination and distribution could be accounted for in part by geographical isolation. What is most interesting about Darwin's method of physical description here is the manner in which he visualizes the profoundly theoretical idea of species origin, fusing the abstract with the palpable, so as to make us *see* the complex variety of physical factors he is taking into consideration:

> Considering the small size of these islands, we feel the more astonished at the number of their aboriginal beings, and at their confined range. Seeing every height crowned with its crater, and the boundaries of most of the lava streams still distinct, we are led to believe that within a period, geologically recent, the unbroken ocean was here spread out. Hence, both in space and time, we seem to be brought somewhere near to that great fact — that mystery of mysteries — the first appearance of new beings on this earth.[48]

This passage is a key to one of the remarkable dimensions of Darwin's prose — his ability to anchor abstract concepts on solid objects. The rhythms of the passage contribute to the tone of mystery and wonder a sense of drama, which we are invited through Darwin's use of "we" to share. There is great economy in this rather unconventional fusing of landscape motifs with theoretical views. It is an economy derived from showing, letting us see for ourselves. In the passage, the words "seeing" and "appearance" emphasize the visual aspect of Darwin's abstractions, Darwin himself, as author, looking down over the whole as over some great sublime map. From a height above the earth and from the theoretical perspective of Lyellian geology, we can see and thus share in Darwin's vision.

But Darwin's mapping activities are highly detailed and quantity-oriented as well. After physically describing the finches of the Galapagos, of which all thirteen species are peculiar to the Archipelago, with their "perfect gradation in the size of the beaks," Darwin observes that of the land birds, twenty-five of twenty-six are new species. He describes the tortoises and lizards, which are also aboriginal species. Of the flowering plants, 100 of 175 are new species, making the Galapagos Archipelago "a distinct botanical province." Locally, the species vary from island to island. Of the thirty-two species found on Chatham Island, sixteen are found in other parts of the world, twelve are exclusive to Chatham Island, and only four of the species unique to the Archipelago are found on more than one island. Most remarkable, Darwin observes, is the "circumstance that several of the islands possess their own species of tortoise, mocking-thrush, finch, and certain plants, these species having the same general habits, occupying analogous situations, and obviously filling the same place in the natural oeconomy of this archipelago . . ." Darwin concludes his distributionary census in great wonder at the sheer "creative force" by which so many new forms are established.

What is perhaps most striking about the landscape of Darwin's Galapagos Archipelago, from the perspective of literary landscape art, is that it constitutes an elaborate and unprecedented discussion of organic position or *place*. The individuals of each species category are located in distinct positions in the general island economy, positions that are far more abstract and elaborately defined than they could possibly be in the texture of an artistic representation based on an impression. It is in the illusive concept of the "species" itself that we can begin to imagine equivalencies of organic position, the actual reality of species becoming less important than the linguistic power of the term to suggest the kinds of spatial equivalence that Darwin needs to describe what he calls the "oeconomy" of the islands. Different species of the same genera replace each other in the economies of different islands, occupying the same places; species are, in this respect, estimates of organic space. They are metaphors of position which have the considerable descriptive power of enabling Darwin to attach highly theoretical abstractions about time and space and organic units of existence to the individuals and events of recognizable places. In the 1845 edition of the *Journal*, Darwin sums up his concept of the Galapagos Islands in combined geological and biological terms, referring to the Archipelago as "a group of satellites, physically similar, organically distinct, yet intimately related to each other, and all related in a marked, though much lesser degree, to the great American continent."[49] Geological and biological space have become far more complexly integrated in Darwin's natural history than in his aesthetic landscape, a feature that after Darwin's *Origin* increasingly distinguished the nature represented in literature and painting from that discovered in the natural sciences.

As Darwin's sense of the general distinctions between geological and biological nature became less pronounced, his concept of biological space became more exact. Howard Gruber has traced a series of theoretical shifts in Darwin's thought from 1831 to 1838, during the voyage years and immediately thereafter, when Darwin shifted from the view that the physical (geological) and organic (biological) worlds were separate, stable, and in perfect harmony with each other, to his evolutionist position that they were continually interacting and evolving.[50] What Gruber demonstrates, among other philosophical points with which he is primarily concerned, is that Darwin's theoretical shift gives increasing emphasis to the physical and quantitative aspects of organic life by merging biology and geology. In the earlier view, the organic world was superimposed, so to speak, on the physical world, their harmony deriving from the fact that they were qualitatively different forms of existence confined to different kinds of activity. Darwin's evolutionism, on the other hand, viewed organic nature as fundamentally integrated with geological nature, the fluctuations in each world conditioning the progressions of the other. As Gruber notes, Darwin by 1835 had clearly begun to link, in his Keeling Island theory of coral reefs, developments in the organic world with the evolution of the geological world. This equating of geological and biological phenomena was encouraged by Herschel's demand for *verae causae* explanations for natural events. One effect such an approach had on Darwin's thinking about landscape was to influence him to reflect on life as a spatial phenomenon in purely, although not exclusively, distributionary terms.[50] It encouraged him to think of life forms on the massive global scale of the geologists.

By 1838, Darwin thus moved from a view consistent with Wordsworth's belief that "man and nature [are] essentially adapted to one another," a view that emphasized the permanence and harmony of all existence, to a view that organic and geological nature were integrated but also in perpetual disequilibrium.[51] Nature was a system of opposing energies — metaphorically, the struggle for existence — leading to continual change and diversification in the familiar landscape. While such a view did not alter the actual appearance of any given landscape — and presumably Darwin could still delight in the splendid descriptive passages of the 1845 edition of the *Journal of Researches* — it did lend an air of illusion to such representations, since the driving force of natural events and of the emergence of life forms was now viewed by Darwin in terms of the elaborate, largely hidden, interactions discovered in the machinery of evolution.

The universal disharmony implied in Darwin's evolutionary views gave new meaning to the concept of *place* in nature and significantly departed from the artistic idea of place in landscape. The physical position occupied by any organism in its natural environment was defined by such prerogatives of space as it could claim by virtue of its superior abilities to survive inter- and intraspecies competition, given certain physical conditions of existence. Hence, the importance of the idea of adaptation — morphological, physiological, and instinctual — in Darwin's concept of biological territory. The beaks of the Galapagos finches were means to their respective niches in the local economy and, in effect, were the mandates of their existence, whether considered collectively, species by species, or individually, bird by bird. To a mind trained in the philosophical discourse of the Academy or, still more, to a mind steeped in the metaphysics of the Transcendentalist tradition, such an idea must have seemed absurd. Yet, by comparison to all previous ideas of space in nature, the niche was a most precise and complex spatial idea; its boundaries for a given species could often be plotted within meters, and the same boundaries were determined as much or more by the biological adaptations that enabled a species successfully to appropriate space as by the inert physical boundaries of geographical topography. The economy of such an idea of biological space, where there is great rigor of competition for place, is breathtaking, especially when compared to the spirit of generosity and abundance of nearly every previous English or European literary ideal of nature. As Darwin observed in a notebook entry of late 1838, "one may say there is a force like a hundred thousand wedges trying to force every kind of adapted structure into the gaps in the oeconomy of nature, or rather forming gaps by thrusting out weaker ones."[52]

While Darwin's developing theories did not alter the physical appearance of landscape, they did ultimately alter what Darwin saw. Once again framing his theoretical ideas in landscapes recollected, Darwin observed in a March, 1838, notebook entry that his impressions of natural settings had a notable aspect of illusion about them, when considered in light of the actual processes taking place.

It is difficult to believe in the dreadful but quiet war of organic beings, going on in the peaceful woods, & smiling fields. — we must recollect the multitude of plants introduced into our gardens . . . which might spread themselves as well as our wild plants; we see how full nature is, how finely each holds its place. — When we hear from authors . . . that in the Pyrenees that the Rhododendron ferrugineum begins at 1600 meters precisely & stops at 2600 & yet know that plant can be cultivated with ease near London — what makes this line, as of trees in the Beagle channel — it is not elements! — we cannot believe in such a line. It is other plants. — a broad border of killed trees would form fringe — but there is a contest & a grain of sand turns the balance.[53]

Beyond the impression of the moment, which is likely to fill one with a feeling of the beauty, the spontaneity and harmony of life in the natural landscape, there is an unsuspected geometry of great preciseness and definition. The geometry of the "smiling field" is the geometry of conflict, a calculable order defined by real biological territories flowing around, through, over, and beneath one another according to the intricate physical relationships among species. The specificity of Darwin's idea of organization in nature is unprecedented and runs directly against the Romantic tendency to seek freedom and release in natural settings, against Darwin's own celebrations in the years of the *Beagle* voyage of transcendent moments of timelessness in the wilderness. The niche of *Rhododendron ferrugineum* amounts to the 1000 meter band along the sides of the Pyrenees. Darwin now understood such recollected landscapes as the *Beagle* Channel mentioned earlier in a new physical sense and was prepared by 1859 with the Publication of the *Origin* to trace the colors, shapes, sounds, and distribution of organic life to wholly physical causes. Physical had eclipsed spiritual law; the "conditions of existence" rather than the *end* of existence were the highest law of development.[54] This naturalistic view cut directly across the Romantic pantheistic impulse to spiritualize nature and more than any other nineteenth-century development served to remove landscape from the category of the Absolute under which it had contributed so immensely to the early stages of the Romantic search for an ultimate.

Darwin's importance for landscape art, then, was probably far less the result of his introducing metaphors of struggle to nature or of his introducing what seemed to many — Herschel included — the element of chance into the causes of existence, than it was the result of his vigorous appropriation, through definition, of natural space. Darwin saw repetitiousness of form, geometries of organization, abstract as such patterns might be, as fundamental natural truths. It would take art fully half a century to follow him into the labyrinth of lines and patterns he had discovered in the familiar landscape. Darwin's appropriation of space was particularly significant for artistic representation of landscape because many of his ideas and demonstrations were mounted on the Romantic images of his early voyage years — the tangled bank image, the tree metaphor, the hundreds of scraps of natural scenes he carried with him during his composition of the *Origin*. He had once moved easily between artistic and naturalist traditions, keeping, for example, during the earliest years of his voyage, separate but parallel notebooks, one for his aesthetic life and another for his scientific life. But as his concept of nature became increasingly intellectual and abstract, his representations became less traditional and emotional, and he drew away from the aesthetic enthusiasm of his *Journal of Researches*.

What Coleridge has considered the blind mnemonic drive of the nomenclaturists to fragment nature and to categorize, Darwin understood as the long, laborious labor of many minds to discover the hidden bonds that described the true community of biological descent. Taxonomical relationships were the geographical and temporal links between life forms, language itself binding together the diverse factors contributing to the history of biological space. In his discussion of classification in the *Origin*, Darwin argued that the natural system, which had attracted the labors of men over the ages, sought not merely to manipulate objects in nature, but rather to discover their true, "hidden" bonds:

> All the foregoing rules and aids and difficulties in classification may be explained, if I do not greatly deceive myself, on the view that the natural system is founded on descent with modification; that the characters which naturalists consider as showing true affinity between any two or more species, are those which have been inherited from a common parent, and, in so far, all true classification is genealogical; that community of descent is the hidden bond which naturalists have been unconsciously seeking, and not some unknown plan of creation, or the enunciation of general propositions, and the mere putting together and separating objects more or less alike.[55]

This elaborate system of metaphors, the Natural System, made species the quantities of organic space; yet it went far beyond the mere quantifying exercise that to Coleridge had seemed an inexcusable trivialization of natural space. True affinity in physical character between two or more species implied a common parent and, hence, a proximity of geographical origin. The thirteen species of finches on the Galapagos Islands were related through their likely common origin in South American species, a

line of thinking that led naturally to the geological association of the
Galapagos with the South American continent. As Gruber has noted,
this geographical branching of species was suggestive of the tree metaphor — the archsymbol of the Romantic movement — which Darwin used
as one of the summary images of his genealogical views.[56] Thus Darwin
was able, remarkably, to unify the Romantic sense of the fecundity and
animation of nature with his own historical and theoretical views of
biological descent.

If by 1838 Darwin's naturalistic view of organic development had
seriously limited the possibility for a traditional metaphysics of organic
form and function, his great map of nature had begun to assemble itself
on a scale that was, even by Percy Shelley's standards, truly heroic. By
1859, Darwin had concluded that theories attributing organic structure
to any force, ideal or material, other than the self-preserving tendency of
the individual were "absolutely" fatal to his theory of descent with modification.[57] So too, he added, were idealistic notions of natural beauty,
beauty for its own sake. Hence, his tree was no longer the mystical symbol of harmony and loveliness that it had been for the Romantic movement; nor was his entangled bank anything like a Coleridgean Xanadu,
"holy and enchanted." Indeed, somewhat on the periphery of that vision
of fecundity sat Darwin, carefully noting its "proportional numbers and
kinds."[58] And yet, even as he methodically connected the kinds and numbers of species in the landscape to the "laws acting around us," with a
result that surprised even Herschel, he was sensitive to and did not
destroy the aesthetic possibilities of his nature.[59] If he, himself, grew
weary of the poetry of his youth, losing, as Wordsworth had, that sense
of the quick immediacy of the experience in nature, Darwin opened up
new possibilities for space in natural landscape. With his concept of the
fluidity of form, he bound the hidden factors of time and space together
in the energy and immensity of organic development. Like the Romantics, Darwin was not a literalist, but saw dimensions in nature that could
only be approximated in language, and while he seized and defined objects and spaces of organic nature that were also the subjects of Romantic
myth and intuition, he also, in some sense, freed them, establishing the
independence of natural form.

NOTES AND REFERENCES

1. Charles Darwin, *Journal of Researches into the Natural History and Geology of the
Countries Visited during the Voyage of H.M.S. Beagle Round the World*, 2 vols. (New

York: Harper, 1846). In this essay, I use the 2nd (American) edition, henceforth to be called *Journal of Researches*.

2. Darwin,[1] vol. 2, p. 310.

3. Samuel Taylor Coleridge, *Poetical Works*, ed. Ernest Hartley Coleridge (New York: Oxford University Press, 1969), p. 94.

4. Darwin,[1] vol. 1, pp. 2, 72, 87, 262, 270.

5. Charles Darwin, "Darwin's Early Unpublished Notebooks," ed. by Paul H. Barrett in *Darwin on Man: A Psychological Study of Scientific Creativity* by Howard E. Gruber (London: Wildwood House, 1974), p. 273. Hereafter cited as "Early Notebooks."

6. Walter F. Cannon, "Darwin's Vision in *On the Origin of Species*," in *The Art of Victorian Prose*, ed. George Levine and William Madden (New York: Oxford University Press, 1968), p. 163.

7. William Wordsworth, "Preface to the Second Edition of . . . *Lyrical Ballads*," in *Wordsworth: Poetical Works*, ed. Thomas Hutchinson, rev. Ernest De Selincourt (New York: Oxford University Press, 1969), p. 735.

8. See, for example, Donald Fleming's "Charles Darwin, the Anaesthetic Man" in *Victorian Studies*, vol. 4 (1961), pp. 219–236; for a more recent treatment of the issue see also John Fowles' "Seeing Nature Whole," in *Harper's*, vol. 259 (Nov. 1979), pp. 49–68.

9. See, for example, Edward Manier's treatment of Darwin and Wordsworth in *The Young Darwin and his Cultural Circle* (Boston, D. Reidel, 1978), pp. 89–96.

10. Charles Darwin, *The Autobiography of Charles Darwin: 1809–1882*, ed. Nora Barlow (New York: W. W. Norton, 1969), p. 138.

11. Samuel Taylor Coleridge, *The Friend* in *The Complete Works of Samuel Taylor Coleridge*, ed. W. G. T. Shedd (New York: Harper and Brothers, 1884), vol. 2, p. 424n.

12. Coleridge,[11] p. 450.

13. Coleridge,[11] pp. 449–450.

14. Coleridge,[11] p. 427.

15. Coleridge,[11] pp. 449–450. See also Trevor Levere, "S. T. Coleridge: A Poet's View of Science," in *Annals of Science*, vol. 35 (1978), p. 35.

16. Samuel Taylor Coleridge, "Hints Towards the Formation of a More Comprehensive Theory of Life" in *Aids to Reflection*, in *Complete Works*, vol. 1, p. 376.

17. Coleridge,[16] p. 381.

18. Coleridge,[11] pp. 425–26.

19. Coleridge,[16] p. 392n. See also Trevor Levere, "Coleridge, Chemistry, and the Philosophy of Nature," in *Studies in Romanticism*, vol. 16 (1977) pp. 349–379.

20. Coleridge,[16] p. 392.

21. John Herschel, *Preliminary Discourse on the Study of Natural Philosophy* (London: Longman . . . and Green, 1830), p. 18.

22. Herschel,[21] p. 135.

23. Herschel,[21] pp. 144–45.

24. Herschel,[21] p. 191.

25. Herschel,[21] p. 215.

26. See, for example, M. H. Abrams, *Natural Supernaturalism: Tradition and Revolution in Romantic Literature* (New York: W. W. Norton, 1973), p. 113.

27. Abrams,[26] pp. 180–82. See also I. A. Richards, *Coleridge on Imagination* (Bloomington: Indiana University Press, 1960), pp. 44–48.

28. Wordsworth,[7] p. 735.

29. William Wordsworth, "Tintern Abbey," *Poetical Works*, p. 164.

30. Herschel,[21] p. 125.
31. Herschel,[21] pp. 4, 43.
32. See also Stanley Edgar Hyman, *The Tangled Bank: Darwin, Marx, Frazer and Freud as Imaginative Writers* (New York: Atheneum, 1962), pp. 14–16. But see also, A. Dwight Culler, "The Darwinian Revolution and Literary Form" in *The Art of Victorian Prose*, p. 232.
33. Darwin,[1] vol. 1, p. 2.
34. Wordsworth,[7] p. 735.
35. Charles Darwin, *Charles Darwin's Diary of the Voyage of H.M.S.* Beagle, ed. Nora Barlow (New York: Macmillan, 1933), p. 39. This work is cited as The *Beagle* Diary.
36. Darwin,[1] vol. 2, p. 308.
37. Darwin,[10] p. 91.
38. Darwin,[35] pp. 292–93.
39. Darwin,[1] vol. 1, p. 216.
40. Wordsworth,[7] p. 734.
41. Wordsworth,[7] p. 735.
42. *Beagle Diary*, pp. 416–17. See also Darwin,[1] vol. 2, pp. 298–99.
43. Darwin,[10] p. 77.
44. Darwin,[5] p. 278. See also Herschel,[21] pp. 144–45.
45. Alexander von Humboldt, *Personal Narrative of Travels to the Equinoctial Regions of the New Continent During the Years 1899–1804*, Trans. Helen Maria Williams (London: Longman, 1814–29), vol. 1, p. xxvi.
46. Humboldt,[45] p. xi.
47. Darwin,[1] vol. 2, p. 310.
48. Darwin,[1] vol. 2, pp. 145–146.
49. Darwin,[1] vol. 2, p. 172.
50. Howard Gruber,[5] *Darwin on Man: A Psychological Study of Scientific Creativity*, p. 127.
51. Wordsworth,[7] p. 738.
52. Darwin,[5] p. 456.
53. Darwin,[5] p. 460.
54. Charles Darwin, *The Origin of Species, . . . a Variorum Text*, ed. Morse Peckham (Philadelphia: University of Pennsylvania Press, 1959), p. 379.
55. Darwin,[54] p. 656.
56. See Gruber,[5] p. 135, for a discussion of this idea.
57. Darwin,[54] pp. 367–368.
58. Darwin,[54] p. 158.
59. Darwin,[54] p. 758.

"Through Science to Despair": Geology and the Victorians

DENNIS R. DEAN

Humanities Division
University of Wisconsin-Parkside
Kenosha, Wisconsin 53141

As a current, useful, easily intelligible science primarily concerned with nature, landscape, environment, and history, geology could hardly have been bettered in appealing to Victorian tastes.[1] Knowledge of the earth had already proven highly relevant to many intellectual and practical endeavors associated with the Industrial Revolution, including mining, manufacture, road and canal building, surveying, cartography, agriculture, hydrology, meteorology, and chemistry, as well as to even less obvious fields like medicine and hygiene.[2] As a leading science of the day, geology set an example that encouraged speculation about the nature of science itself. By revolutionizing depictions of landscape and natural processes in both science and the arts, geology allowed the Victorians to apprehend the deeply impressive, increasingly materialistic world that surrounded them. A ubiquitous and emblematic science, geology gave new concepts, images, and habits of observation to the age.

For many Victorians, geology had a further use, revealing to them (as it seemed) the nature and manner of God's creative intentions in fashioning the earth as a fit abode for man. However, when increasing geological evidence made this comfortable assumption suspect, anxious Victorians began to wonder what further unexpected, even loathsome, revelations of divine neglect geology would next bring forth. References to their continuing distress are prominent throughout the Victorian literary record, and a general pattern of challenge and response is discernible if traced with care. While several previous scholars have alluded to portions of this important conflict, no one has yet attempted to be comprehensive.[3] The present essay, therefore, traces those developments within Victorian geology that most influenced contemporary religious affirmations and those related developments within Victorian theology that so fully disillusioned a once optimistic century. While mentioning the necessary titles, both in science and in religion, I also note some responses to these debates and moods in literary works, which are, as usual, chronicles of their times.

Geology, its historians agree, became a modern science during the final dozen years of the eighteenth century, when James Hutton, ignoring Genesis, first definitively recognized the ability of slow geological processes to create, destroy, and re-form whole continents if given sufficient time. Because Hutton insisted that present-day causes are the same as those that have always been operative, his position was later called Uniformitarian. Although his theory was freely attacked for its religious implications (no Creation and no Flood), it gradually gained adherents.[4] Yet Catastrophists, who were therefore more sympathetic to Genesis, continued to argue upon reasonable grounds that geological forces were once (and perhaps still) both stronger and more rapid than those we customarily see. Accordingly, they rejected the immense age that Hutton had attributed to the earth. Hutton's emphasis upon volcanic forces was also rejected by an older school of geologists called Wernerians (after Abraham Gottlob Werner, the German mineralogist), who believed that almost all the world's rocks had been deposited as sediments from a universal ocean. Both the Huttonian and the Wernerian theories preceded William Smith's discovery that strata are characterized by their fossil content, an insight that became particularly important after 1800, when competent prehistoric reconstructions by the French comparative anatomist Georges Cuvier gave the British their first real glimpse of vertebrate life older than the mammoth. Cuvier's *Essay on the Theory of the Earth* (Edinburgh, 1813) made Catastrophism the leading school of British geological thought, a position it maintained well into the Victorian period.

The first British geologist of central importance to the Victorians was William Buckland (1784–1856), whose lectures to the undergraduates at Oxford (from 1813) were formative. His Inaugural Address as reader of geology, *Vindiciae Geologicae: or, The Connexion of Geology with Religion Explained* (1820), affirms Creation and the Flood while rejecting the efficacy of present-day causes and Huttonian attempts to divorce geology from the Bible. A delightfully uninhibited lecturer, Buckland was immensely popular with his students, among whom were Thomas Arnold, John Ruskin, Charles Lyell, and John Henry Newman. Of these, Newman alone thought geology, in 1821, far too tentative to be considered a science. As promoted by Buckland, however, it was "most entertaining, and opens an amazing field to imagination and to poetry."[5]

Perhaps Newman was thinking of Shelley's *Prometheus Unbound* (1820) and Byron's *Cain* (1821). Throughout the decade poetic versions of the Flood were especially popular, with Byron's drama *Heaven and Earth* (1822) the best-known example. A number of Victorian readers first learned of geology through such literary intermediaries.

Buckland himself defended the Flood on paleontological grounds in *Reliquiae Diluvianae* (1823), but his was the last respectable attempt to do so, since the historicity of the deluge swiftly became an issue. By around 1825 the situation was almost as Samuel Butler was later to remember in his well-known novel *The Way of All Flesh* (1903; written in the 1870s): Theobald Pontifex, his bumbling cleric, "had taken a great deal of pains with his sermon, which was on the subject of geology — then coming to the fore as a theological bugbear. He showed that so far as geology was worth anything at all — and he was too liberal entirely to pooh-pooh it — it confirmed the absolutely historical character of the Mosaic account of the Creation, as given in Genesis" (chap. X). Speaking as president of the Geological Society of London in 1826, Buckland could still honestly praise "the resident clergy" for their contributions to geology. But his personal belief in the Flood as a unique geological agent was fading; irrefutable stratigraphical evidence now demanded a succession of deluges rather than one. By 1830, therefore, Buckland could no longer affirm the theory he had published only seven years previously.[6] That same year Thomas Carlyle began *Sartor Resartus*, a spiritual autobiography that is often regarded as the first masterpiece of Victorian literature. It utilizes metaphors derived from the Huttonian-Wernerian debates so prominent in Edinburgh during the Regency, when he had been a student there, but ignores Buckland and the Deluge controversy. Curiously, then, this foremost Victorian literary prophecy bases its call for transcendental thought — the giving of spiritual significance to natural facts — upon an obsolete dispute. Although deeply interested in geology at one time, Carlyle read little pertaining to it after his spiritual regeneration in 1822. While remaining jauntily confident that "Of Geology and Geognosy we know enough" (indeed, "to many a Royal Society, the nature of a world is little more mysterious than the cooking of a dumpling") he actually resolves the problem of geology and religious values by ignoring it. Nevertheless, Carlyle was a major influence upon almost every subsequent nineteenth-century writer who confronted the science more directly.[7]

Victorian geology can be said to have begun in June 1830, when the first of three volumes of the *Principles of Geology* by Charles Lyell

(1797-1875) appeared.[8] Eventually there would be eleven subsequent editions, some further titles, and a series of important papers. Being one of the most gifted, articulate, and reliably decent men of science in his day, Lyell had a great deal to do with maintaining the essential civility that generally characterized Victorian discussions of geology almost until his death. His arguments were always moderate, never shrill, and Lyell was at pains to commend the work of those with whom he disagreed. Furthermore, he used successive editions of the *Principles of Geology* to record changes in his own thinking (most famously in the tenth, with regard to natural selection). Candor (often called "manliness"), rationality, generosity, and tact were all admired qualities throughout most of the Victorian years, and while Lyell was by no means alone in embodying them, his was clearly a superior example. To one familiar with the values of this period, therefore, it might well appear that Lyell's greatest contribution to Victorian geology was the quality of his character.

A second feature of Lyell's argument was, of course, its content. Like Darwin's *Origin of Species* later on, the *Principles of Geology* is full of fascinating examples, lucidly set forth. As its subtitle specifies, this famous book was "an attempt to explain the former changes of the earth's surface, by reference to causes now in operation." Whatever one thought of Lyell's explanation, there was no better popular encyclopedia of geological information than his; Victorians bought and read his book as if it were a novel. Regarding the principles themselves, we now distinguish — as Lyell did not — between the endurance of physical laws (Actualism) and the sameness of geological causes and results (Uniformitarianism). Although Lyell had an immense array of geological, paleontological, historical, and ecological facts at his command, he failed nevertheless to see that change is, in some respects, directional. Despite occasional appeals within his work to Providence and the Author of Nature, Lyell emphasizes a nonprogressive, if not quite clockwork, universe of natural law. While this aspect of his argument influenced many readers, others were equally disturbed by his abandonment of Scripture and preferred geological theorists who sought to reconcile the two. Most of the real opposition to Lyell, however, came from fellow geologists who found some variety of Catastrophism more convincing than his steady-state, Uniformitarian vision of the globe.[9]

As a former student of Buckland's, Lyell was very much aware that geological theorizing had theological repercussions. Being convinced that Biblical authority could only hamper science, he was determined to effect their separation. This, then, became a third significant feature of Lyell's work, and he achieved it in two ways: first, by restricting his proofs to purely geological evidence, without appeals to natural theology; and, second, by covertly attacking theologically derived cosmologies in a cleverly slanted "history" of geological thought with which the *Principles* in all of its editions begins. Although this same history later became something of a model for those controversialists who, during the last three decades of the nineteenth century, more openly disparaged the influence of religion upon science, Lyell would not have sanctioned the animosities that followed. His own tactful approach to this delicate issue is apparent from surviving accounts of his lectures, as, for example, this one by Henry Crabb Robinson:

> May 4th [1832] — I continued at home till it was time to go to the King's College [London], where Lyell delivered his introductory lecture on Geology, of which I understood scarcely anything, — but I liked what I did understand. . . . He decorously and boldly maintained the propriety of pursuing the study without any reference to the Scriptures; and dexterously obviated the objection to the doctrine of the eternity of the world being hostile to the idea of a God, by remarking that the idea of a world which carries in itself the seeds of its own destruction is not that of the work of an all-wise and powerful Being. And geology suggests as little the idea of an end as of a beginning to the world.[10]

Through Lyell's work and John Herschel's, however, Victorian England during the 1830s came face to face with what Tennyson would later call its "terrible Muses," geology and astronomy, or, more precisely, with the immensities of time and space, which wonderfully and fearfully revealed both the powers of man's mind and the seeming insignificance of his being.

At first, perhaps, the powers of his mind were more apparent, largely because they were thought to reveal (in the transcendental manner) new evidence of God's grandeur. So, for Herschel in 1830, man is a speculative being who contemplates the world around him, admiring "the harmony of its parts, the skill and efficiency of its contrivances":

> Thus he is led to the conception of a Power and an Intelligence superior to his own, and adequate to the production and maintenance of all that he sees in nature — a Power and Intelligence to which he may well apply the term infinite, since he not only sees no actual limit to the instances in which they are manifested, but finds, on the contrary, that the farther he enquires, and the wider his sphere of observation extends, they continually open upon him in increasing abundance; and that as the study of one prepares him to understand and appreciate another, refinement follows on refinement, wonder on wonder, till his faculties become bewildered in ad-

miration, and his intellect falls back on itself in utter hopelessness of arriving at an end.[11]

Geology, he tells us later on, "in the magnitude and sublimity of the
objects of which it treats, undoubtedly ranks, in the scale of the sciences, next to astronomy," a comparison that would echo through
the century.[11] Meanwhile, Lyell's remarkable *Principles*, already a
best-seller, would soon foster a widespread, sophisticated enthusiasm for earth science among those who read it. "I find in Geology
a never-failing interest," wrote the young Charles Darwin from
Chile in 1834; "it creates the same grand ideas respecting this world
which Astronomy does for the universe."[12] And this enthusiasm
was not restricted to budding scientists. Thus, John Constable, the
artist, wrote in November 1835 of his fondness for geology, a
science "which, more than any other, seems to satisfy my mind."[13]
For many at this time geology was novel, exciting, and appealing — a stimulus to the imaginative, and a solace to the thoughtful.

During the 1830s, then, writers often found geological observations,
theories, and terminology relevant to their endeavors. Besides Carlyle,
others whose novels, essays, and poems manifest some considerable
awareness of the science include Thomas Love Peacock, his novel
Crotchet Castle (1831) deriving its analysis of British landscape from a
geologic map; Goethe, whose *Faust*, part II (1832) reflects the departed
Neptunist-Vulcanist controversy; Edward Bulwer-Lytton, whose *Eugene
Aram* (1832), is a novel full of anachronistic geological sensitivity and
whose *The Last Days of Pompeii* (1834) depicts a famous geological
catastrophe; Robert Browning, whose *Paracelsus* (1835) is a long poem
in which alchemy and volcanic geology interestingly combine; Alfred
Tennyson, part of whose work "The Epic';' (1837) is quoted below; John
Stuart Mill, who in his essay on Bentham (1838) incorporates an
elaborate allusion to Buckland; and John Sterling, whose interest in
geology during the 1830s, as shown in his poem "The Ages" (1839), was
later to be recalled by Carlyle.[14] John Ruskin's early poem "A Tour
through France to Chamonix" (1835) reliably indicates the bent that his
mature genius would take. Although its theories on the Deluge remind us
that Ruskin was Buckland's pupil, most professional geologists were now
convinced that the Flood is untenable as science. Accordingly, they one
by one abjured: Adam Sedgwick in 1831, G. B. Greenough in 1834, Edward Hitchcock (the American) in 1835, and Buckland himself in 1836,
only a year after Ruskin's poem. Lyell, who had never affirmed Deluge

geology, spoke out against it in 1833.[15] Among the leaders, therefore, Deluge geology soon ceased to be an issue.

The scriptural and scientific implications of its failure continued to be of general interest, however, as we have already seen in Herschel's book and Lyell's lecture. Also in 1832, Thomas Arnold, Victorian England's most famous schoolmaster, dissociated "the conclusions of geology" from "our belief in revelation" in an important, forward-looking sermon. Another educator, the geologist and clergyman Adam Sedgwick, concurred. In a December lecture (published as *Discourse on the Studies of the University*; Cambridge, 1833) Sedgwick defended a vague form of natural theology, but outspokenly denounced the whole endeavor of scriptural or Mosaic geology. Despite his recent abjuration of the Flood, Sedgwick was (with Buckland) the most eminent Christian geologist in Britain. In response to Sedgwick's disavowal, therefore, consternation soon erupted among the orthodox, much of it preserved for us within the pages of *The Christian Observer*, a periodical edited by the Rev. Samuel C. Wilks, who was strongly interested in the whole question. ("A large number of geologists," he wrote ominously, "are, we fear, infidels — or at least sceptics. . . .") Wilks also concocted a good-natured poem, "The Fossil Shell," which descants with orthographical license about "Strange icthyosaurus, or inguanidon,/With many more I cannot verse upon — ." As for Wilks' contributors, one proclaimed in his title "Popular Geology Subversive of Divine Revelation!", while others preferred moderation. W.D. Conybeare, for example, another distinguished clergyman-geologist, insisted that the Bible advised men only about their relations with God; it was not intended to teach physical science. Because of Sedgwick's prestige, his *Discourse* angered the literalists and significantly afflicted their cause.[16]

A second stimulus to less temperate debate was the appearance in 1836 of William Buckland's *Geology and Mineralogy Considered with Reference to Natural Theology*, in which his virtual abandonment of Deluge geology was made public. For those marginal to the scientific community, there was an immense feeling of betrayal. Thus, a parson characterized in Tennyson's poem "The Epic" (1837) reacts, probably to Buckland, by "hawking at Geology and schism" and the "general decay of faith." Books like George Fairholme's on the Mosaic Deluge (1837) and William Cockburn's *Letter to Professor Buckland Concerning the Origin of the World* (1838) show how typical he was. Even before reactions to *Geology and Mineralogy* had subsided, however, Buckland spoke out again, this time in a sermon of 27 January 1839 on "The Sentence of

Death." It may be difficult for us to sympathize with, or even to understand, the resulting disillusionment when Buckland proved that death had entered the biological world long before the advent of man and could not, therefore, have been a consequence of his Fall, as Genesis has held. During this further controversy, the word "anti-geologist" first entered our language.[17]

The latter 1830s also saw a number of book-length contributions to the science and Scripture debate, including Gideon Mantell's *Thoughts on a Pebble* (1836), Nicholas Wiseman's *Twelve Lectures on the Connexion between Science and Revealed Religion* (1836), Thomas Gisborn's *Consideration on the Modern Theory of Geology* (1836), the already-mentioned works of Fairholme (1837) and Cockburn (1838), Mantell's *The Wonders of Geology* (1838), John Murray's *A Portrait of Geology* (1838), the Rev. John Pye Smith's *The Relation Between the Holy Scriptures and Some Parts of Geological Science* (1839), and William Sidney Gibson's *The Certainties of Geology* (1840). Of the writers just named, Mantell was second only to Lyell in popular influence, his several books having numerous editions. Although this transitional literature has generally been slighted, and even maligned, its most successful authors were moderate, enlightened men attempting to mitigate religious objections to geology. The obscurantists were never popular.

Meanwhile, geology itself was rising to new heights of professionalism and independence. Its practitioners, for example, soon dominated the British Association for the Advancement of Science, which had been founded in 1831. At each of the Association's annual meetings, there would be an important presidential address regarding science, and many of the individual papers also received wide publicity. Although the government-sponsored Geological Survey (1835) had less popular appeal, its work was valued and funded by the same laymen for whom Lyell's wholly secular *Elements of Geology* (1838) and Roderick Murchison's great *Silurian System* (1838) were written, the latter attesting to current efforts toward standardizing the stratigraphic sequence.[18]

The book of Genesis was rapidly becoming irrelevant to this more professional geology since both scientists and clergy saw the need for separation. Thus, William Whewell argued in 1837 that "Although it may not be possible to arrive at a right conviction regarding the origin of the world, . . . extraneous considerations, and extraneous evidence . . . must never be allowed to influence our physics or our geology." Similarly, the Rev. William Vernon Harcourt, speaking as president of the British Association in 1839, recalled the case of Galileo (as Sedgwick's

Discourse had done seven years before) in suggesting on behalf of geology that it was sheer folly "to interpret the laws of nature by the expressions of Scripture." Yet, even geology's most sympathetic defenders seriously underestimated its intellectual vigor. Thus, to Whewell, president of the Geological Society in 1839, science seemed to be entering a quieter, primarily descriptive phase; he could scarcely have been more wrong.[19]

In 1840 Louis Agassiz formally announced one of the most revolutionary concepts in Victorian geology — that there had once been an immense Ice Age, by which many geological and paleontological effects formerly attributed to the Deluge could easily be explained. Another momentous event of the same year, as hindsight allows us to ascertain, was the belated publication of Samuel Taylor Coleridge's *Confessions of an Inquiring Spirit*, which undermined a decade of attempts to reconcile geology and the Bible by introducing Germanic textual criticism into England. We should be guided by the *essential* message of our sacred writings, he argued, and not by the "imperfections of knowledge" in some less inspired passages. Real Christianity, said Coleridge, in no way requires uncritical acceptance of "the hydrography and natural philosophy of the Patriarchal ages."[20] Within ten years even *The Christian Observer* was echoing this position. It is fair to say, then, that after 1840 there were no longer rational grounds in either geological or Biblical studies to support what had previously been put forward as "scriptural" geology.

A particularly characteristic figure of this transitional time was Thomas Hawkins, the geologist and poet, who in 1840 published both a Biblical epic called *The Last Angel and the History of the Old Adamites* and also a work of popular science, his *Book of the Great Sea-Dragons, Ichthyosauri and Plesiosauri*. The existence of such monsters had been known since the beginning of the century, through the work of Cuvier, and the first dinosaurs were described and named during the 1820s by Buckland and Mantell. Only in 1841, however, did the word "dinosaur" appear, coined then by Richard Owen, the comparative anatomist (or "English Cuvier," as he was often called, to his disgust), whose work and personality influenced popular conceptions of geology throughout the 1840s and '50s. Both Carlyle (1842) and Charles Dickens (1843) went out of their way to meet him, for scientists had become the heroes of the hour.[21] In contrast, there were the increasingly defeated English clergy, who, already thrown into turmoil by the geological confessions of William Buckland, were now stunned by the evidently Romish bent of John Henry Newman's *Tract XC*, which effectively ended the Anglican

Church's most promising years of doctrinal initiative. Meanwhile, its author's brother, Francis, was among those who were finding traditional theology obsolete. For a time, it became unpopular to say too much about religion, although one fresh voice among the geological apologists, Hugh Miller's in *The Old Red Sandstone* (1841), charmed everyone with the beauty and vigor of its knowledgeable creationist assertions.

For geology in general, the early 1840s were years of triumph. Besides Owen, Agassiz, and Miller (all of whom were firmly pious), there were new, more secular heroes like Charles Darwin of the *Beagle*, whose unpretentious *Journal of Researches* (1839), detailing marvellous adventures, was followed by other of his geological publications in 1842, 1844, and 1846. Owen contributed important paleontological works throughout the decade, was active in social circles, and even had some theological influence. Charles Lyell, too, was very much a force, his books and lectures remaining popular both in England and America. Scientific analyses of scenery were in such demand with travelers that Adam Sedgwick's appended geological letters noticeably helped the sales of William Wordsworth's revised *Guide to the Lakes* (1842). In *Modern Painters* (1843ff.), similarly, John Ruskin insisted on the relevance of geology to landscape art. Contrarily, Dickens' anti-American novel *Martin Chuzzlewit* (1843) and several of his later ones include nostalgic reminiscences of now-departed Biblical geology and the security of Eden.[22] W.G. Ward's Catholicizing *Ideal of a Christian Church* and Robert Chambers' *Vestiges of the Natural History of Creation* (both 1844) — strange counterparts — then precipitated other furors, although the latter's impact was blunted somewhat by Mantell's *Medals of Creation* (1844), Whewell's *Indications of the Creator* (1845), Miller's *Footsteps of the Creator* (1847), and a grotesquely expanded fifth edition of Sedgwick's *Discourse* (1850).[23] In 1845, meanwhile, Buckland was appointed Dean of Westminster, Newman joined the Roman Catholic church, and the Great Famine began.

Orthodox religion suffered yet another major blow in 1846, when George Eliot's translation of D.F. Strauss' destructively critical *Life of Christ* appeared. Extraordinary waves of agnosticism and outright disbelief now swept across England, affecting those in the universities (like A.H. Clough) especially. This mood was soon reflected in two novels by J.A. Froude, *Shadows of the Clouds* (1847) and its more explicit sequel, *The Nemesis of Faith* (1849), which includes "Confessions of a Sceptic." Two nonfictional works by Francis W. Newman, *The Soul* (1849) and *Phases of Faith* (1850), are only slightly more positive. From now on, Higher Criticism (as it was called) would often discountenance traditional belief.

Emily Brontë's *Wuthering Heights* (1847) is the first major British novel to be named for a geological process (erosion). Other literary indications of the same year suggest more obviously how influential the science was becoming: enthusiastic praise from Edward FitzGerald; a geological fieldtrip and much proudly displayed scientific jargon in Tennyson's *The Princess* (Book III); and Benjamin Disraeli's satire upon the geology and transmutation theory of Chambers' *Vestiges*.[24] Meanwhile, Ruskin (having become secretary of the British Association's geology section) could not exclude his favorite science even in writing a book about architecture, while William Makepeace Thackeray's novel *Vanity Fair* (1848) refers more prissily to a "ladylike knowledge of botany and geology" in chapter 12. The science was regularly accorded great prestige. "So at this moment," wrote a prominent American visitor to Britain, "every ambitious young man studies geology; so members of Parliament are made, and churchmen."[25] In 1849 Prince Albert joined the Geological Society of London, after having seen to a knighthood for Lyell the year before. Geology had become the fashionable, almost the indispensable knowledge of the times.

What with famine, Chartism, political revolutions, religious defections, paleontological discoveries, and evolutionary theories, however, the 1840s were disillusioning many as to the benevolence of either history or nature. "All creation is equally insane," wrote Emily Brontë in 1842: "Nature is an inexplicable puzzle, life exists on a principle of destruction; every creature must be the relentless instrument of death to the others, or himself cease to live." A new and unsettling sensitivity toward the problem of biological survival was emerging, one with which only evolutionists like Chambers, Darwin, and Wallace could be comfortable. As the world of nature now appeared more and more hellish, Hell itself being in disrepute, natural theology — the inference of God's attributes from His works — was soon to become a deeply troubled field. Early poems by Matthew Arnold (the son of Thomas), including "Quiet Work," "In Harmony with Nature," and "In Utrumque Paratus" (all 1849) are indicative of the new alienation, which John Stuart Mill then fully articulated in his essay *Nature* (1850: published in 1874). God was now, for some, an unlikely possibility, and nature's real message to man was not Wordsworthian benevolence but vicious disregard.[26]

The indifference of nature was of primary concern to Alfred Tennyson, a poet whose knowledgeable interest in geology manifests itself throughout his long career.[27] After having studied the science under Adam Sedgwick at Cambridge, Tennyson then turned to Lyell, reading the *Principles* thoughtfully and with some horror in 1837. During these same years Tennyson was deeply afflicted with a continuing grief for the death in 1833 of his most promising undergraduate acquaintance, Arthur Henry Hallam. Tennyson wrote a series of lyric poems describing his varied thoughts and emotions during this prolonged crisis and later synthesized the whole into a single work, *In Memoriam, A.H.H.* (1850), which was highly esteemed upon publication and won Tennyson the laureateship. Read carefully, it vividly details his intellectual struggles with Lyellian geology. What is the significance of even a superior individual's death, he wonders, in a miniscule world, dwarfed by space and time, in which whole species perish routinely? Faced with the evidence of geology, it seemed almost impossible to retain a belief in Providence:

> Are God and Nature then at strife,
>   That Nature lends such evil dreams?
>   So careful of the type she seems,
> So careless of the single life . . . ?
>
> 'So careful of the type?' but no.
>   From scarped cliff and quarried stone
>   She cries, 'A thousand types are gone:
> I care for nothing, all shall go. . . .'
>   (poems 55, 56)

Tennyson can resolve the issue only by rejecting natural, and even rational, theology altogether. But this fideistic escape of his was widely acclaimed in its time by scientists and clergy alike, for whom Tennyson appeared to be the intellectual salvation of his age.

Another student of Adam Sedgwick's going on to literary fame was Charles Kingsley, who underwent a spiritual crisis in 1839 (probably owing to the Buckland controversy) and was rescued from despair by the optimistic fervor of Thomas Carlyle, with geological aid from Buckland and Hugh Miller. By 1848 Kinglsey felt ready to speak out on behalf of his generation and did so in a provocative first novel, *Yeast*. It is the story of Lancelot Smith, geologist, who, like Kingsley himself, searches to find something beyond materialism, despite the inadequacy of traditional creeds. In a second novel, *Alton Locke* (1850), published the same year as Tennyson's *In Memoriam*, Kingsley again discusses the literary and theological implications of contemporary science. Locke's reaffirmation of Christianity is preceded by a remarkable "Dreamland" chapter (36), which traces the evolutionary history of terrestrial life, from madrepore to man. Kingsley's other novels, books, and poems throughout the 1850s also reveal his interest in geology; by 1858, he was even giving lectures on the science.[28] Like Tennyson, Kingsley was a well-read amateur, and his knowledge of geology was sound.

Although both men were convinced that geology and traditional religion could be reconciled, others were far less assured. For a third major Victorian writer strongly committed to the science, this was again a time of crisis, probably because of *In Memoriam:* in May 1851 John Ruskin wrote to a friend that his faith was "being beaten into mere gold leaf, and flutters in weak rags from the letter of its old forms. . . ." "If only the Geologists would let me alone," he protested, "I could do very well, but those dreadful Hammers! I hear the clink of them at the end of every cadence of the Bible verses. . . ."[29] Matthew Arnold, too, was hearing ominous sounds from the world of nature. In his famous "Dover Beach," probably written the same year, there is only the "melancholy, long, withdrawing roar" of a once-full "Sea of Faith" to console us now. An example of such withdrawal was William Whewell's *Of the Plurality of Worlds* (1853), which utilizes morphological arguments from Richard Owen to deny both the existence of life on other planets and the teleological tradition within natural theology; although Hugh Miller (*Geology Versus Astronomy;* Glasgow, 1855) and Owen himself fumed with disagreement, the tide ignored them both.

Major literary statements like Arnold's "Stanzas from the Grand Chartreuse" and Browning's "Bishop Bloughram's Apology" (both 1855) now regard traditional Christianity as unquestionably passé: in the latter poem, Bloughram urbanely (and, for Browning, hypocritically) disdains an intellectual regression of six hundred years that would blot out "Geology, ethnology, whatnot" and get us "square with Genesis again" (lines 680, 683). Creation, the Flood, Adam, Eve, and Noah had irretrievably lost their literal meaning. Although some minor writers disputed assertions by Baden Powell (*The Unity of Worlds*, 1855) and others that the attempted reconciliations of science and Scripture had failed, these rearguard defenders (P.H. Gosse; *Omphalos* [1857], in particular) were merely candidates for Martin Tupper's innocuous "Cabinet of Fossils," a contemporary poem. In general, even the argument from design was no longer credible, and theological writers were therefore reduced to justifying their associations of God and nature through increasingly vaguer references to consistency, order, symmetry, and harmony — in short, by postulating a world view in which divinity was no more than natural law.[30]

The doctrinal weakness of his church at this time was particularly apparent to Anthony Trollope. In his novel *Barchester Towers* (1857) a man is nominated for the episcopacy who is "an eminent naturalist, a gentleman most completely versed in the knowledge of rocks and minerals, but supposed by many to hold on religious subjects no special doctrines whatever" (ch. 1). Trollope probably had William Buckland in mind. In 1848, three years after being appointed Dean of Westminster, Buckland had made his last important public appearance — to receive a medal. In defiance of that year's devastating scepticism, Buckland steadfastly reaffirmed his belief in natural theology. Shortly thereafter, as if his world had collapsed about him, this once affable apologist fell into a gloom-ridden dementia, which lasted until his death in 1856. Nor was he the only victim of the times. That December, Hugh Miller corrected in proof his forthcoming *Testimony of the Rocks* (1857), wrote an affectionate note to his wife, and shot himself to death. With Buckland and Miller, an era of facile accommodation died as well. Henry Mansel's influential Bampton Lectures (*The Limits of Religious Thought*, 1858) emphasized that man can learn of God only through supernatural revelation, not through reason or empiricism. Although natural theology had buffered increasing hostilities between science and religion for more than a century, its influence was now exhausted. Sensing this, Charles Kingsley (in his preface to the fourth edition of *Yeast*, February 1859) described the failure of Neo-Anglicanism, particularly in regard to natural fact; and with Edward FitzGerald's translation of the *Rubaiyat*, appearing a few months later, the new Victorian creed arrived.

On 24 November 1859 the publication of Darwin's *Origin of Species* brought Victorian naturalism to the fore. Like Tennyson and Kingsley, Darwin had studied geology under Adam Sedgwick, but there were few concessions anywhere within the *Origin* to supernaturalism. Although Darwin's book was essentially a triumph for Victorian geology, his insistence "on the imperfection of the geological record" (chap. IX) swiftly diminished geology's prestige. "The noble science of Geology," Darwin concluded, "loses glory from the extreme imperfection of the record. The crust of the earth with its embedded remains must not be looked at as a well-filled museum, but as a poor collection made at hazard and at rare intervals" (p. 487). This theme soon became so dominant that even scientists began to stress the subjectivity of geological analysis. Meanwhile, a different kind of intellectual weather had blown in. During the British

Association meeting at Oxford in June 1860, Thomas Henry Huxley defended Darwin and our ostensibly simian ancestry against an intemperate sneer made by Bishop Wilberforce, and with this famous confrontation between science and religion a new period of increased polarization and animosity began.[31]

Although evolution soon became the watchword of the day, geology had by no means exhausted its destructive influence upon traditional theology. The next hammer blow was *Essays and Reviews* (1860), which followed Darwin by four months and was all the more alarming in that six of its seven authors were clergy — spokesmen for the Broad Church that Coleridge inspired and Clough named. The goal of this movement was to reconcile Christianity and the Bible with their scientific and textual critics so as to retain educated persons (many of whom were drifting toward agnosticism) within the fold. Six of the book's seven essays, accordingly, include some notice of geology, and at least five accept its findings as truths to which enlightened religion must conform. In particular, "On the Mosaic Cosmogony," by C.W. Goodwin, the one layman, attacks Buckland, Miller, and all who would attempt to harmonize modern science with a Bible having quite other concerns. Bishop Wilberforce and (by petition) 11,000 other Church of England clergy steadfastly denounced such opinions, the entire volume being eventually condemned. The *Essays and Reviews* controversy, together with its Colenso aftermath of 1862 (which again involved geology), outshone Darwin's for a time.[32]

A new aspect of the general debate, precipitated by the discovery of paleolithic flints and Neanderthal remains, rose to prominence in 1863, with Lyell's *The Antiquity of Man* and Huxley's *Man's Place in Nature*.. Biblical literalism being no longer seriously in contention, the search now was for a scientific genesis. Throughout the 1860s a particularly energetic voice in this and other debates regarding the cultural role of science was that of Thomas Henry Huxley. For John Henry Newman, in *The Tamworth Reading Room* (sect. 3, 1863 version), it was presently apparent that pursuing science freely, unsubservient to Genesis, gave one "an uppish, supercilious temper, much inclined to scepticism." Meanwhile, Huxley was writing to Kingsley that "Whether astronomy and geology can or cannot be made to agree with . . . Genesis — whether the Gospels are historically true or not — are matters of comparatively small moment in the face of the impassable gulf between the anthropomorphism (however refined) of theology and the passionless impersonality of the unknown and unknowable which science shows everywhere underlying the thin veil of phenomena." Kingsley himself stood nearly alone that year in maintaining his resolute belief that "science and the creeds will [like Victorian gentlemen] shake hands at last."[33] Browning, for example, rebuked any such reconciliations in "Caliban upon Setebos" (1864), a poetic satire upon the natural theology of fallen, Darwinian man. As for some of the others, William Whewell regretted seeing "the clear definite line, which used to mark the commencement of the human period of the earth's history, made obscure and doubtful"; Benjamin Disraeli, who had already ridiculed one theory of evolution in *Tancred*, chose publicly to regard man as an angel, not an ape; Newman defended his religious beliefs against the incursions of science in his famous *Apologia*; and Pope Pius IX spoke out similarly on behalf of all Catholics with *Syllabus Errorum*, a sweeping condemnation of naturalism and other modern trends (all 1864). *Geological Magazine*, a rather different product of the same year, generally eschewed theology but encouraged contributions from amateurs; several were by John Ruskin, who also published a curious introduction to crystallography called *Ethics in the Dust* (1866). Meanwhile, William Thomson (later Lord Kelvin) was seriously challenging the now-dominant Uniformitarian hypothesis with plausible scientific arguments that severely limited the possible age of the earth; geology continued to lose both credibility and prestige as a result.[34]

The next important year was 1869, when the British Museum opened its mineral collection to the public; an irresolute "Metaphysical Society" was founded; Huxley coined the word "agnostic"; and Matthew Arnold (who was one) resigned himself to believing that truths will come when they may. Accordingly, he revised one stanza of his 20-year-old poem "In Utrumque Paratus" to allow for the possibility of human evolution, but the lines disappeared again in subsequent editions.[35] Broadly speaking, geology was of less cultural importance now, and what issues there were within it (the Eozoon question, for instance) had become so technical that amateurs could no longer contribute or even understand. Its influence as a specific science, therefore, although apparent from continuing literary references, is less noticeable during the 1870s than is a now general initiative by increasingly professionalized scientists to free themselves entirely of theological constraints. In 1871, for example, Darwin in his *Descent of Man* made no significant attempt to placate devotees of Genesis; religious tests were abolished at both Oxford and Cambridge; and Huxley suggested in print that one could be either a clergyman or a scientist but not both. According to one of Swinburne's poems, "God is buried and dead to us."[36] Perhaps no one before had ever said it with such appalling certitude; in any case, teleological arguments and references to the Creator were disappearing rapidly from scientific prose.

Many authors now, and not only scientists, were advocating or depicting situations favorable to the autonomy of science. In his *English Men of Science* (1872), for example, one of several such biographical tributes to scientists currently appearing, Francis Galton also declared scientific pursuits "uncongenial to the priestly character"; he then published "A Statistical Enquiry into the Efficacy of Prayer" in John Morley's *Fortnightly Review*, a periodical that would blast traditional theology throughout the 1870s. For British freethinkers, it seemed to be a decade of impending victory. "We are in the midst of a gigantic movement greater than that which preceded and produced the Reformation," wrote Huxley in 1873, "nor is any reconcilement possible between free thought and traditional authority."[37] Although Huxley cautioned in the same letter that his campaign on behalf of skeptical materialism would require years of effort, with Leslie Stephen and Charles Bradlaugh the heterodox London described by C.M. Davies in 1874 (*Heterodox London*) seemed already to have arrived.

While James Thomson's Shelleyan poem "City of Dreadful Night" (published in 1874 with Bradlaugh's assistance) portrays the agonies of unbelief, cooler minds like Arnold's, Bradlaugh's, Stephen's, and (posthumously) Mill's were affirming its necessity. An especially notorious advocate of scientific independence was John Tyndall, whose Belfast Address at the British Association meeting of 1874 ("We claim, and we shall wrest, from theology the entire domain of cosmological theory") enraged the Irish and conservative opinion generally.[38] Forthright historical attacks upon religion, as an impediment to science, characterize the 1870s. When Lyell died in 1875, with burial in Westminister Abbey following, even Dr. Arnold Stanley's generous memorial sermon straightforwardly rejected attempts to regulate science by Scripture and vice versa.[39] George Smith (*The Chaldean Account of Genesis*, 1876) then exposed the Babylonian origins of Noah's flood, and it was the end of an era in popular geology. Henceforth, noteworthy books by amateurs, like Kingsley's *Town Geology* (1877) and Ruskin's *Deucalion* (1879) would be, in spirit, almost wholly secular.

While the geologists themselves were writing popularized "geological sketches," introductions to the science, histories of the field's accomplishments, and autobiographies — or having their lives and letters compiled posthumously — some of their spiritual (or at least social) problems remained. George Gissing's novel *Born in Exile* (1892) attempts to show us what such problems were. The Church of England, it now seemed, was as obsolete as William Buckland's Deluge geology.[40] During the latter half of the 1880s, when none other than Prime Minister Gladstone tried repeatedly to defend the old Biblical literalism, his ingenious but empty arguments promptly collapsed under a dual attack—scientific, from Huxley; and theological, from the Rev. Samuel Rolles Driver of Oxford. A long fight was over at last.[41]

If victory for the scientific outlook seemed now assured, human happiness was not. General pessimism prevailed throughout the latter third of the Victorians' century, darkening both their science and their literature. Although several writers founded a genuine optimism upon the extremely evident technological progress of their own time (Mathilda Blind's geological epic *The Ascent of Man* [1889], for example), others were deeply impressed by the amoral viciousness of Darwinian nature and the failure of revealed religion. Human existence seemed little more than a perpetual struggle against the indifference of nature and the hostility of one's fellow men. Whether one chose Hardy's stoic resignation, Conrad's heroic resistance, or Pater's solipsistic hedonism, scant salvation availed outside the self. For still-productive older writers like Tennyson, Browning, and Ruskin, only despondency was left, all three vehemently opposing major aspects of late nineteenth-century science. Neither they nor anyone else derived much consolation from the recent doctrines of geology, which were fostering a deeper gloom.

The lessons of geology were four. First, time is immensely superior to man. Accordingly, Swinburne versifies "The Triumph of Time" as early as 1866 and reaffirms the concept with characteristic marine imagery in "A Forsaken Garden" (1876) and "By the North Sea" (1880). During the nineteenth century's closing years other poets also acknowledge time's power; for example, Hardy ("Rome", 1887), Browning ("Bad Dreams II", 1889), William Sharp, ("High Noon at Midsummer in the Campagna", 1891), and A.E. Housman ("On Wenlock Edge"; 1896). Even more powerfully, the characters in late Victorian fiction (and its readers) learn of their insignificance through direct exposure to geological time. Thus, in Thomas Hardy's novel *A Pair of Blue Eyes* (1873), Henry Knight (an amateur geologist) is walking along the edge of a coastal cliff, when it gives way, and he slides precariously downward, only to save himself by grasping at a tuft of grass. The significance of this near-destruction at nature's hands is emphasized for Knight when he scans the cliff face and finds a fossil trilobite looking back. He then contemplates the long vista of geological time, the development and extinction of life forms, and his

own wretched insignificance: "The immense lapses of time each formation represented had known nothing of the dignity of man. They were
grand times, but they were mean times too, and mean were their relics.
He was to be with the small at his death" (chap. 22). Hardy returns to the
same theme in *A Group of Noble Dames* (1891), when chattering, in-
consequential humanity is juxtaposed with skulls of ichthyosaurs and
bones of iguanodons. Of what significance are such lives?

Passages dealing with the same theme also appear in other late Victorian novels. For example, William Hale White's *Autobiography of
Mark Rutherford* (1881) includes a religious crisis reminiscent of *In
Memoriam*, Rutherford's "greatest difficulty" being his "inability to
believe that the Almighty intended to preserve all the mass of human beings, all the countless millions of barbaric half-bestial forms which, since
the appearance of man, have wandered upon the earth, savage or civilized." "Is it like Nature's way," he asks, "to be so careful about individuals, and is it to be supposed that, having produced, millions of years
ago, a creature scarcely nobler than the animals he tore with his fingers,
she should take pains to maintain him in existence for evermore? (chap.
VI). Human evolution, then, was raising theological problems of its
own. A sequel, *Mark Rutherford's Deliverance* (1885) revolts against
hopelessness, but the once consoling study of geology is first taken up
and then abandoned in an accompanying short story. A third novel, Gissing's *Born in Exile* (1892), echoes Hardy's *Pair of Blue Eyes* in a closely
related but briefer scene. While "geologizing" one Sunday, its hero, Godwin Peak, confronts an impressively stratified outcrop and promptly
loses himself "in something like nirvana, [growing] so subject to the idea
of vastness in geological time that all human desires and purposes shriv-
elled to ridiculous unimportance." The geological record, then — as Tennyson had seen and rejected so much earlier — reduces humanity to tran-
sient insignificance in the history of a godless world.[42] This, for many
Victorians, was the primary import of geology.

A second lesson was that prehistoric creatures vastly bigger and
stronger than ourselves once ruled the earth and now have disappeared.
Dinosaurs or gigantic mammals therefore have roles — albeit sometimes
minor ones — in Tennyson's *In Memoriam* (1850) and *Maud* (1855);
Dickens' *Bleak House* (1852); Thomas Love Peacock's *Gryll Grange*
(1860); Disraeli's *Lothair* (1870) and *The Coming Race* (1871); George
Eliot's *Middlemarch* (1871–72); Swinburne's "The Higher Pantheism in a
Nutshell" (1880); H.G. Wells' *A Vision of the Past* (1887); George Du
Maurier's *Peter Ibbetson* (1891); Hardy's *Noble Dames* (1891); and Oscar

Wilde's "The Sphinx" (1894), not to mention a vast popular literature. In several instances, allusions to extinct giants emphasize the obsolescence of British aristocracy (Eliot's Lord Megatherium, for example) and are for their writers hopeful, but if man as such is any improvement upon past life, it has to be his mental or moral qualities that matter, as Lyell had argued in 1863.[43] By century's end, however, Victorians were by no means sanguine that excellence of any kind would ensure human survival, although few British authors were thoroughly naturalistic.

The third and most bitter lesson was that human civilization, our species, and even life itself are in decline, like faith. Tennyson sensed it in writing both "Despair" (1881) and "Vastness" (1885). William Sharp clearly saw the possibility of human extinction in "The Last Aboriginal" (1884) and Richard Jefferies brought it home in *After London* (1885). For Mark Rutherford, the same year, "Our civilisation seemed nothing but a thin film or crust lying over a volcanic pit," and he often wondered "whether some day the pit would not break up through it and destroy us all" (*Deliverance*, ch. V). Ruskin spoke dejectedly in 1886 of our entering "a new glacial age," while May Kendall's darkly humorous "Lay of the Trilobite" (1887), probably influenced by Hardy, likewise hints at our species' demise.[44] Characteristic of the times were such titles as *Twilight of the Gods* (Richard Garnett) and *Looking Backward* (Edward Bellamy), while Mrs Humphrey Ward's novel of the same year (1888), *Robert Elsmere*, once more brought loss of faith before the public. No writer more resolutely embodied such loss than Thomas Hardy, whose "Nature's Questioning" (1898) is as far from natural theology as one can go. Nor may we forget H. G. Wells and his "On Extinction" (1893), which makes special reference to us. Like its century, everything that Victorian humanism revered seemed to be coming to an end.

The fourth and most comprehensive truth impressed upon the late Victorians by geology is that the earth itself is dying (as William Thomson, Lord Kelvin, had proposed). An unusual number of *fin de siècle* poems, therefore, deal with (often mythological) representations of the earth — for example, Swinburne's "Hertha" (1871); Meredith's "The Appeasement of Demeter" (1887), and Tennysons "Demeter and Persephone" (1889) — and while the idea of decline is sometimes opposed, it is nonetheless in evidence. A number of literary works are perfectly explicit about it: "Obermann Once More" (Matthew Arnold, 1867); "Love Is Enough" (William Morris, 1872, line 1); "Locksley Hall Sixty Years After" (Tennyson, 1886); "A Nympholept" (Swinburne, 1891, lines 155–56); "Envoy: When Earth's Last Picture Is Painted" (Rudyard Kipling, 1892);

and especially *The Time Machine* (H.G. Wells, 1894, with later versions). Christina Rossetti's poem "Advent" (1893) gives to the dying earth those personal applications that so many greying Victorians felt. It was not only *their* world that was running down, but the world as a whole. "Is Christianity Played Out?" asked the London *Daily Chronicle* in 1893, and there were about 2,000 replies, most of them negative. Even so, Victorian attitudes had moved away from religion and, as the poet Meredith said a year later, "through Science to despair."[45]

For us, as opposed to the Victorians, the earth is immensely older but far more vigorous, life is likely to go on, the grandeur of prehistory is our own creation, and time (like space) is merely new material for the human intellect. Our world belongs to man, not God, technology and science having replaced theology for us. Like the Victorians, we too fear the decline of the earth, but unlike them we fear that it may result from depletions made by man himself. We not only accept the science of geology, but depend upon it to sustain us as well. In this sense alone, perhaps, we can appreciate the fact of Victorian anxiety about geology, if not its substance. Although geology may well foster a sense of wonder in us, and that sense may have religious applications, most of us do not see the close association between geology and religion that the Victorians originally did, and perhaps both fields are better off as a result. The despair, I think, was worth it.

NOTES AND REFERENCES

1. The most adequate history of Victorian geology is still Karl Alfred Von Zittel's international *History of Geology and Paleontology to the End of the Nineteenth Century*, trans. Maria M. Ogilvie-Gordon (London, 1901; rpt. Weinheim: J. Cramer, 1962); Sir Archibald Geikie's history, *The Founders of Geology* (1897; 2nd ed., 1905, rpt. New York: Dover, 1962) ends around 1840. *A Source Book in Geology, 1400–1900*, ed. Kirtley F. Mather and Shirley L. Mason (Cambridge: Harvard University Press, 1970) offers well-chosen excerpts from significant papers. Two helpful specialized treatments are R.J. Chorley, A.J. Dunn, and R.P. Beckinsale, *The History of the Study of Landforms*, vol. I (Frome and London: Wiley, 1964) and Peter J. Bowler, *Fossils and Progress* (New York: Science History, 1976).
2. Geology's utility was frequently defended in essays and talks. See, for example, D.T. Ansted, *The Application of Geology to the Arts and Manufactures. Being Six Lectures on Practical Geology* (London, 1865). There is not much modern scholarship, but *Images of the Earth: Essays in the History of the Environmental Sciences*, eds. L.J. Jordanova and Roy S. Porter, (Chalfont St. Giles: The British Society for the History of Science, 1979) emphasizes the interdisciplinary applications.
3. Charles C. Gillispie's *Genesis and Geology: A Study in the Relations of Scientific Thought, Natural Theology, and Social Opinion in Great Britain, 1790–1850* (1951 rpt.

New York, Evanston, and London: Harper and Row, 1959) was pioneering and remains helpful. H.H. Thomas, "The Rise of Geology and Its Influence on Contemporary Thought," *Annals of Science*, vol. 5 (1941–47), pp. 325–341; and F. Sherwood Taylor, "Geology Changes the Outlook," in Harman Grisewood, ed., *Ideas and Beliefs of the Victorians* (New York: Dutton, 1966), pp. 189–196, now seem superficial.

For Victorian religious belief and its literature, Basil Willey's *Nineteenth Century Studies* (London: Chatto and Windus, 1964) and *More Nineteenth Century Studies* (New York: Columbia University Press, 1956) were most useful to me. A.O.J. Cockshut in *The Unbelievers: English Agnostic Thought, 1840–1890* (London: Collins, 1964) takes less notice of geology. Other studies include M.A. Crowther, *Church Embattled: Religious Controversy in Mid-Victorian England* (Hamden, Conn: Archon, 1970); Desmond Bowen, *The Idea of the Victorian Church* (Montreal: McGill University Press, 1968), esp. pt. II, chap. iv; P.T. Marsh, *The Victorian Church in Decline: Archbishop Tait and the Church of England, 1868–1882* (Pittsburgh: University of Pittsburgh Press, 1969), esp. chap. 2, "Theological Paralysis"; and R. P. Flindall, ed., *The Church of England, 1815–1948: A Documentary History* (London: Society for Promoting Christian Knowledge, 1972), esp. for *Essays and Reviews* and Colenso.

Following the publication of R. Hooykaas's, *Natural Law and Divine Miracle* (Leiden: Brill, 1963), it has become more common to stress the compatibility of scientific and theological beliefs. Three recent articles are: John Hedley Brooke, "The Natural Theology of the Geologists: Some Theological Strata," in Jordanova and Porter[2]; Frank M. Turner, "The Victorian Conflict between Science and Religion: A Professional Dimension," *Isis*, vol. 69 (1978), pp. 493–516 (with extensive notice of further scholarship); and Richard Yeo, "William Whewell, Natural Theology and the Philosophy of Science in Mid Nineteenth Century Britain," *Annals of Science*, vol. 36 (1979), pp. 493–516.

The book-length works I cite often had numerous editions (sometimes importantly revised). For poems and essays, consult the authors' collected works and standard Victorian anthologies.

4. See Dennis R. Dean, "James Hutton and His Public, 1785–1802," *Annals of Science*, vol. 30 (1973), pp. 89–105; and "James Hutton on Religion and Geology," *Annals of Science*, vol. 32 (1975), pp. 187–193.

5. For Buckland, see the essay by Susan F. Cannon in the *Dictionary of Scientific Biography* and Anna B. Gordon, ed., *Life and Correspondence of William Buckland* (London, 1894). I quote Newman from Geoffrey Faber, *Oxford Apostles* (London: Faber & Faber, 1933; rpt. New York, 1976), p. 60, as do Chorley *et al.*[1] Disliking the tentativeness of geological hypotheses, Newman sought "a consistent and luminous theory of certainties" [ibid.] His fruitless study of the Rev. John Bird Sumner's *Treatise on the Records of the Creation* (1816) in 1823 (Faber, p. 122) probably alienated him from any attempt to reconcile geology and Scripture. There are further allusions to geology by him in *Newman: Prose and Poetry*, ed. Geoffrey Tillotson (Cambridge: Harvard University Press, 1970), pp. 129 (last line), 429, 552ff., 819.

6. See Leroy E. Page, "Diluvium and Its Critics in Great Britain in the Early Nineteenth Century" in Cecil J. Schneer, ed., *Toward a History of Geology* (Cambridge: The M.I.T. Press, 1969), pp. 257–271; and Buckland's address in *Proceedings of the Geological Society of London*, vol. 1 (1827), pp. 55 and 60. For Buckland's change of opinion by 1830, see Katherine M. Lyell, ed., *The Life, Letters, and Journals of Sir Charles Lyell, Bart.*, 2 vols. (London, 1881; rpt. Westmead, England: Gregg, 1970), vol. I, p. 276.

7. Although Carlyle began *Sartor Resartus* in 1830, it was not published until 1833–34 and

not in book form until 1836. The most fully annotated edition of *Sartor Resartus*, a difficult work, is that edited by Charles F. Harrold (New York: Odyssey, 1937), from which I quote p. 4. For other geological passages, see that edition, pp. 18, 90, 149, 286, etc., and *The French Revolution* (1837), vol. I, bk. III, chap. IV. Later Victorians influenced by Carlyle include Sterling, Kingsley, Dickens, Froude, Mill, Ruskin, Huxley, Tyndall, and Gissing.

8.   Lyell's *Life, Letters and Journals*[6] is extremely helpful. Leonard G. Wilson's as yet incomplete biography, *Charles Lyell*, vol. I (New Haven and London: Yale University Press, 1972), includes some useful background. We owe much of our current understanding of Lyell to Martin J.S. Rudwick, who has summarized his point of view in his book *The Meaning of Fossils*, 2nd ed. (New York: Science History, 1976), chap. IV. "The "Lyell Centenary Issue" of *The British Journal for the History of Science*, vol. 9 (1976), pp. 91–242, is an interesting collection of papers, although I take issue with several of them in *Annals of Science*, vol. 34 (1977), pp. 607–611.

9.   For an overstated account, see Philip Lawrence, "Charles Lyell versus the Theory of Central Heat: A Reappraisal of Lyell's Place in the History of Geology," *Journal of the History of Biology*, vol. 11 (1978), pp. 101–128.

10.   Thomas Sadler, ed., *Diary, Reminiscences, and Correspondence of Henry Crabb Robinson*, 3 vols. (London, 1869), vol. III, p. 8. See also Lyell, *Principles of Geology*, 1st ed., vol. III (1833), pp. 382–385.

11.   John F.W. Herschel, *A Preliminary Discourse on the Study of Natural Philosophy* (London, 1830), pp. 4–5 (quoted) and 281ff., quoting p. 287.

12.   Francis Darwin, ed., *The Life and Letters of Charles Darwin*, 2 vols. (New York, 1887), vol. I, p. 227. Compare this statement with his earlier rejection of geology (thanks to the Wernerian teachings of Robert Jameson) in the *Autobiography*, sec. 1.

13.   C.R. Leslie, *Memoirs of the Life of John Constable* (London: Phaidon, 1951), p. 272 (see also p. 226). Charles Kingsley expressed similar views during the 1840s and was opposed by John Henry Newman in *The Tamworth Reading Room* (1841), sec. 2.

14.   James C. Simmons establishes "the topicality of *The Last Days of Pompeii*" in *Nineteenth Century Fiction*, vol. 24 (1969), pp. 103–105, but it is not true that the August 1834 eruptions of Vesuvius were "the most destructive . . . in modern centuries" (p. 103). Browning's imagery in *Paracelsus* may have been influenced by the great attention given to Graham Island, a temporary submarine volcano in the Mediterranean (1831ff.); see my forthcoming essay in *Isis*. I owe the Mill reference and two others later in the present essay to Susan F. Cannon, *Science in Culture: The Early Victorian Period* (New York: Science History Publications, 1978), chap. 1. Carlyle's *Life of Sterling* appeared in 1851; see esp. pt. II, chap. V and pt. III, chap. II.

15.   Lyell, *Principles*, 1st ed., vol. III (1833), pp. 270–273.

16.   Thomas Arnold, *Sermons*, 3 vols. (London, 1829–1834), vol. II (1832), p. 482; Arnold was strongly influenced by Coleridge. Sedgwick's *Discourse* (rpt. New York: Humanities Press, 1969) attacks Scriptural geology in note (F); see esp. p. 106. I quote Wilks from *The Christian Observer*, vol. 34 (1834), pp. 207–208n; his poem is at pp. 219–229, quoting p. 224.

17.   See *The Christian Observer*, vol. 39 (1839), esp. pp. 400ff.; and John Wallis Clark and Thomas McKenny Hughes, eds., *The Life and Letters of the Reverend Adam Sedgwick*, 2 vols. (Cambridge, 1890), vol. I, 469–470, and vol. II, 76–80.

18.   A useful source that should be returned to print is George Basalla, William Coleman, and Robert H. Kargon, eds., *Victorian Science: A Self-Portrait from the Presidential Addresses to the British Association for the Advancement of Science* (Garden City, N.Y.:

Doubleday, 1970). See also Arthur P. Stanley, ed., *The Life and Correspondence of Thomas Arnold, D.D.*, 12th ed., 2 vols. (London, 1881; rpt. Farnborough: Gregg, 1970), vol. II, p. 142, for the Devonian controversy of 1839.

19. William Whewell, *History of the Inductive Sciences*, 3rd ed., 3 vols. (London, 1857, rpt. London; Cass, 1967), vol. III, p. 485. I have the 1839 opinions of Harcourt and Whewell from John Pye Smith, *The Relation Between the Holy Scriptures and Some Parts of Geological Science*, 5th ed. (London, 1852), pp. 380 and 381.

20. For the development of Agassiz's concept, see John and Katherine Imbrie, *Ice Ages: Solving the Mystery* (Short Hills, N.J.: Enslow, 1979), chaps. 1 & 2. S.T. Coleridge, *Confessions*, 3rd ed. (London, 1853; rpt. Stanford: Stanford University Press, 1957), quoting letter IV (p. 61); see also letter VI.

21. Dinosaurs were first named in Richard Owen's "Report on British Fossil Reptiles," *Report of the British Association for the Advancement of Science* (1841), p. 103. Chapter one of Adrien J. Desmond, *The Hot-Blooded Dinosaurs* (New York: Warner, 1977) emphasizes Owen, while the Rev. Richard Owen's *Life of Richard Owen* (2 vols., London, 1894) includes a lengthy tribute by Huxley; for Carlyle, Dickens, and Tennyson, see the index in Owen. The famous dinosaur reconstructions at Syndenham, and the Iguanodon dinner that took place within one of them on 31 December 1853, dramatized the work of both Owen and Hawkins.

For the religious situation, see F.E. Kingsley, ed. *Charles Kingsley: His Letters and Memories of His Life*, abridged (New York, 1884) esp. chaps. III and VI; and Ralph Waldo Emerson, *English Traits* (1857), chap. XIII.

22. Sedgwick contributed three letters on the geology of the Lake District to John Hudson, ed., *A Complete Guide to the Lakes* (with text by Wordsworth) in 1842, adding a fourth in 1846 and a fifth in 1853. See also Clark and Hughes,[17] vol. I, pp. 246-252 and vol. II, pp. 39-41. Ruskin's fullest treatment of geology is in vols. 3 and 4 (both 1856). For my remarks on Dickens I am indebted to Bert G. Hornback's, *"Noah's Arkitecture": A Study of Dickens' Mythology* (Athens, Ohio: Ohio University Press, 1972), esp. pp. 162-163 and 169-173.

23. Milton Milhauser's, *Just Before Darwin: Robert Chambers and Vestiges* (Middletown, Wesleyan University Press, 1959; chap. 6 esp.) has now been supplemented by the work of Peter J. Bowler,[1] pp. 53ff.

24. *Letters of Edward FitzGerald*, ed. J.M. Cohen (Carbondale, Ill.: Southern Illinois University Press, 1960), pp. 66-69. John Kilham, *Tennyson and The Princess* (London, 1958), discusses geology in chap. XI. In Disraeli's novel *Tancred, or The New Crusade* (1847), the title character dramatizes his dissatisfaction with Western creeds by making a superficial pilgrimage to Jerusalem. The satire on Chambers is in bk. II, chap. IX; in consecutively numbered editions, this is chap. XV ("Disenchantment").

25. John Ruskin, *The Seven Lamps of Architecture* (1849), chap. II, par. 13; Ralph Waldo Emerson, *English Traits* (1857), chap. XIV.

26. Emily Brontë, *Five Essays Written in French*, trans. Lorine White Nagel (Austin, Texas: University of Texas Press, 1948), p. 17. For analysis and application, see especially chap. 4 in J. Hillis Miller, *The Disappearance of God* (Cambridge: Harvard University Press, 1963). *Wuthering Heights* was written in 1845 and published in December 1847. Lionel Trilling in *Matthew Arnold* (Cleveland and New York: Meridian, 1963) discusses Arnold's view of nature in chap. 3. See also F.E. Kingsley,[21] pp. 245-246.

27. Tennyson's geology has been reviewed by several scholars, including Willey (*More Nineteenth Century Studies*[3]) and Eleanor B. Mattes, *In Memoriam: The Way of a Soul* (New York: Exposition Press, 1951), but more remains to be done.

28. *Yeast*, published serially, was not in book form till 1851. *Letters and Memories*[21] contains much on Kingsley's geology, and there is an especially characteristic passage in *Glaucus* (1855), pp. 10–13. Kingsley's lectures became *Town Geology* (1877), mentioned below.

29. E.T. Cook and Alexander Wedderburn, eds., *The Works of Ruskin*, 39 vols. (London: G. Allen, 1909), vol. XXXVI, p. 115.

30. See the Rev. Baden Powell, *Essays on the Spirit of the Inductive Philosophy, the Unity of Worlds, and the Philosophy of Creation* (London, 1855; rpt. Farnborough: Gregg, 1969), p. 135; "The instances in which we can trace a *use* and a *purpose* in nature, striking as they are, after all constitute but a very small and subordinate portion of the vast scheme of universal order and harmony of design which pervades and connects the whole." Yeo[3] has found similar passages in other writers, and eventually there would be geologist Henry Drummond's *Natural Law in the Spiritual World* (1883). Like Kingsley in *Glaucus* the same year, Powell rejects "the hopeless chimera of erecting theories of geology on the Mosaic narrative" (p. 303ff.).

31. For the impact of Darwin, see *Victorian Science*,[18] sec. 11. Although the present essay does not extend to Darwinian controversy, some overlap is inevitable. On Darwin and religion see Alvar Ellegard, *Darwin and the General Reader* (Göteborg: Göteborg University Press, 1958); Michael Ruse, *The Darwinian Revolution* (Chicago: University of Chicago Press, 1979); and Neal C. Gillispie, *Charles Darwin and the Problem of Creation* (Chicago: University of Chicago Press, 1979). Darwin's geological writings aroused no significant theological controversies, but see *Victorian Science*,[18] p. 424, regarding the charge of subjectivity. Studies of Darwin's literary influence include those of Lionel Stevenson (*Darwin Among the Poets* [1932; rpt. New York: Russell and Russell, 1963]) and Georg Roppen (*Evolution and Poetic Belief* [Oslo: Oslo University Press, 1956). For the Wilberforce episode, see Leonard Huxley, ed., *Life and Letters of Thomas Henry Huxley*, 2 vols. (London, 1900; rpt. Westmead: Gregg, 1969), chap. XIV, from which many later accounts derive.

32. Besides Goodwin, the other six authors are Frederick Temple, Rowland Williams, Baden Powell, Henry B. Wilson, Mark Pattison, and Benjamin Jowett; all but Pattison at least mention geology. For Colenso and geology, see esp. Peter Hinchliff, *John William Colenso, Bishop of Natal* (London: Nelson, 1964), chap. 4; and Trilling,[26] pp. 190–196.

33. T.H. Huxley, *Collected Essays*, 9 vols. (London, 1893; rpt. New York: Greenwood, 1968), *passim*; *Life and Letters*,[31] vol. I, pp. 238–244 (three important letters), quoting p. 239. F.E. Kinglsey,[21] chap. XXII, quoting p. 344. I next quote I. Todhunter, ed., *William Whewell, D.D.: An Account of His Writings, with Selections from His Literary and Scientific Correspondence*, 2 vols. (London, 1876; rpt. Farnborough: Gregg, 1970), vol. II, p. 435. (A "reconciliation of the scientific with the religious view is still possible," Whewell believes, "but it is not so clear and striking as it was.")

34. Joe D. Burchfield, *Lord Kelvin and the Age of the Earth* (New York: Science History, 1975) presents the scientific controversy, and its literary influence is alluded to in *Victorian Science*,[18] pp. 485–486. Although Huxley had himself criticized the reliability of geological chronology (in his essay "Geological Contemporaneity and Persistent Types of Life" [1862]), he was an outspoken opponent of Kelvin's.

35. For Arnold, see David J. DeLaura, *Hebrew and Hellene in Victorian England: Newman, Arnold, and Pater* (Austin and London: University of Texas Press, 1969), p. 88 (to which I allude) and p. 108, and chap. 5 and 6 *passim*. His revision of "In Utrumque Paratus" is discussed by Miller,[26] p. 226n.

36. T.H. Huxley, *Collected Essays*,[33] vol. II, p. 149. Swinburne, "To Walt Whitman in America," line 106; for Victorian antecedents, see A.H. Clough, *Dipsychus* (1850; pub. 1865) and D.G. Rossetti, "The Burden of Nineveh" (1870). Nietzsche was slightly later. For the Eozoon question, see *Victorian Science*[18] sec. 8, esp. p. 210; and George P. Merrill, *The First Hundred Years of American Geology* (1924; rpt. New York: Hafner, 1969), chap. X.

37. Galton, *English Men of Science* (1872), p. 24, as quoted by Turner,[3] p. 365. Huxley, *Life and Letters*,[31] vol. I, p. 397.

38. For the text of Tyndall's Belfast Address, see *Victorian Science*,[18] sec. 15. Two contemporary reactions are to be found in C. L. Cline, ed., *The Letters of George Meredith*, 3 vols. (Oxford: Oxford University Press, 1970), vol. I, p. 493; and K.M. Lyell,[6] vol. II, p. 455.

39. I rely upon the excerpt appearing in A.D. White, *A History of the Warfare of Science with Theology in Christendom*, 2 vols. (London: MacMillan, 1896), vol. I, p. 247. See also K.M. Lyell,[6] vol. II, pp. 459–463. Following Stanley's initiative, burial in Westminister Abbey was no longer contingent upon orthodoxy.

40. The novel includes five geologists. Martin Warricombe was such a lifetime devotee of the science that he named his son Buckland. "Mr. Warricombe's attainments were respectable, but what could be said of a man who had devoted his life to geology, and still (in the year 1884) remained an orthodox member of the Church of England?" (p. 269). The other three geologists are Godwin Peak and two of his geological tutors, Professor Thomas Gale and a Mr. Gunnery, who arouses Godwin's contempt for religious interpretations of geological phenomena ("Deluge?" growled Mr. Gunnery. "*What* deluge? *Which* deluge?"). Soon, Godwin leaves college and begins to write essays for an atheistic periodical, attacking those who still attempt to reconcile science and religion. He falls in love, of course, with Martin Warricombe's daughter.

41. Huxley, *Life and Letters*,[31] vol. II, pp. 114–118; 269–271, 276. See also A.D. White,[39] vol. I, pp. 243–246.

42. Gissing, p. 180. See also J.O. Bailey, "Hardy's 'Imbedded Fossil,' " *Studies in Philology*, vol. 42, (1945), pp. 663–674.

43. *In Memoriam*, poem 56; *Maud*, sec. IV, poem VI; *Bleak House*, the first paragraph; *Gryll Grange*, chap. XXI; *Lothair*, chap. 26; *Middlemarch*, the character Lord Megatherium; and the others *passim*. For other instances of prehistoricized aristocrats or Tories, see K.M. Lyell,[6] vol. I, p. 473; and T.L. Peacock, *Works* (Halliford Edition), vol. VIII, p. 254; Lyell, *The Antiquity of Man* (London, 1863), chap. XXIV.

44. We must not overlook the influential Krakatoa eruption of 1883; see Richard D. Altick, "Four Victorian Poets and an Exploding Island," *Victorian Studies*, vol. 3 (1960), pp. 245–260. The four poets are Tennyson, Bridges, Swinburne, and Hopkins; Ruskin is also involved. I quote Joan Evans and John Howard Whitehorse, eds., *The Diaries of John Ruskin*, 3 vols. (Oxford: Clarendon Press, 1956), vol. III, p. 1132. May Kendall's poem is from *Dreams to Sell* (London, 1887). According to William Hale White the same year, "We all have too vast a conception of the duty which Providence has imposed upon us; and one great service which modern geology and astronomy have rendered is the abatement of the fever by which earnest people are often consumed" (*Revolution in Tanner's Lane* [London, 1887], pp. 355–356).

45. George Meredith, "Foresight and Patience," line 90. For the *Daily Chronicle* episode, see John A. Lester, Jr., *Journey Through Despair: 1880–1914* (Princeton: Princeton University Press, 1968), pp. 39–40.

Astronomical Imagery in Victorian Poetry

JACOB KORG

Department of English
University of Washington
Seattle, Washington 98195

T HE PROGRESS the Victorian poets made in reconciling the scientific
and humanist traditions with each other, moderate as it was,
represents a significant achievement. These poets faced great obstacles in
accommodating their art to the growing dominance of science. Especially
prominent in the nineteenth-century debate between science and
theology was the dualism of spirit and matter formulated by Descartes,
and this created a particularly intractable problem for the Victorians
because it conflicted with their religious convictions and with the
worldview they had inherited from the Romantics. Yet the Victorian
poets managed at least to begin the task of bringing new insights about
the nature of the physical universe into the circle of significant moral and
spiritual experience. Nothing illustrates this better than their treatment of
astronomy.

There was no crisis of astronomy for poets in the eighteenth century
because Newton's principles defined an orderly and intelligible system.
But Romanticism, with its emphasis on the primacy of subjective reality,
made the idea of a universe that could be fully explained by the action of
physical forces seem spiritually sterile. The cosmic metaphor in the title
of Pierre-Simon de Laplace's great exposition of the Newtonian solar
system, *La Mécanique Céleste*, was a triumphant formulation of the
eighteenth-century view; but considered within the Romantic framework
of the unity of man and nature, it reduced man to a machine as well. In
his ironic, "The New Sinai," the Victorian poet, Arthur Hugh Clough,
needed no more than Laplace's title to expose the inadequacy of his
metaphor:

Earth goes by chemic forces; Heaven's
A Mécanique Céleste!
A heart and mind of human kind
A watch-work as the rest!

137

The advance of astronomy after Newton posed an even more direct threat to nineteenth-century poetry and its traditions. Since the earliest times, poetry had employed such concepts as the fixity of the stars, the music of the spheres, and the symbolism attached to the sun, the moon, the planets, constellations and eclipses, as powerful sources of order and authority. Scientific astronomy gradually but persistently dismantled the mythopoetic approach to the heavens, reducing all such ideas to the status of mere intellectual toys. This development reached a certain culmination with Sir William Herschel's investigations of space beyond the planets and established the fact that the solar system is no more than an infinitesimal part of a vast sidereal universe. These findings, much more revolutionary and disturbing than any that had gone before, confronted the Victorian imagination with serious cosmological and spiritual dilemmas.

Herschel first achieved fame when he sighted Uranus in 1779, becoming the original of Keats's "watcher of the skies" who feels a sense of exultation "when a new planet swims into his ken." This discovery had no revolutionary implications, for it only added a feature to the Newtonian account of the solar system, but Herschel's later work, his exploration of the heavens with an improved telescope, literally transformed the shape and dimensions of the universe, opening unimaginable depths of outer space, and dispelling forever the lingering notion that the sky was a dome studded with fixed stars. He showed that many of the nebulae are resolvable into individual stars resembling the sun, and that the sky is dotted with three-dimensional formations of these bodies. Furthermore, all is in a state of constant change, for the stars move and undergo internal transformations, and the varying conditions of the nebulae represent different stages of a developmental process.

In a paper written for the Royal Society in 1811, Herschel offered evidence that the nebulae could be transformed into stars through condensation brought about by gravity, and he divided the nebulae into 31 classes that could be regarded as the successive stages of a continuous development, none differing from the adjacent ones, as he said, more than the yearly stages in the growth of the human body. In this way the heavens acquired an evolutionary significance.

> They are now seen to resemble a luxuriant garden, which contains the greatest variety of productions, in different flourishing beds; and one advantage we may at least reap from it is, that we can, as it were, extend the range of our experience to an immense duration. For, to continue the simile I have borrowed from the vegetable kingdom, is it not almost the same

thing, whether we live successively to witness the germination, blooming, foliage, fecundity, fading, withering and corruption of a plant, or whether a vast number of specimens, selected from every stage through which the plant passes in the course of its existence, be brought at once to our view?[1]

Looking back at these contributions to astronomy and their influence on nineteenth-century thought about fifty years later, the intellectual historian W.E.H. Lecky reported that the "morphological conception" of the universe strongly encouraged the idea that natural law produced a progression from simple to more complex forms.[2]

The Victorians reacted to these discoveries in a variety of ways. Some found information about the stars and planets irrelevant to human concerns; some optimistically interpreted the new facts, in accordance with the principles of natural religion, as evidence of the design of a benign Creator; and others simply accepted these findings of the natural law that governed terrestrial life. But there was also the profound suspicion that Herschel's thrusts into unthinkable depths of space and time had reduced man to an unimportant accident in the vastness of an indifferent cosmos. Herschel's work carried forward the tradition initiated in the sixteenth century by Giordano Bruno, who contended not only that the universe is changing, but also that it is infinite, without a center, and filled with an infinite number of worlds. Bruno himself regarded this plenitude as a manifestation of the greatness of God, but the displacement of man and his earth from the position they had held in earlier cosmologies led to cultural episodes of pessimism and anxiety in the Victorian and modern periods.[3]

In George Eliot's *Daniel Deronda* (1876), the egotistic heroine is overtaken by a sense of alienation when she studies astronomy at school. Only when she returns to the human dimension of ordinary social life can she regain her self-confidence. The depressing effect astronomy could have on the Victorian mind is vividly dramatized in Thomas Hardy's *Two on a Tower* (1882), whose youthful astronomer-hero, Swithin St. Cleeve, mentions two sources of disturbance that preyed on the minds of imaginative people. Speaking of the dimensions astronomers encounter, he says, "There is a size at which dignity begins; further on, a size at which awfulness begins; further on, a size at which ghastliness begins." He regrets the passing of the old idea that the sky is a dome. The actual sky is "a horror" that contains "monsters." And he echoes Pascal's famous remark about the spaces between the stars by saying that "Until a person has thought out the stars and their interspaces, he has hardly learnt that there are things much more terrible than monsters of shape, namely, monsters of magnitude without known shape. Such monsters are the voids and waste places of the sky."

But this is not all. Herschel's discovery of the evolution of stars, together with the growing knowledge of novae and variable stars, had shown that the universe was not fixed, but unstable, and that the sun, together with the earth, might some day come to an end. In the seventeenth century John Donne had been disturbed by the discovery of new heavenly bodies because they disrupted the accepted understanding of the cosmos.[4] In the nineteenth century, however, the realization that existing stars might disappear seemed even more threatening. Consequently, St. Cleeve complains, "For all the wonder of the everlasting stars, eternal spheres, and whatnot, they are not everlasting, they are not eternal; they burn out like candles." He advises anyone who wants to remain cheerful to avoid astronomy, for "it alone deserves the character of the terrible."[5] We can see that a sidereal universe totally alien to human intelligence and incompatible with human existence provided Hardy with a perfect vehicle for his conviction that man lives in a hostile or indifferent world.

This desperate vision of the cosmos generated a critical problem within poetry itself. The Victorian situation is perhaps better understood if it is compared with that of the Metaphysical Poets, who lived in a time of intense spiritual crisis precipitated by the destruction of the geocentric model of the universe in the work of Copernicus and Galileo; however, the Renaissance doctrine of correspondences and belief in the unity of man and the cosmos under God survived this revolution. It was still possible to find in the creation affinities and analogues that disclosed the consistency of the Divine Plan, and the discoveries of science fell into place as examples of these perceptions. The reshaped universe remained within the reach of human feeling and intelligence, and the poets were able to base figures of speech embodying strong emotional experiences on concepts, terms of art, and even instruments drawn from science. All of this prepared the way for the eighteenth-century view that the scheme of nature revealed by science was a "great chain of being," in which man had a specified place, and the orderliness and justice of the creation became a common poetic theme.[6]

In the Romantic period, however, a decisive bifurcation in thought occurred, as figures like Wordsworth and Coleridge urged the claims of a newly discovered faculty, the Romantic imagination, against the rationality upheld by Godwin and Bentham. In spite of their differences, the opposing parties agreed that each mode of thought had its own powers and was sharply divided from the other.[7] Wordsworth was comparatively generous to science, for he thought that scientists were impelled by the same motives as poets, and he predicted that science and poetry would eventually harmonize with each other; but he also warned that mere rational analysis could have a destructive effect on the moral and aesthetic sensibilities and identified poetry as its antithesis. Twenty years later, Thomas Love Peacock, in an essay called "The Four Ages of Poetry," attacked poetry as an infantile and archaic holdover from primitive times that interfered with serious thinking and that was bound to degenerate because it did not address itself to the "scientific and philosophical part of the community." His friend, Shelley, replied in his "Defence of Poetry" (1821) by defining poetry as the direct expression of the imagination, which is the source of cultural forms that "reasoners and mechanists" only develop and fill in. In the essay, "What is Poetry?" written in 1833, John Stuart Mill pursued the distinction by saying that poetry speaks to the feelings through images that arouse emotion, while science attempts to command belief through logical propositions.[8] And a later and less familiar critic, George Brimley, in his "Poetry and Criticism" observed that poetry and science might deal with the same subjects, but that science seeks to reduce all the features of an object to cause and effect, whereas poetry develops the rich complexity science is likely to set aside and presents objects as concretions organized into units by subjective thought.[9]

In this way, the Victorian poets faced an absolute division that the Metaphysical Poets of the seventeenth century had been spared. They were prevented from joining intuitive and intellectual faculties, as the Metaphysical Poets had, by the famous dissociation of sensibility, the conviction that the truths accessible to each were irrelevant or opposed to the other. The dilemma of the Victorian poets is reflected in the problem of metaphor. The ground of Romantic metaphor is the premise that man and nature share common moral and spiritual values, and that the external world, approached imaginatively, can be seen as a symbolic embodiment of human feelings. This continuum and the universe of discourse based on it were threatened from two directions by science in the Victorian period. First, science could often validate rationally the insights Romanticism attributed to the imagination. Blake's proverb, "What is now proved was once only imagin'd," gives the imagination priority, but it also leaves open the inference, which came to seem very plausible in an age dominated by developmental views, that imagination is no more than a provisional form of reason. Such ideas as the kinship of men and animals and the rebirth of nature in the spring, which had originated as metaphors expressing profound religious beliefs, were shown to be physically and verifiably true and thereby stripped of their mystique. Even worse, the conceptions of imagination were easily dwarfed by the awesome facts of the nature that science was discovering, so that reportorial and referential uses of language threatened to displace poetic ones. Edward FitzGerald clearly saw that science was capable of disabling poetic language when he wrote:

> Yes, as I often think, it is not the poetical imagination, but bare Science that every day more and more unrolls a greater Epic than the Iliad; the history of the World, the infinitudes of Space and Time! I never take up a book of Geology or Astronomy but this strikes me. And when we think that Man must go on to discover in the same plodding way, one fancies that the Poet of to-day may as well fold his hands, or turn them to dig and delve, considering how soon the march of discovery will distance all his imaginations, [and] dissolve the language in which they are uttered.[10]

A passage from Arthur Hugh Clough's poem, "Easter Day," vividly illustrates the power of rational thought to unravel metaphor and reduce language to its literal function. The skeptical Clough, finding little in the fallen world to justify a belief in the Resurrection, demystifies the Christian myth by taking the metaphor in Christ's words to his Apostles, "I will make you to become fishers of men," and poetically urging them to follow only its literal sense, "And catch, not men, but fish."

Science also threatened to destroy the axis of similitude on which Romantic metaphor turns. Among the Metaphysical Poets, who thought of logic and rhetoric as cooperative rather than rival disciplines, the two parts of the metaphor were linked by logical or abstract relationships compatible with scientific modes of thought. Because it had an intellectual basis, the Renaissance principle of correspondence between human and nonhuman terms found in the metaphysical conceit was able to survive radical changes in the conception of the universe, and could even generate new relationships in accordance with changing worldviews. Something of this kind takes place in John Donne's "First Anniversarie," which treats the irregularities discovered in the movements of heavenly bodies by "the new philosophy" as counterparts of the corruption of society. The Romantic image, on the other hand, depended on an intuitive, general sense of affinity between man and nature that was increasingly threatened by the scientific investigation of nature. As the reactions of Hardy's astronomer show, the scientific discoveries that confronted the Victorian poets made it difficult to regard the physical universe as a symbolic embodiment of mental processes or a theater of human moral striving.

The issue became critical with the advent of Darwinism, but well before that time the revelations of astronomy and geology—"terrible Muses," as Tennyson called them in his poem, "Parnassus"—had been troubling thoughtful people who were committed to the Wordsworthian vision of nature. Against the backdrop of a cosmos that had been in existence eons before the coming of man, one that extended through reaches of space and time that exceeded man's conceptual powers, the human episode seemed no more than a brief and insignificant anomaly. The man-nature image and the Romantic vision as a whole were in danger of seeming frivolous and irresponsible, for Herschel's disquieting explorations of the stars had cancelled the second term of the Romantic metaphor.

In spite of this, and perhaps because they sensed that the evidence of astronomy was particularly threatening to the Romantic tradition, some of the Victorian poets made serious efforts to integrate astronomical facts into their work, especially through metaphors based on them. I want to examine astronomical images in the work of three Victorian poets who employed the nonscientific logic of metaphor to create new similarities between the terrestrial and celestial worlds. By this effort they attempted to reweave man into the fabric of the cosmos. I shall begin with Browning, who accepted the new facts, but not the ideas they implied, move on to Tennyson, who was more receptive to astronomy than any other poet of his time, and conclude with Francis Thompson, a lesser figure whose mysticism makes his acceptance of astronomical knowledge especially significant.

◅§   §►

Robert Browning was a man of powerful independent intellect who adhered faithfully to the traditional body of Christian beliefs. While he employed many scientific ideas, and often took up the debate between reason and intuition, he never changed his view that ultimate truth is open to intuition alone. Nevertheless, he possessed a strongly empiric temperament, was capable of vigorous logical reasoning, and sometimes found that scientific discoveries corresponded to intellectual patterns he had already formulated. Many years before *The Origin of Species* appeared, he conceived a general developmental view that became a prominent theme in his poetry, and his metaphysical ideas often sound like ingenious transpositions of evolutionary theory.[11]

While he rejected the claim that science had definitively changed man's concept of himself and his world, Browning found it possible to accept scientific facts without surrendering his conviction that the physical universe was controlled by the wisdom and goodness of its Creator. "No star bursts heaven's dome," says a passage in the late poem, "Fust and His Friends," "but Thy finger impels it." He approved of scientific discovery because he felt that science and its instruments exposed both the nature and the limitations of the cognitive faculty and showed the universe to be the work of a remarkable Intelligence beyond human understanding.

Although the heavenly bodies, both as sources of insight and as reflections of *a priori* ideas, are often found in Browning's thinking, most of his allusions to them owe little or nothing to science.[12] The numerous astronomical images in "Abt Vogler," a dramatic monologue spoken by a composer, are preNewtonian, if not entirely prescientific, and seem to have been suggested by the Pythagorean linkage of music and the planets. One such image interestingly shows how astronomy, like science in general, made it impossible to accept traditional ideas about man's relation to the universe. Abt Vogler compares the imperfection of man and the perfection of God: "On the earth the broken arcs; in the heaven a perfect round." But it had been known since Kepler's time that the planetary orbits are elliptical, not round, and cannot serve as symbols of the human idea of perfection.

On the other hand, a small but important group of Browning's poetic images employs specific astronomical facts of the kind generally available to well-informed laymen. These images are significant in two ways: historically, because they show that a powerful mind could use the new findings to express traditional moral and spiritual ideas, effectively linking intuitive and empirical thinking; and rhetorically, because they create a new expressive synthesis, combining rational elements with a poetic that remains basically Romantic.

These effects appear in the astronomical images of *Paracelsus* (1833), a drama written early in Browning's career, about a scientist who learns the danger of allowing intellectual ambition to replace human affection. The poet in the play, Aprile, expresses his ambition by saying,

> Even as a luminous haze links star to star,
> I would supply all chasms with music . . .

joining the traditional Pythagorean myth about the music of the spheres with Herschel's discovery that some nebulae contain a cloud-like "shining fluid." Through a parallel with the familiar communicative power of music, the image assigns an intelligible function to the newly discovered and puzzling nebulae. At the end of the play, Paracelsus compares himself and Aprile to two stars,

> . . . the over-radiant star too mad
> To drink the life-springs, beamless thence itself —
> And the dark orb which borders the abyss,
> Ingulfed in icy night . . . ,

This comparison employs the varying stages of stellar evolution pointed out by Herschel to dramatize human destinies. But the most significant star image in *Paracelsus* is perhaps the most incidental one. In his prefatory note to the play, Browning acknowledged its apparent lack of unity and said that he hoped the reader would make the necessary connections among its elements, adding, "were my scenes stars, it must be [the reader's] cooperating fancy which, supplying all chasms, shall collect the scattered lights into one constellation — a Lyre or a Crown." The premise behind this analogy between the play and the stars is perhaps no more than a sophisticated variant of the idea that the work follows nature. But it contains the significant implication that the sidereal universe can serve as a model for a structure that does not conform to recognized patterns, but is based on unfamiliar principles that must be grasped if the structure is to be understood.

A balance between rationality and intuition is especially apparent in Browning's later work, written when evolution and its repercussions had entered the religious controversy over the Higher Criticism. Browning's "Epilogue" to his volume of poems, *Dramatis Personae*, consists of statements on the possibility of faith by three speakers; the second uses the image of a dying star to lament the disappearance of faith. Joining traditional imagery with recent astronomical knowledge, he asks what has become of the star of Bethlehem that guided the Magi — and by implication, all men — to Christ, thus identifying its disappearance with the decline of faith. The imagery reflects the perception that the stars are constantly changing. Did not everyone lament, he asks

> When a first shadow showed the star addressed
> Itself to motion, and on either side
> The rims contracted as the rays retired. . . .

The dim stars left in the sky are

> a mist
> Of multitudinous points, yet suns, men say . . .

But the star that was formerly bright cannot be found, and men cannot be assured that it will burn brightly in response to their needs. In this way, the disappearance of stars implied in the developmental view of the universe casts doubt on the traditional affinity between the heavens and human affairs.

In fact, the continued survival of the nearest star, the sun, remained an insoluble problem throughout the nineteenth century. When it was understood that the sun could not contain enough fuel to sustain ordinary combustion for more than a few centuries, various hypotheses were offered to explain its stability. One of these, supported by Tennyson's friend, Sir Norman Lockyer, proposed that the sun's heat was renewed by the impact of meteorites. More plausible was Von Helmholtz's theory that the sun, in accordance with the nebular hypothesis, was still shrinking, and that its heat was caused by the centripetal motion of its particles. This meant that it would shrink to half its size in 5 million years, and would ultimately disappear or grow dark.[13]

This possibility is often considered in nineteenth-century poetry; it is dramatized in Byron's "Darkness" (1816), a powerful fantasy about events on the earth after

> The bright sun was extinguish'd, and the stars
> Did wander darkling in the eternal space,

and a generation later Tennyson, in *In Memoriam*, speaks of the "dying sun" as part of his vision of a pointless universe. A still later Victorian poem, James Thomson's "City of Dreadful Night" (1870–74), attributes the spiritual sterility of urban life to science and technology, and sets the scene with the image of a sun approaching exhaustion that rose over the horizon but then

> stopped and burned black, except a rim,
> A bleeding, eyeless socket, red and dim,

a symbol of despair that recalls the dying star of Bethlehem in Browning's poem.

The speaker who takes up this theme in "Epilogue" does not speak for Browning, for he is identified with Ernest Renan, the author of the agnostic *Life of Christ*, which Browning did not approve. On the contrary, the sun is the hero of Browning's vital and pulsating universe, and one of its associated wonders, the spectrum, is often used in his imagery. Both of these appear in "Parleying with Bernard de Mandeville" (1887) in an argument against the protest that the universe fails to exhibit the goodness of God.

In this poem, Browning uses a stellar image which suggests that the physical properties of the stars are less significant than their potential for symbolizing the Romantic view that knowledge of the world is incomplete without the contribution of the imagination. To one who complains that the constellation of Orion cannot be a man because it lacks a body, Browning answers, "Look through the sign to the thing signified." His longer parable on a related theme concerns the sun. After a wonderful description of the response of nature to the sunrise, Browning tells us that all things rejoice in the sun except Man, who grumbles because it fails to satisfy his intellectual needs. He cannot manipulate its energies as the Mind behind it can, or share the intelligence embodied in it. To placate this malcontent, Browning offers the example of Prometheus, who caught the minor emanations of the sun with a burning-glass and put them to use without attempting to capture its stupendous fires themselves. This, says Browning, is the right principle: use nature for practical ends, interpreting its signals without trying to fathom the essential nature of things. Don't try to follow the sunbeam to its source,

> Rather ask aid from optics. Sense, descry
> The spectrum — mind, infer immensity!

Science is capable of revealing the facts of nature. Reflection on what they mean cannot produce further knowledge, but only wonder.

The limitation of the intellect when it is confronted by the spectacle of the universe is also the subject of one of Browning's most majestic passages on astronomy. It appears in the Pope's monologue in *The Ring and the Book*, in which the seventeenth century Pope, Innocent XII, conscious of his fallibility, explains why he nevertheless feels that he must agree to judge the case before him. Man, he says, cannot expect to share the insights of God. When he tries to comprehend the moral cosmos, man is capable only of arriving at a small, if intelligible, image of a reality that is altogether beyond his understanding or power of expression.

> Man's mind, what is it but a convex glass
> Wherein are gathered all the scattered points
> Picked out of the immensity of sky,
> To reunite there, be our heaven for earth,
> Our known unknown, our God revealed to Man?
> Existent somewhere, somehow, as a whole;
> Here, as a whole proportioned to our sense, —
> There (which is nowhere, speech must babble thus!)
> In the absolute immensity, the whole
> Appreciable solely by Thyself . . .

There is another inaccessible universe, he adds, in the microscopic world, and the cosmos, seen as a whole, exhibits the order and plenitude of the Great Chain of Being. Within this hierarchy, says the Pope, he, like every other creature, has a place that is exclusively his, that of the representative of God in the seat of judgment. The point is made clear through an analogy with the greatly expanded universe that was still a novelty in the Pope's time (the end of the seventeenth century):

> Just as, if new philosophy know aught,
> This one earth, out of all the multitude
> Of peopled worlds, as stars are now supposed, —
> Was chosen, and no sun-star of the swarm,
> For stage and scene of Thy transcendent act . . .

an allusion to the Incarnation. Shelley thought the sidereal universe revealed by Herschel dwarfed the doctrine of the Virgin Birth. He wrote, in his notes to "Queen Mab": "It is impossible to believe that the Spirit that pervades this infinite machine begat a son upon the body of a Jewish woman . . . "[14] The Pope, on the other hand, considers the coming of Christ greater than the universe, for it is

> Thy transcendent act
> Beside which even the creation fades
> Into a puny exercise of power.

In the example as a whole, the new worlds revealed by astronomy and the appearance of Christ on earth are combined into an argument that draws upon the resources of both science and religion.

◄§ §►

   Alfred Tennyson was the only Victorian poet who seriously undertook to reconcile traditional values with the philosophy of science. At Tennyson's death, Huxley said that he was the only poet since Lucretius who understood science, and declared at another time that the "insight into scientific method" displayed in *In Memoriam* was "quite equal to that of the greatest experts."[15] Tennyson's poem, one of the central spiritual documents of the period, is usually understood in relation to biological evolution, but it was published nine years before *The Origin of Species* appeared, and, like much of Tennyson's poetry, reflects his lifelong interest in astronomy. He had learned about Sir William Herschel's work on nebulae at an early age; some time before he was eighteen, he said to a brother who felt too shy to attend a dinner party. "Fred, think of

Herschel's great star-patches, and you will soon get over all that."[16] When he went to Cambridge, Tennyson had as his tutor William Whewell, the mathematician and philosopher of science who was soon to publish the Bridgewater Treatise on astronomy that gained him an enviable reputation. Tennyson was also a friend of the astronomer, Sir Norman Lockyer, who said of him, "His mind is saturated with astronomy."[17] In 1910 Lockyer, in collaboration with his daughter, published *Tennyson as a Student and Poet of Nature*, a volume consisting of numerous quotations with some commentary. The first 84 pages are devoted to astronomy under such chapter headings as "Cosmogony and Evolution," "The Evolution of Stellar Systems," and "The Starry Heavens." Lockyer reported that Tennyson was "often in the observatory" and that he displayed a knowledge of current developments in astronomy during talks that took place in the 1870s and '80s. There was a telescope at Tennyson's home at Aldworth, and on one occasion he and Lockyer used it to observe binary stars. The effect of astronomy on Tennyson's sense of perspective is beautifully suggested in the remark he made after Lockyer had shown him some nebulae through the telescope: "I cannot think much of the county families after that."[18]

Tennyson was conscious of change as a principle of nature, and the evidence suggests, as William R. Rutland has observed, that for him, as for many others, the source of this idea was astronomy.[19] One of the scientific works that shaped the state of mind expressed in *In Memoriam*, Robert Chambers' *Vestiges of the Natural History of Creation*, began its account of the differentiation of the forms of life with a description of the solar system that reflected the universality of natural law. Chambers held that the mathematical relationships that govern the solar system were produced by an external intelligence. "We come, in short, to a Being beyond nature—its author, its God. . ."[20] This was familiar doctrine; what was new and alarming, and contrary to accepted ideas about the creation, was Chambers' contention that God had not created a world of fixed forms resembling those visible today, but a universe whose laws kept the world in a state of constant change.

> The Creator, then, is seen to have formed our earth, and effected upon it a long and complicated series of changes, in the same manner in which we find that he conducts the affairs of nature before our living eyes; that is, *in the manner of natural law.* This is no rash or unauthorized affirmation. It is what we deduce from the calculations of a Newton and a Laplace, on the one hand; and from the industrious observation of facts by a Murchison and a Lyell on the other.[21]

Tennyson maintained his Christian faith because he was able to see purposeful and orderly change — "eternal process moving on" — both as a principle embodied in the physical universe revealed by science and as evidence of the intentions of God. No spectacle in nature dramatized this more impressively than nebular evolution as Herschel envisioned it.

Tennyson is an especially valuable witness to the Victorian spiritual crisis because he was fully exposed to the doubts the new science could inspire. His ultimate confidence in Divine purpose was earned against the resistance of despair, an aspect of his feelings that is perhaps more representative of the nineteenth century's state of mind than his Christian optimism. In his early poetry, astronomical imagery is generally found in optimistic or neutral contexts. In "Armageddon," a youthful effort that was unpublished until after his death, a fantastic heavenly cosmos looms mysteriously over the landscape. The sun, moon, and stars enact a lurid drama at the scene of the battle that takes place before the Last Judgment. Finally, the poet is left facing the night sky, and feeling a pulsation in the atmosphere,

> As if the great soul of the Universe
> Heaved with tumultuous throbbings on the vast
> Suspense of some great issue,[22]

an image that attributes a vague, if portentous human identity to the cosmos. In "The Palace of Art," a much more finished work, astronomical study is one of the temptations that lead the "soul" who is the heroine of the poem to enter a life of seclusion. The three stanzas "expressive of the joy wherewith the soul contemplated the results of astronomical experiment" describe the soul visiting an observatory, where she views the active universe described by Herschel, which is described in imagery that can only be called decorative:

> Regions of lucid matter taking forms,
>   Brushes of life, hazy gleams,
> Clusters and beds of worlds, and bee-like swarms
>   Of suns and starry streams.[23]

The other astronomical image is sufficiently remarkable. In her isolation, says the soul, she will be

> Still as, while Saturn whirls, his steadfast shade
>   Sleeps on his luminous ring,

a marvelous rendering, possibly drawn from Herschel's studies of Saturn, of the deceptive immobility of the outer universe.

At first, then, Tennyson enjoyed astronomy. But in *In Memoriam*, which expressed the spiritual doubts he felt at the death of his friend, Arthur Hallam, he shows that he felt that the heavens might project the inhuman "ghastliness" felt by Hardy's astronomer. Tennyson questions the meaning of both life and death, and sees in the cosmos images embodying the purposelessness of existence. He imagines Sorrow telling a terrible secret:

> "The stars," she whispers, "blindly run;
> A web is woven across the sky;
> From out waste places comes a cry,
> And murmurs from the dying sun."

As Whitehead has observed, the line about the stars denies the possibility of free will and declares that all actions, including those of human beings, are fully determined by mechanical forces.[24] Tennyson adds to this vision of pointless determination the nightmare of the dying star, reading from the heavens a message that confirms his private despair.

In Memoriam does not resolve the debate between science and traditional religion, but it does give poetic form to knowledge from fields as diverse as astronomy, geology, embryology and anatomy, while formulating the profound spiritual questions posed by these fields. The poem extends the traditional world in two ways. Tennyson is especially conscious of the symbolism of sun, moon, and stars as they shape the seasons, define natural forms by the lights and shadows they cast, and intimate spiritual messages in their movements. But he is also capable of standing outside of space and time and of viewing the universe disclosed by the astronomers, "star and system rolling past," as the scene of man's moral and spiritual career. The developmental view that Tennyson derived through Chambers and Lyell from Herschel—and that taught him that

> The solid earth whereon we tread
> In tracts of fluent heat began . . .

and led from the evolution of inorganic matter to biological evolution—this view provided him with the sense in which the poem finally comes to rest. This is the characteristic and distinctively Victorian view that God's work is to be found, not in particular creations, miraculous as they may seem,

> I found Him not in world or sun,
> Or eagle's wing or insect's eye;

but in an intuitive sense of progress and development whose material manifestations science was capable of disclosing.

Throughout his career, Tennyson continued to bring a consciousness of astronomy to his assessments of the human condition. Sometimes it seemed necessary to set the nonhuman universe aside. Man's concerns may seem insignificant within the universe his intellect is capable of revealing, but he must nevertheless devote himself to his own values. The hero of *Maud* reflects that "Our planet is one, the suns are many," and makes this a basis for the reclusive Stoicism he embraces. Even Tennyson's Laureate production on the death of the Duke of Wellington describes the astronomically inspired "eternity" to which the dead hero is travelling as an irrelevance:

> Through world on world in myriad myriads roll
> Round us, each with different powers,
> And other forms of life than ours,
> What know we greater than the soul?

That this attitude became difficult to maintain in the face of the growing authority of science is shown by such poems as "Despair," in which a speaker who has attempted suicide blames his spiritual condition on the nihilism of popularized science, which has preached Doubt until "The Sun and Moon of our science are both of them turned into blood." The speaker regarded the light of the stars as deceptive, for "the dark little worlds running round them were worlds of woe like our own," and when he abandoned religion, he felt that

> . . . the homeless planet at length will be wheeled
> through the silence of space,
> Motherless evermore of an ever-vanishing race,

visions of despair clearly generated by astronomy.

The burden of astronomy described by Hardy's hero in *Two on a Tower* appears in the short dialogue called "Epilogue to the Charge of the Heavy Brigade," in which a poet says that earlier poets had an easier time than contemporary ones, who see man "in Space and Time," and realize that even Homer's fame will ultimately vanish in a universe that forms only a small part of a huge galaxy. His stanza in this scene uses terms drawn from Herschel's accounts of the nebulae:

> The fires that arch this dusky dot —
>   Yon myriad worlded way —
> The vast sun-clusters' gathering blaze
>   World-isles in lonely skies
> Whole heavens within themselves, amaze
>   Our brief humanities.

In "Vastness," Tennyson, speaking in his own voice, reviews the futility of the whole range of human efforts and ideals in light of the speculation that "Many a planet by many a sun may roll with the dust of a vanished race," and he asks what the use of it all may be when everything will be "Swallowed in Vastness, lost in Silence, drowned in the deeps of a meaningless Past?" The only answer, an anguished assertion that the dead live on, hardly withstands the crushing nihilism of the major part of the poem.

Francis Golffing has characterized Tennyson as the only major Victorian figure who refused to participate in the debate between science and humanism because he anticipated a time when the two would be reconciled in a system based on "humanized science." In a group of later poems Tennyson expressed a serene confidence in the vision of a utopian collaboration between technological and moral energies. "The key terms" of this vision, says Golffing, "become increasingly those of astrophysics. . . "[25] He presents as perhaps his strongest piece of evidence, "The Dreamer," which is described as the last poem Tennyson actually completed. In it a man hears the Voice of the Earth complaining that it is tormented by dissension, but counsels it not to despair,

> For moans will have grown sphere-music
> Or ever your race be run!
> And all's well that ends well,
> Whirl, and follow the Sun!

The right course is resignation to a scheme of things identified metaphorically with the order of the solar system.

Francis Thompson was born in 1859, the year *The Origin of Species* was published, so that he belonged to a generation familiar with evolution. He had some awareness of science, for he spent six years as a half-hearted medical student, and evidently gave some attention to astronomy, for in 1903 he published in the *Academy* a review of Agnes M. Clerke's *Problems in Astro-Physics*. He was, perhaps more than any figure in English literature since Blake, a cosmic poet, who undertook to read the truths of human life from the sky, and he mentioned as one of the objects of man's ambition, "all starry majesties, / And dim transstellar things." Although he dealt with the heavens as a religious mystic, he integrated science and mysticism by employing images in which astronomical facts are essential elements. "Carmen Genesis," for example, is a retelling of the story of Creation as it is found in the Bible, but Thompson follows the nebular hypothesis in his account, providing dynamic causes for the formation of the heavenly bodies. A force he calls "dividual splendour" is responsible for "spinning" them through a process of "globing" and something like gravity is present when the movements of the solar system are described in this way:

> With interspheral counterdance
> Consenting contraries advance
> And plan is hid for plan.

In Thompson's most famous poem, "The Hound of Heaven," the sinner flees God through the reaches of a universe of pre-Ptolemaic design, a place with arches, windows, and gateways. Yet astronomy is at the heart of Thompson's image of the spiritual pursuit. To locate it, we must turn to "Orient Ode," a song that seems thoroughly unscientific, for it presents the Day as a priest lifting the sacrament of the sun in a ceremonial that closes when he sets it down "Within the flaming monstrance of the West." Later in the poem, however, the heavenly bodies are involved in a different metaphor, one that could not have been conceived without a knowledge of the effects of gravity described by Newton and Laplace. Here the cosmic scene is that of a hunt, where the sun, as a hunter, urges his hounds, the planets, forward with fear, but holds them back with the leash of love so that the centrifugal and centripetal forces that determine planetary orbits symbolize the sinner's conflicting feelings toward the source of his salvation. The poet addresses the sun:

> Before thy terrible hunt thy planets run;
> Each in his frighted orbit wheels,
> Each flies through inassuageable chase,
> Since the hunt o' the world begun,
> The puissant approaches of thy face,
> And yet thy radiant leash he feels.
> Since the hunt o' the world begun,
> Lashed with terror, leashed with longing,
> The mighty course is ever run. . . .

In "The Hound of Heaven," the hunting image appears without any suggestion that it is based on the Newtonian solar system. But "Orient Ode" enables us to see that the war of emotions that afflicts the sinner as he flees God is a counterpart of the interplay between gravitational and inertial forces, and that the dynamics of the soul replicate the dynamics of the universe. Thompson apparently did not feel that science threatened his mystical convictions or that astronomy challenged his image of the heavens. He met them by using physical laws as metaphoric terms for the spiritual forces that govern his vision of life.

❦ ❧

Not all of the Victorian poets accepted scientific knowledge in general, or astronomical facts in particular, as these three did. Coventry Patmore, in a poem called "The Two Deserts" (1868), specifically rejects the worlds revealed by the telescope and the microscope as spiritually barren, but it is significant that his defense of traditional views reflects a keen consciousness of science. He begins by declaring

> Not greatly moved with awe am I
> To learn that we may spy
> Five thousand firmaments beyond our own,

and goes on to say that he is less antagonized by the microscopic dimension, because it is near at hand and alive, but prefers the middle region to which man has been assigned, where wonder and beauty are found. His condemnation of the moon as ugly and the sun as a kind of hell is no doubt narrow-minded, but it leads him to the engaging notion that the nonterrestrial universe, with all its horrors, is a kind of joke on the Creator's part, a project intended "to make dirt cheap," so that it would contrast with the spiritual values of man.

Finally, it is important to observe that scientific astronomy can be seen extending the resources of poetic language and the field of poetic reference even when no overt allusions to astronomy are made. The effective final image of George Meredith's 1883 sonnet, "Lucifer in Starlight," is a brilliant example of this use. The fallen angel is described as soaring over the earth with evil intent, but retreating when he sees the stars in their courses:

> Around the ancient track marched rank on rank
> The army of unalterable law.

In the Victorian context, the word "law" has a double force here, both mythic and scientific. Some explanation is needed to clarify this effect.

John Stuart Mill, in his essay, "Nature," posthumously published in 1874, about nine years before Meredith's poem appeared, had explored the complexities of the concept of "law," and thought it necessary to caution his readers against confusing the word's familiar meaning as a rule of conduct or morality, and its scientific one, in the sense of a regular natural phenomenon. The concept of natural law was one of the irresistible intellectual forces of the nineteenth century. It was extended beyond the natural sciences, generated new fields of study wherever reliable generalizations could be formulated on the basis of evidence, and promised — or threatened — to transform all studies into sciences. Astronomy

156 ANNALS NEW YORK ACADEMY OF SCIENCES

was one of its major sources. The discoveries of regular motion in the solar system by Kepler, Newton and Laplace, and Herschel's extension of their principles to outer space, which showed that the distant stars are controlled by the same gravitational force that operates when a child drops a penny, impressively demonstrated the uniformity of nature. Throughout the nineteenth century, new observations and methods of measurement showed that the established astronomical principles had wider and more refined applications. Beginning in 1838, the measurement of the parallaxes of fixed stars made it possible to determine their distances, and the sighting in 1846 of a new planet, Neptune, whose existence had been predicted on the basis of eccentricities in the orbit of Uranus, was a triumphant vindication of Newtonian ideas. In their adherence to physical and mathematical laws, the heavens seemed to be nothing less than an intellectual system, and Meredith, in his sonnet, calls the stars "the brain of heaven." His "unalterable law" identifies the regular movements of the stars with divine authority, making the point, not through the confusion Mill warned against, but through clear and intelligible metaphor, that moral law is as binding in the spiritual realm as natural law is in the physical universe.

The perfect meshing of scientific and spiritual traditions seen in Meredith's sonnet never became a general cultural premise among the Victorians, as Wordsworth's identification of man and nature had in the Romantic period. All that can be said is that some of the Victorian poets believed, in varying degrees, that science and poetry stood in a complementary relation to each other, and that they tried to overcome the disjunction that had arisen between the two by devising a language capable of accommodating both thought and feeling, fact and symbol, the truths of nature and the truths of the imagination. Their use of scientific astronomy is perhaps one of the best illustrations of their limited, but significant effort to reshape the Romantic tradition into a poetic idiom fully contemporary with Victorian civilization.

1. Quoted from Herschel in *A Short History of Astronomy* by Arthur Berry (London: John Murray, 1898), p. 340.
2. W. E. H. Lecky, *History of the Rise and Influence of the Spirit of Rationalism in Europe* (1865; rpt. New York: George Braziller, 1955), p. 294.
3. Alexander Koyré, *From the Closed World to the Infinite Universe* (1957; rpt. New York: Harper and Brothers, 1958), pp. 43–44. It should not be forgotten that Bruno's ideas and many other modern insights were anticipated in the work that remains the supreme union of science and poetry, Lucretius's *De Rerum Natura*. In Book V, which is devoted to

cosmology, Lucretius argues that while the cosmos is immortal, like the elements of which it is made, its particular features will disappear, as man does. He attributes the origin of the earth and sky to random combinations of atoms, not to an act of intelligence.

4. On the other hand, many regarded Galileo's revelation of new stars and satellites as a cause for optimism, and there was some speculation and fantasy about the possibility of life in other worlds. See Marjorie Nicolson, "The 'New Astronomy' and English Imagination" in *Science and Imagination* (Ithaca, N.Y.: Great Seal Books, Cornell University Press, 1956).

5. The quotations from Hardy in this and the preceding paragraph are from *Two on a Tower* (London: Macmillan, 1912), chap. 4, pp. 34–35.

6. See William Powell Jones, *The Rhetoric of Science* (London: Routledge and Kegan Paul, 1966).

7. For a survey of Romantic and post-Romantic opinion on this subject, see M.H. Abrams, *The Mirror and the Lamp*, chap. XI, "Science and Poetry in Romantic Criticism," (1953; rpt. New York: W.W. Norton, 1958).

8. First published in the *Monthly Repository* in 1833, reprinted in *Mill's Literary Essays*, ed. Edward Alexander (Indianapolis: Bobbs-Merrill, 1967), pp. 49–63.

9. George Brimley, *Essays*, ed. William George Clark, 3rd ed. (London: Macmillan, 1882), pp. 195–196. The separation between the two modes of expression has, of course, survived and deepened, and enjoyed continual theoretical support in the twentieth century. It is a recurrent theme in the work of I.A. Richards, one of whose earliest publications was a statement entitled "Art and Science" (*Athenaeum*, June 27, 1919, pp. 534–535), which attempted to formulate a distinction between the two. Richards' influential restatement of the Victorian view, *Science and Poetry* (1926), maintains that poetry deals with truths and beliefs that are independent of scientific standards, that there is no conflict between science and poetry, and that it is a serious mistake for poetry to challenge the authority of science or to pretend to offer verifiable truths of the kind only science can produce. He says, "In its use of words poetry is just the reverse of science." In the reprint of *Science and Poetry* published in 1970 as *Poetries and Sciences,* Richards' views on this subject are mainly unchanged, although he softens his statement by inserting "most" before "poetry." For a later examination of the subject see *The Identity of Man* (1965; rtp. Garden , N.Y. City: Doubleday, 1971) by Jacob Bronowksi. Here poetry and science are sharply differentiated, but their values are considered complementary to each other, and Bronowski regards the modern failure to grasp their relationship as "the visible sign of a loss of confidence in the identity of man."

10. Quoted (not quite accurately) from a letter written to E. B. Cowell [1847] in Christopher Ricks' edition of *The Poems of Tennyson* (London, 1969), p. 1410, in the headnote to "Parnassus." The original text appears in *The Letters and Remains of Edward Fitz-Gerald,* ed. William Aldis Wright (London: Macmillan, 1903), vol. I, p. 262.

11. A discussion of Browning in relation to evolution appears in Lionel Stevenson's *Darwin Among the Poets*, chap. 3 (New York: Russell & Russell, 1963).

12. The book-length study on this subject, *Browning's Star-Imagery* by C. Willard Smith (Princeton: Princeton University Press, 1941) pays no attention whatever to the influence scientific astronomy may have had on Browning.

13. However, Von Helmholtz's theory was quickly found to be unsatisfactory. For one thing, the time it allowed for the previous existence of the sun fell far short of the age of terrestrial fossils that bore evidence of the sun's radiation, amounting to less than 1 percent of the needed length of time.

14. *Complete Poetical Works of Percy Bysshe Shelley*, ed. Thomas Hutchinson (London: Oxford University Press, 1945), p. 801.

15. *Life and Letters of Thomas Henry Huxley*, ed. Leonard Huxley (New York: Appleton, 1901), vol. II, p. 359; and Wilfrid Ward, "Thomas Henry Huxley," *Nineteenth Century* (August, 1896), p. 286.

16. Hallam Tennyson, *Alfred Lord Tennyson: A Memoir* (New York: Macmillan, 1897), vol. I, p. 20.

17. Tennyson,[16] vol. II, p. 381. For Lockyer's friendship with Tennyson, see A.J. Meadows, *Science and Controversy: A Biography of Sir Norman Lockyer* (Cambridge, Mass.: The M.I.T. Press, 1972).

18. Sir Joseph Norman Lockyer and Winifred L.Lockyer, *Tennyson as a Student and Poet of Nature* (London: Macmillan, 1910), pp. 2-4.

19. William R. Rutland, "Tennyson and the Theory of Evolution," *Essays and Studies*, vol. 26 (1940), p. 9.

20. Robert Chambers, *Vestiges of the Natural History of Creation*, 11th ed. (London: J. Churchill, 1860), p. 9.

21. Chambers,[20] p. 102.

22. *The Poems of Tennyson*, ed. Christopher Ricks (London: Longmans, 1969), p. 74. Tennyson later incorporated much of this imagery into his Chancellor's Prize poem, "Timbuctoo" (1828).

23. This and the other two stanzas about astronomy were not included in the body of the poem, but appeared in a note to the 1832 version, which said that they would have been in the text itself "if the Poem were not already too long." See Ricks' edition, pp. 412-413. The stanzas do not, of course, form part of the standard text of the poem. Ricks points out (p. 413) that "lucid," "matter," and "forms" were scientific terms.

24. Alfred North Whitehead, *Science and the Modern World* (1925; rpt. New York: Mentor Books, 1964), pp. 74-75. Whitehead adds that Tennyson refused to face this issue, which was current at the time.

25. Francis Golffing, "Tennyson's Last Phase: The Poet as Seer," *Southern Review*, 2 n.s. (Spring, 1966), pp. 264-285. The quotation is from p. 269.

The "Aims and Intentions" of *Nature*

DAVID A. ROOS
Department of English
Northwestern University
Evanston, Illinois 60201

DURING the past ten years, scholars from various fields have focused upon the founding, in 1869, of the weekly scientific journal *Nature* as a turning point in the history of Victorian culture. Historians of education have looked upon *Nature* as one of the central forums wherein schemes for "the promotion of scientific research and education" could be articulated and debated.[1] Social historians interested in the "emergence of the professionalized society in the West" and sociologists of science interested more specifically in the professionalization of modern science have regarded the creation of *Nature* as one of the landmarks that designate the self-conscious awareness of distinct, specialized, professional interests among the previously rather ill-defined group of "new scientists."[2] Scholars interested in Victorian periodicals have interpreted the appearance of *Nature* as one of the most significant moments in the "fragmentation of the common context" of nineteenth-century intellectual discourse; such analyses argue that the tendency towards intellectual stratification and topical specialization in periodical literature was both contributing cause and partial effect of the larger cultural fragmentation that seems to have occurred during the late nineteenth century.[3] The establishment of *Nature*, in other words, seems to have coincided with the decisive recognition among scientists of the impossibility of maintaining a single norm of truth; *Nature* thus marks what another scholar has called "the demise of the truth-complex" and the (sometimes reluctant) acceptance of a "multi-normative," relativistic attitude towards modern culture.[4]

All of these analyses take for granted that some decisive divergence between two or more cultures did in fact occur near the end of the last century. All of these analyses also assume that, on the whole, the intentions of the founders, editors, and contributors of *Nature* produced their

159

desired effects. Or, to be more precise, these analyses assume that the effects produced by *Nature* are an accurate and complete indication of the intentions of the founders, editors, and contributors. Thus, scholars can look backward from the point of view of a modern professional scientist and recognize the intention of *Nature* to solidify the nascent professional impulse among late-Victorian scientists. Similarly, since the signs of modern cultural fragmentation seem so obvious, the attribution of either aggressive or reluctant intentions to the men who helped produce that cultural divergence seems unavoidable. Either those new professional scientists were aggressively trying to carve out for themselves a socioeconomic niche within modern, commercial society, or those same professional scientists were reluctantly being forced to admit that they could no longer share their main interests with the public at large. All of these analyses are well documented and all of them seem to me to be partially true. But all of them ignore in various ways the crucial evidence of the text of *Nature* itself.

The extreme alternative to the anachronistic and Whiggish tendencies of such analyses is best expressed by John Maddox, who took advantage of his position as the editor of *Nature* to review the dramatic changes in the journal which occurred between 1869 and 1973.

> To begin with, the journal was a gossip sheet. . . It was almost an accident that *Nature* stumbled into being a learned journal as well as a weekly vehicle for news and comment. After decades during which readers would write in with observations hardly more profound than that they had heard the cuckoo for the first time that spring, they began to use the journal for saying that they had found an electron, or a neutron, or were able to describe why DNA should be central to an understanding of the life process.[5]

Maddox's hyperbolic rhetoric begs to be discounted, but his comments draw attention to a rather different perspective on *Nature* than that discussed above. Maddox simply insists that *Nature* was not created in its modern form, but that it only gradually became a prestigious professional journal. In other words, by insisting that "It was almost an accident that *Nature* stumbled into being a learned journal," Maddox asserts that this development was not entirely the result of the intentions of the original founders, editors, and contributors. Considered by itself, Maddox's analysis is even more misleading than those he contradicts; *Nature* was never "a kind of gossip sheet." Nonetheless, by forcing scholars to differentiate between the development or evolution of *Nature* and what the first editor, Norman Lockyer, explicitly called the "aims and intentions" of *Nature*, Maddox has suggested how to supplement the partial perspectives of previous analyses.

In this essay I attempt to articulate a more complex analysis of the "aims and intentions" of *Nature* by examining the evidence that previous analyses have either ignored entirely or glanced over lightly. Specifically, I argue that at least one of the intentions of *Nature* was to mediate between increasingly diverse and sometimes antagonistic segments of late Victorian society — between scientists and artists, between professional scientists and interested amateurs, between scientific generalists and specialists, and between specialists in different fields.[6] Although *Nature* may, in fact, have been one of the divisive forces that led to the fragmentation of late nineteenth-century culture, my analysis attempts to prove that the founders, editors, and contributors to *Nature* understood their "aims and intentions" to be cohesive — and, cohesive not simply in relation to science, but also in relation to the arts and humanities. Finally, since I agree with John Maddox that *Nature* became something substantially different from what it was originally intended to be, I attempt to identify several causes for what I shall call the "unintended effects" that transformed *Nature* into a modern, professional journal.

In order to explain fully the "aims and intentions" of *Nature*, part of its prehistory must first be summarized. During the early nineteenth century, when the English word "science" still had the general sense of the German word *wissenschaft*, members of the British reading public who were interested in one or more of the specific "sciences" could depend upon quarterly journals such as the *Edinburgh* or *Quarterly Review* to provide adequate scientific information to satisfy both amateur and professional, generalist and specialist. When Charles Lyell first began writing the geological essays that eventually culminated in his monumental, three-volume *Principles of Geology*, his intention was to "write something that would be read by the public," and the countless reviews and responses to Lyell's work that appeared in the major quarterly journals during succeeding years were written by men of similar intentions and similar intellectual qualifications.[7] Early Victorian science existed within — indeed, depended upon — an explicitly public, accessible, and sharable context, and even in the case of Charles Darwin's theory of natural selection, causes for major theoretical shifts can be traced to the pages of the *Edinburgh Review* as well as to the pages of the *Philosophical Transactions of the Royal Society*.[8] But in 1834 William Whewell coined the new English word "scientist" to designate with one collective name all "the students of the knowledge of the material world." In his *Philosophy of the Inductive Sciences,* Whewell later wrote: "Thus we might say that as an Artist is a Musician, Painter, or Poet, a Scientist is a Mathematician, Physicist, or Naturalist." Although the division between two cultures embodied in Whewell's definition may seem all too obvious to the modern scholar, the full impact of this simple linguistic act was not to be felt for more than fifty years. Many Victorians, including even Thomas Henry Huxley, resisted the radical innovation and the no less radical specialization that they thought was implied by the new word "scientist."[9] There is no doubt that William Whewell had some intention of stabilizing what we would now call the "professional" identity of those special sort of men who "cultivate science in general." But it is equally without doubt that Whewell had no intention of fragmenting that public space upon which sharable human knowledge depends. Whewell's consistent and, indeed, *a priori* commitment to an idealistic, unitary theory of knowledge and reality made it impossible for him to conceive of any irreconcilable conflict between science and the humanities. [10] Nonetheless, as years passed and the unintended effects of Whewell's linguistic innovation became part and parcel of the modern consciousness, it became less and less possible to imagine and recoup a sense of what it felt like to live, to think, and to talk in a more cohesive, public, nonspecialist, nonprofessional environment. By the end of the nineteenth century, the chances of a major scientific achievement being influenced by an article in the *Edinburgh Review*, or any other generalist, nonprofessional journal, were almost nil.

The founding of *Nature* in 1869 was the culmination of a decade-long attempt to create a higher-level, but still general and accessible forum for the discussion of scientific matters than the old quarterly journals provided. More frequent and more extensive coverage of scientific matters was necessitated by institutional changes within the scientific community itself. Each of the many specialized societies that were established during the early decades of the century naturally wanted a specialized journal to facilitate communication among its members; thus, by 1860, the Astronomical Society, the Geological Society, the Chemical Society, and the Botanical Society, to name just four, each had at least one specialized journal that focused upon its own particular field of interest. Frequently, membership in a specialized society was a prerequisite for publication in such journals; therefore, since professional status during the first half of the century was often identified by publication in a specialized journal, specialization and professionalization were mutually reinforcing. Even in those cases in which a specialized journal was not formally connected with a specialized scientific society, the attempt to maintain a "popular," public perspective gave way, seemingly inevitably, to a more professional approach. To cite just one example, Edward Newman founded

The Zoologist in 1843 as "A Popular Miscellany of Natural History," but changed the subtitle in 1877 to "A Monthly Journal of Natural History" in order to reflect the loss of his "popular" audience. Less specialized and more frequent scientific information was published in such highbrow intellectual newspapers as *The Athenaeum*, the *Saturday Review*, and the *Spectator*, as well as in the numerous "current science" articles that appeared in the newer, ever-proliferating bimonthly and monthly journals such as *Macmillan's Magazine* or the *Fortnightly Review*. Nonetheless, despite the apparent expansion of scientific publication of every sort on every intellectual level, men of science still felt that their full needs were not being met.

Thomas Henry Huxley best expressed the anxiety that many of his colleagues felt as they surveyed the increasingly diversified field of scientific publications. "Everything in Science is becoming too much divided into compartments," Huxley complained, urging that some high level yet still public and accessible forum for sharing scientific information be created in order to rejoin these recently divided and separated "compartments."[11] Huxley recognized the need to prevent not only the excessive specialization of science, but also the excessive professionalization of science. If science is to be a cohesive rather than a divisive force in modern culture, it must remain public, sharable, and, in some sense of the word, popular. Therefore, from the beginning until the end of his career, Huxley insisted upon echoing Sir Humphry Davy's reassuring claim that "Science is in fact nothing more than the refinement of common sense making use of facts already known to acquire new facts."[12] Huxley, like Davy, thought that the needs of a professional man of science could not in the final analysis conflict with the needs of the public, and therefore in 1861 he tried to transplant the Dublin *Natural History Review* to London in order to meet part of those needs. Unfortunately, the *Natural History Review* was a quarterly and, like its close competitor the *Quarterly Journal of Science*, proved unable to provide sufficient fresh scientific information to a broad enough audience in order to succeed. Therefore, in 1863, Huxley joined with Norman Lockyer, Joseph Dalton Hooker, John Tyndall, Charles Darwin, and more than thirty other men of science from different fields to support and staff a new weekly "Review of Literature, Science, and the Arts," named, appropriately, *The Reader*. These distinguished men of science worked along with equally distinguished theologians and men of letters such as F.D. Maurice, Emanuel Deutsch, F.W. Farrar, Thomas Hughes, William Michael Rossetti, Charles Kingsley, and John Ruskin. The credentials of a "professional" were respected, but not required of the contributors to *The Reader*, thus allowing a nonprofessional student of geology like John Ruskin to engage in a long, heated public controversy with a professional geologist like John Tyndall.[13]

In terms of content, however, it was difficult to distinguish the reviews of current novels and theological and scientific literature that *The Reader* contained from those found in, say, *The Cornhill* or *Macmillan's Magazine*. Therefore, *The Reader* was financially unable to survive for more than four years, despite the outstanding quality of its contributors. Nonetheless, some of the distinctively liberal, scientific, and nonprofessional flavor of *The Reader* can be appreciated when one recalls that among his other duties, Huxley shared part of the responsibilities for editing the weekly "Theology" section.[14] Contrary to the recent scholarly view that Huxley and many other Victorian men of science were struggling to establish their own "profession" by destroying or undermining "the intellectual legitimacy" of the theological "profession,"[15] *The Reader* offers abundant evidence that Huxley and other admittedly "unorthodox" men of science were in fact still struggling to maintain a broad, public intellectual context that would replace the faltering natural theology of earlier generations and transcend any divisive, "professional" points of view. Huxley spoke for himself and for *The Reader* when he wrote in a lead article entitled "Science and 'Church Policy'," that all "professional prepossessions" must be set aside in order to understand the true relationship between science, religion, and theology. In a manner reminiscent of William Whewell, Huxley emphasized the "unity of all phenomena" and the indivisibility of man's emotional and intellectual perception of reality. But unlike Whewell, Huxley refused to accept an isolating "professional" notion of "science" as only "knowledge of the material world." Instead, Huxley reasserted his own and *The Reader's* conception of "science" as the totality of sharable human knowledge, including knowledge about God.

> Religion has her unshakable throne in those deeps of man's nature which lie around and below the intellect, but not in it. But Theology is a simple branch of Science, or it is nought; and that "Church Policy" which sets it up against Science is about as reasonable as would be the advocacy of the claims of the rule of three to superiority over arithmetic in general.[16]

In the wake of the great "Oxford Debate" and the publication of *Essays and Reviews*, Huxley and the other contributors to *The Reader* were perfectly aware of the increasing tensions between the narrowly defined "professions" of "science" and "theology," but that is exactly why in passages such as the one quoted above, *The Reader* attempted to avoid unnecessary, divisive conflict by articulating a conception of "science" that included "theology." In other words, *The Reader* represented a characteristically broadminded Victorian attempt to integrate man's knowledge of himself and the world he lives in from an authoritative and scientific point of view.

The Reader expired in 1867, and the first issue of *Nature* appeared on Thursday, November 4, 1869. All modern scholars agree that *Nature* is the direct descendant of *The Reader*, and most explanations of the intentions of *Nature* either simply repeat or else expand without qualifying the account given by Norman Lockyer, the first editor. In the Jubilee edition of *Nature* on November 6, 1919, Lockyer succinctly summarized the circumstances that led to the creation of *Nature*: "When *The Reader* ceased publication the idea occurred to me of starting a general scientific journal of a more comprehensive scope than the *Natural History Review*, which, like other specialised scientific periodicals, had failed for want of circulation." Lockyer then reprinted the following prospectus, which had been distributed in late 1869 "to bring the aims and intentions of the journal before scientific readers and others."[17]

The object which it is proposed to attain by this periodical may be broadly stated as follows. It is intended:

First, to place before the general public the grand results of Scientific Work and Scientific Discovery, and to urge the claims of Science to a more general recognition in Education and in Daily Life; and

Secondly, to aid Scientific men themselves, by giving early information of all advances made in any branch of Natural Knowledge throughout the world, and by affording them an opportunity of discussing the various Scientific questions which arise from time to time.

To accomplish this twofold object, the following plan is followed as closely as possible.

Those portions of the paper more especially devoted to the discussion of matters interesting to the public at large contain:

I. Articles written by men eminent in Science on subjects connected with the various points of contact of Natural Knowledge with practical affairs, the public health, and material progress; and on the advancement of Science, and its educational and civilising functions.

II. Full accounts, illustrated when necessary, of Scientific discoveries of general interest.

III. Records of all efforts made for the encouragement of Natural Knowledge in our Colleges and Schools, and notices of Aids to Science-teaching.

IV. Full Reviews of Scientific Works, especially directed to the exact Scientific ground gone over, and the contributions to knowledge, whether in the shape of new facts, maps, illustrations, tables, and the like, which they may contain.

In those portions of *Nature* more especially interesting to Scientific men are given:

V. Abstracts of important papers communicated to British, American, and Continental Scientific societies and periodicals.

VI. Reports of the meetings of Scientific bodies at home and abroad.

In addition to the above, there are columns devoted to Correspondence.

The "aims and intentions" of *Nature* as Lockyer described them in this prospectus were obviously broader, more generalist, and less professional than those of the *Natural History Review*, but narrower, more specialist, and more professional than those of *The Reader*. The range of "science" to be discussed was expanded beyond the botany, biology, and geology that constituted most of the subject matter of the *Natural History Review*, but it was reduced from that broad conception of *wissenschaft* that had allowed regular discussions of history, literature, philosophy, and theology in *The Reader*. What were the causes of this shift in intentions? Did the shift occur because Lockyer, Huxley, and other men of science lost faith in their ability to articulate a sharable, cohesive intellectual context for science such as *The Reader* had offered?[18] Or, did the shift occur because these men of science recognized the "professional" advantages of developing their own separate, group identity apart from the rest of society?[19] Or, were there any other, more practical, and less culturally profound causes for this shift in intentions?

Lockyer's Jubilee article identified at least one major practical cause for this shift in intentions: the economic failure of the *Natural History Review* and *The Reader* caused the founders, editors, and contributors of *Nature* to redefine their range of interests in order to increase the chances for survival within the increasingly competitive, professionalized, and specialized field of periodical journalism. This interpretation is strengthened when one recalls that *Nature* was not the only new journal founded in 1869 that claimed to be descended from *The Reader*.

While Lockyer was still negotiating with Alexander Macmillan to publish *Nature* and even before he had managed to recruit enough regular contributors to staff his new journal, he discovered that Charles Appleton and an influential group of research-oriented, educational reformers were already planning to create a new journal to fill the intellectual gap left by the demise of *The Reader*. Appleton circulated a prospectus for his new *Monthly Journal of Science* in April 1869 and soon thereafter managed to secure the support of John Murray, who published the works of Charles Darwin and other men of science. Although Appleton's new journal was to include the same broad range of topics as *The Reader*, natural science was originally intended to receive more emphasis than history, literature, philosophy, or theology, and the entire range of topics was to be treated on a more learned, more professional level. Luckily for Lockyer, two sets of circumstances combined to help him avoid a competitive clash between *Nature* and Appleton's new journal. First, John Murray insisted upon lowering the scholarly and professional intentions of Appleton's new journal in order to assure its financial success. Alexander Macmillan, on the other hand, was so committed to *Nature* that he absorbed substantial financial losses for the first thirty years of its publication. Second, since Lockyer attempted to recruit many of the same eminent men of science that Appleton did, both editors could reformulate and, in effect, negotiate the intentions of their respective journals by depending upon figures like Huxley and Hooker to warn against and to mediate any competitive struggle. In other words, Lockyer could literally "afford" to define the content and the audience of *Nature* in a more specialized, more professional manner than *The Reader* in order not to overlap too much with Appleton's new journal, and Appleton could correspondingly de-emphasize natural science and lower his professional standards in order not to compete financially with *Nature*. When Appleton's new journal finally appeared in October 1869, it had been renamed *The Academy* and it duplicated almost perfectly the format and the intentions of *The Reader*.[20] When *Nature* appeared less than a month later, its format and its stated intentions revealed little overlap or similarity with either *The Reader* or *The Academy*, and an unnecessary economic contest was thereby avoided.[21]

Nonetheless, despite the fact that *Nature* was clearly more specialist and more professional than *The Reader*, the "aims and intentions" outlined in Lockyer's prospectus were not *only* or even *primarily* "professional." *Nature* was intended, "First, to place before the general public the grand results of Scientific Work and Scientific Discovery," and only "Secondly, to aid Scientific men themselves." Even when Lockyer attempted to categorize the six portions of *Nature* that were intended to fulfill either public or professional functions, the public portions outnumbered the professional portions by two to one. And if one were to count the actual number of pages devoted to these public and professional portions in any of the early issues of *Nature*, the statistics would weigh even more heavily in favor of the nonprofessional, public portions. Now it could be argued that even these ostensibly nonprofessional, public portions of *Nature* in fact fulfilled the eminently professional function of increasing the prestige of scientists in the eyes of the public.[22] Such an argument would entail the conclusion that Lockyer's clear distinction between public and professional functions was illusory and fallacious. In other words, such an argument, athough technically accurate and acceptable from a modern point of view, would be anachronistic and inaccurate insofar as it ignored and obfuscated the explicit intentions of *Nature's* founders, editors, and contributors. Lockyer may, indeed, have miscalculated the ability of interested amateurs among the public to comprehend even the nonprofessional portions of *Nature,* and therefore the actual readership may have been and probably was more professional than Lockyer ever intended.[23] But, again, the important point to keep in mind is that such unintended effects cannot be properly understood unless the intended effects are made clear first.

An examination of the various articles, editorials, and controversial correspondence that appeared between 1869 and 1919 makes it clear that the founders, editors, and contributors attempted to preserve the journal's public character despite the gradual, unintended elevation of professional standards that continually reduced the public's access to *Nature.*

Thomas Henry Huxley, who was just as much a founding father of *Nature* as Lockyer himself, exercised his paternal prerogative by opening the first issue with a typically controversial lead article. His translation of Goethe's romantic "Aphorisms on Nature" was perfectly designed to achieve the new journal's twofold objective of stimulating communication between scientists and the general public as well as between scientists of different fields. In much the same way that Huxley startled the readers of the first issue of *The Academy* by boldly asserting that "The teleological and the mechanical views of nature are not, necessarily, mutually exclusive,"[24] he now startled the readers of *Nature* by praising Goethe's all-encompassing pantheistic rhapsody. What could Darwin's Bulldog possibly mean by digging up and displaying with such prominence the antiquated notions of teleology and *naturphilosophie*? Huxley knew that both his public and his professional audience would respond with some such bewildered question, and with consummate rhetorical skill his final paragraphs turned the question back to his audience by forcing them to consider whether their own "scientific" conceptions of "nature" would outlast those of Goethe, the poet. Although Huxley concluded his essay by quoting Goethe's recantation of his early naïve pantheism, he also made it clear that even Goethe's palinode suggested that those "Aphorisms" remained symbolically true.

"If we consider the high achievements by which all the phenomena of Nature have been gradually linked together in the human mind; and then, once more, thoughtfully peruse the above [Aphorisms] from which we started, we shall, not without a smile, compare that comparative, as I called it, with the superlative which we have now reached, and rejoice in the progress of fifty years."

Forty years have passed since these words were written [by Goethe], and we look again, "not without a smile," on Goethe's superlative. But the road which led from his comparative to his superlative, has been diligently followed, until the notions which represented Goethe's superlative are now the commonplaces of science — and we have a super superlative of our own.

When another half-century has passed, curious readers of the back numbers of NATURE will probably look on our best, "not without a smile;" and, it may be, that long after the theories of the philosophers whose achievements are recorded in these pages, are obsolete, the vision of the poet will remain as a truthful and efficient symbol of the wonder and the mystery of Nature.

Huxley stated explicitly that he wanted to focus on Goethe's "Aphorisms" because "It seemed to me that no more fitting preface could be put before a Journal, which aims to mirror the progress of that fashioning by Nature of a picture of herself, in the mind of man, which we call the progress of science."[25] He wished to convince the public and to remind his colleagues that although science often progressed by being methodologically reductive, it should never, indeed, could never be epistemologically or ontologically reductive. The symbolic truth that Goethe's "Aphorisms" revealed was that "nature" always transcends and eludes man's attempt to reduce it to "science." Huxley was, in effect, reminding his readers of the Kantian lesson that science can never fully explain "the wonder and the mystery of Nature," and that, therefore, some intuitive and affective response to nature must always supplement any method of explanation.[26] Even by his simple decision to call men of science "philosophers" rather than "scientists," Huxley indicated his firm intention to preserve a nonreductive, nonspecialist conception of science.

Huxley shared this intention to express a broad, accessible conception of modern science with Norman Lockyer, the editor of *Nature*. Lockyer was one of the most famous astronomers of the nineteenth century, but his real forte was to act as a public spokesman within the controversial arena of late Victorian science. Despite the competitive pressures mentioned earlier that encouraged Lockyer to abandon the broad format of *The Reader*, he struggled constantly during the fifty years of his editorship to maintain a clear perspective on the relationship between the increasingly specialized, professionalized, and divergent areas of human knowledge. In one of the first of a long series of articles on "Government Aid to Science," Lockyer rhetorically emphasized the grand, synthetic view of science and culture that *Nature* stood for when he asked: "Is it really necessary to tell any educated man of the nineteenth century that science, art, and literature, with one or two other matters, are simply civilisation; and that civilisation affects not particular classes, but whole communities?"[27] Government aid to science *alone* was not what was necessary, Lockyer argued; rather it should be extended to every major area of public education and culture. Lockyer had no intention of fragmenting "civilisation" by destroying the bonds between "science, art, and literature." Indeed, in his many editorials on educational reform, Lockyer sometimes explicitly echoes Matthew Arnold's humanistic educational ideal of "knowing ourselves and knowing the world," and when Charles Appleton and *The Academy* began to editorialize about the need to restructure university endowments in order to subsidize research in all fields of knowledge, Lockyer and *Nature* were quick to add a strong voice of support.[28] It is undoubtedly true that such efforts to promote the "endowment of research" only hastened the specialization, professionalization, and hence fragmentation of modern education. But, although Lockyer was not blind to such unintended effects, he nonetheless maintained an explicit faith that scientific and humanistic knowledge could and should be integrated through education.

Less than one month after his first editorial leader on government aid to science, Lockyer took the opportunity provided by the introduction in Parliament of W.E. Forster's Education Act of 1870 to promote a well-considered plan to appoint a Minister of Public Instruction, and again *Nature* revealed what I believe are her true colors when Lockyer stated, "We understand the comprehensive term natural knowledge to include Education, Science, the Fine Arts, and Music." Lockyer was perfectly willing to suggest that it was preferable to appoint separate subministers for each area, "not only with a view to division of labour, but to the special efficiency of each." But he insisted that a single Minister of Public Instruction should "take charge of the whole range of natural knowledge in all matters in which the state in any way intervenes to advance such knowledge."[29] When the "division of labour" between science and other fields of knowledge was pushed too far beyond the reasonable requirements of efficiency, Lockyer and *Nature* protested vehemently. For example, in a lead article entitled "The Inauguration of the Yorkshire College of Science," Lockyer boldly stated,

> We do not want a Yorkshire College of Science, but a Yorkshire College in which science will be found in its proper place. It must be remembered that the whole duty of these local colleges is not limited to the instruction in the particular sciences which more directly relate to the manufacturing industries of the districts in which they are placed; they must be made to act as *nuclei* for higher culture by the establishment of chairs of Art and Literature.[30]

When the countless articles in *Nature* demanded that modern science be integrated into the educational curriculum, the consistent intention was to emphasize *integration*, not merely science. If an individual Victorian scientist were pushed in the midst of an educational controversy to the rhetorical extreme of choosing *either* science *or* the humanities, he might, like Thomas Henry Huxley, respond that "the free employment of reason, in accordance with scientific method, is the sole method of reaching truth."[31] But *Nature* during Lockyer's tenure as editor was always a more humane and less exclusive spokesman for the scientific community as a whole, and even when the famous "Two Cultures" debate broke out between Huxley and Matthew Arnold, Lockyer was unafraid to enter the argument in order to mediate and alleviate the tensions between science and the humanities.[32]

No, *Nature* was neither written for nor by "scientists" in any narrow sense of that word. Science was obviously the superstructure that integrated the diverse articles, letters, and editorials that appeared in *Nature* during the Victorian era, but it was a more comprehensive and less reductive science than many of us can possibly recognize today. Because Lockyer, like Huxley, lamented the increasing specialization and compartmentalization of modern science, he recruited among his regular contributors men like John Brett, an astronomer and painter, who, because he was a member of both the Astronomical Society and the Royal Academy of Arts, was especially able to bridge any gaps between apparently divergent professions. In a long and controversial lead article entitled "Natural Science at the Royal Academy," Brett acknowledged that there were increasing tensions between scientists and artists in late Victorian England, but by chiding both groups, he made it clear that he felt a fruitful union and interchange between science and art could still be maintained. Brett's article concludes:

> What is the moral of all this? Simply that the scientific men pay too little attention to the broader aspects of the visible world; while artists on their part pass by the clear fountain of natural beauty and content themselves

with dreamily sipping lukewarm water from the corroded vessels of their forefathers, the one group of doers standing apart from the other; whereas, if either would go to school with the other, they would, in my opinion, each stimulate and aid the labours of the other and divide between them a far larger share of the spoils of the world.[33]

Brett's exhortation may sound strained and somewhat hopeless to our modern ears, but his intentions and his expectations were typically Victorian — idealistic and earnest. In order to achieve such goals, *Nature* sometimes included reviews of such works as Louis Viardot's *Wonders of Italian Art*, which even the reviewer admitted could hardly be considered "scientific" in any narrow sense, but which deserved discussion exactly because it indicated one of those areas where, as Brett had put it, scientists and artists could "go to school" together.[34] At other times, a more purely "scientific" article would make a passing reference to some topic such as the "correlation of colour and music," and a long, heated series of letters would follow that argued the finer points of artistic and scientific significance.[35]

Lockyer most clearly demonstrated his refusal to allow professional standards or specialist interests to dominate *Nature* when he himself reviewed the Royal Academy of Arts exhibitions in 1878, in 1883, and in 1887. No doubt encouraged by the previous articles of Brett, Lockyer wrote such an extensive analysis and criticism of the Royal Academy exhibition in 1878 — six articles, extending over two months — that he actually considered expanding them into a book to be called "The Scientific Aspects of Art."[36] As his daughter later testified, "Lockyer's experience of the pictorial art of his time . . . was by no means a narrow one," and in addition to his criticism of "moons in impossible phases, rainbows painted in perspective, and similar atrocities," these reviews of the Royal Academy exhibitions contained some sensitive and accurate judgments of the paintings he discussed.[37] Some of the Royal Academy members whose works Lockyer had criticized were not entirely pleased with all of his comments, however, and numerous other men of science also wanted to have their say about the scientific and artistic questions under dispute. The correspondence columns of *Nature* for the next four months were filled with rejoinders, additional criticism, additional rejoinders, and sometimes even disinterested scientific and artistic commentary.[38] Enthusiasm never lagged once arguments like this were opened in the pages of *Nature*. Professional artists and professional men of science argued right along with those interested "amateurs" who read *Nature* because they still believed in the possibility of attaining a general, nonspecialist understanding of the relationship between humanistic and scientific knowledge.

Perhaps the longest such argument began as a dispute concerning the significance of certain unusual solar phenomena, but during the course of more than two years it eventually expanded to include discussions of the Krakatoa volcano eruptions and the literal and symbolic pollution of the natural environment by modern society.[39] Professionals and amateurs shared equally in this complicated controversy, offering new observations as evidence or making new interpretations of previously established evidence. Although in a few instances a professional meteorologist like W. Clement Ley might criticize the scientific expertise of an amateur like John Ruskin, the dominant tone of the articles and correspondence was genuinely open and inquisitive, indicating, I think, the continuing accessibility of *Nature* to all interested readers. Criticisms like Ley's were undoubtedly responsible for the gradual elevation of professional standards that eventually diminished the public's access to *Nature*. But the fact that the artist Robert Leslie and the poet-priest Gerard Manley Hopkins could repeatedly contribute to this long controversy without being criticized for their lack of professional credentials is a more accurate indication of the intention of *Nature* to maintain a free and open discussion of science and its relationship to all realms of human knowledge.[40]

It would be possible to list many more articles and arguments that demonstrate the intentions of *Nature* to act as a cohesive force between various scientific, artistic, and humanistic factions. But it would be absurd to suggest that *Nature* was able to maintain those intentions against the increasing pressures to specialize and professionalize. However, without going into elaborate detail, I would suggest that even in the midst of highly specialized and almost entirely professionalized arguments about epistemology and methodology, *Nature* never intentionally championed "the positivist epistemology" of "vigilant verification," as some scholars imply.[41] Even when the impulse to specialize and professionalize became irresistible, *Nature* attempted to mediate any divisive arguments and to encourage the inclusive, rather than the exclusive solution. I will cite only one last example to substantiate my point.

At the 1869 meeting of the British Association for the Advancement of Science at Exeter, Professor J.J. Sylvester, the eminent Cambridge mathematician, presented a paper in which he attempted to refute Thomas Henry Huxley's assertion that mathematics was not a very important tool within the modern educational curriculum. Huxley had argued that since mathematics was primarily deductive, it tended to encourage dependence upon premises accepted only on authority or faith. Although Sylvester's address would have automatically appeared in the published proceedings of the Exeter meeting, Lockyer hurriedly solicited an expanded version of the paper for *Nature* and thus initiated a four-month-long controversy in which letters would appear weekly from prominent mathematicians, astronomers, physiologists, and philosophers. In the expanded version of his address, entitled "A Plea for Mathematicians," Sylvester included a 500-word footnote in which he defended a reference that he had made to Kant's conception of space and time. Sylvester wrote:

> It is very common, not to say universal, with English writers, even such authorised ones as Whewell, Lewes, or Herbert Spencer, to refer to Kant's doctrine as affirming space "to be a form of thought," or "of the understanding." This is putting into Kant's mouth (as pointed out to me by Dr. C.M. Ingleby), words which he would have been the first to disclaim, and is as inaccurate a form of expression as to speak of "the plane of a sphere," meaning its surface or a superficial layer, as not long ago I heard a famous naturalist do at a meeting of the Royal Society.[42]

Whether space is a form of thought or a form of intuition according to Kant, is a highly specialized question which most amateurs would be perfectly willing to allow the professionals to decide, and the long series of letters that followed Sylvester's essay indicates the seriousness with which Victorian men of science handled such professional questions.[43] But Sylvester was not merely concerned with pointing out the professional mistakes of "Whewell, Lewes, or Herbert Spencer."[44] Sylvester's elaborate discussion of the methodologies and epistemologies of modern mathematics was also an explicit defense of the function of faith and authority in both science and religion.

In another long, 1200-word footnote to his essay, Sylvester stated explicitly for the readers of *Nature* what the ultimate consequences of such professional disputes about methodology and epistemology were. Despite the fact that the footnote ostensibly concerned the development of theories of four-dimensional geometry, none of the readers of *Nature* could have misunderstood Sylvester's clear intention to define this narrow, professional problem within the widest possible public context.

> In philosophy, as in aesthetic, the highest knowledge comes by faith. I know (from personal experience of the fact) that Mr. Linnell can distinguish purple tints in clouds where my untutored eye and unpurged vision can perceive only confused grey. If an Aristotle, or Descartes, or Kant assures me that he recognizes God in the conscience, I accuse my own blindness if I fail to see with him. If Gauss, Cayley, Riemann, Schalfli, Salmon, Clifford, Kronecker, have an inner assurance of the reality of

transcendental space, I strive to bring my faculties of mental vision into accordance with theirs. The positive evidence in such cases is more worthy than the negative, and actuality is not cancelled or balanced by privation, and matter plus space is none the less matter. I acknowledge two separate sources of authority — the collective sense of mankind, and the illumination of privileged intellects.[45]

Sylvester refused to allow his audience to ignore the theological implications of his argument, despite the fact that his was not an orthodox defense of revealed religion even by Victorian standards. His essay makes it clear that the attempt to articulate a coherent, "scientific" understanding of man's relationship with God had not suddenly ended after the publication of *The Origin of Species*. And the fact that *Nature* published Sylvester's essay is an equally clear sign that the intentions that had informed *The Reader's* editorial policy had not entirely disappeared either. Although science was daily becoming more specialized and professionalized, the editors, founders, and contributors of *Nature* attempted to maintain that earlier cohesive vision of science's relationship to the rest of human knowledge for as long as possible.

Unfortunately, the unintended effects of publishing articles like Sylvester's may have been more divisive than cohesive. The general public could hardly be consoled to read that it was just as difficult to "recognize God in the conscience" as it was to arrive at "an inner assurance of the reality of transcendental space." As Robert M. Young has so cogently argued, the strategies whereby many late-Victorian men of science reconciled their theological and scientific beliefs became so abstract, that is, so specialized and so professionalized, that the general public could no longer fully comprehend or evaluate them.[46] Therefore, despite the intentions of *Nature* to discuss such problems when appropriate, the gradual elevation of the level of discourse up to "professional" standards made it more and more difficult for the interested "amateur" or the general public to share the intellectual context of such discussions. Indeed, it seems to be the ironic fact that Lockyer's intention to recruit the most "eminent" men of science to write for *Nature* eventually undermined his intention "to place before the general public the grand results of Scientific Work and Scientific Discovery." Knowing something by no means implies the ability to communicate it to a broad and diverse audience, and as the "eminent" men of science who wrote for *Nature* became more used to the new, higher level of discourse that such a journal made possible, they became less and less used to understanding and discussing science within a broad, accessible, public context.

Pinpointing the exact moment when the content of *Nature* became more "professional" than "public" is probably impossible. But I think it is possible to pinpoint a moment when that shift had become so evident that even Thomas Henry Huxley had to admit reluctantly that the intentions of *Nature* had changed. When Huxley was asked to write the lead article for the twenty-fifth anniversary issue of *Nature*, he decided to reflect back upon the translation of Goethe's "Aphorisms on Nature" with which he had opened the first issue. But although Huxley was still, in 1894, willing to state in private that Goethe's romantic poem was "the best expression of the modern aspect of Nature known to me,"[47] he also knew that neither the public nor his professional colleagues were likely any longer to understand fully what Goethe's beautiful words meant. Therefore, rather than once again being accused of madness, Huxley restricted his retrospective article on "Past and Present" to a much narrower, professional analysis of modern science.[48] Huxley still lamented the compartmentalization and specialization of modern science which he saw taking place around him, and he tried once again to suggest how future advances in scientific theories might be achieved by transcending narrow, specialist boundaries. But he knew that *Nature* had changed and that the readers of *Nature* would no longer be usefully startled, but only slightly embarrassed if he once again tried to revive Goethe's inspirational vision of modern science. After retiring from the Presidency of the Royal Society in 1885, Huxley spent almost all of his efforts during the last ten years of his life not on writing about the new professional science, but on writing about science and religion within the old, broad, public context of the earlier Victorian period. None of these essays on science and religion appeared in *Nature*. Instead, Huxley returned to the medium of general periodical literature from which *Nature* had intentionally diverged in 1869. Although the development that *Nature* had undergone during the years since 1869 had undoubtedly constituted progress from a professional point of view, Huxley knew that something had been lost at the same time, and I am sure he was regretful. Perhaps we should be too.

NOTES AND REFERENCES

1. George Haines, *Essays on German Influence upon English Education and Science, 1850–1919*, Connecticut College Monographs, No. 9 (Hamden: Archon, 1969), pp. 50–53.
2. See Frank M. Turner, "The Victorian Conflict between Science and Religion: A Professional Dimension," *Isis*, vol. 69 (1978), pp. 363, 376; A. J. Meadows, *Science and Controversy: A Biography of Sir Norman Lockyer* (Cambridge, Mass.: Massachusetts Institute of Technology Press, 1972), pp. 16–38, esp. pp. 28–29; Roy M. MacLeod, "A Note on *Nature* and the Social Significance of Scientific Publishing, 1850–1914, "*Victorian*

Periodicals Newsletter, No. 3 (November, 1968), pp. 16–17; and the series of short articles by MacLeod published in the centenary edition of *Nature*, vol. 224 (1969), pp. 423–461. MacLeod's discussion of the prehistory and evolution of *Nature* is the most comprehensive in existence, and I rely heavily upon his account in the analysis that follows. But, despite the fact that MacLeod clearly recognizes the multiplicity of intentions that gave rise to *Nature*, he consistently emphasizes the "professional" aspect; see his "Science and the Treasury: Principles, Personalities, and Policies, 1870–1875," and W.H. Brock's "The Spectrum of Science Patronage," in *The Patronage of Science in the Nineteenth Century*, edited by G. L'E. Turner (Leyden: Noordhoff, 1976), pp. 123, 188.

3. Robert M. Young, "Natural Theology, Victorian Periodicals, and the Fragmentation of the Common Context," paper presented to the King's College Research Centre Seminar on Science and History (Spring, 1969), pp. 10, 33.

4. See Susan Faye Cannon, *Science in Culture: The Early Victorian Period* (New York: Science History Publications, 1978), especially chap. 9, "The Truth-Complex and Its Demise." Although Cannon does not discuss *Nature* explicitly, her analysis of the general cultural and scientific trends of the period parallels that of Turner, MacLeod, and Young, and thus deserves to be compared with them in this context.

5. John Maddox, "Scientific Journalism: John Maddox on the Nature of *Nature*," *The Listener*, vol. 90 (1973), p. 762.

6. The classic description of the "mediatorial" function of Victorian nonfiction prose is given in A. Dwight Culler's "Method in the Study of Victorian Prose," *Victorian Newsletter*, no. 9 (Spring, 1956), pp. 1–4. Although Culler's original formulation did not seem to include scientific prose such as that found in *Nature*, other scholars have more recently suggested that nineteenth-century men of science wrote the same type of "mediatorial" prose that Culler described. See Michael Timko, "The Victorianism of Victorian Literature," *New Literary History*, vol. 6 (1975), pp. 607–627; and Eugene Goodheart, "The Failure of Criticism," *New Literary History*, vol. 7 (1976), pp. 390–391, note 3.

7. Cannon,[4] pp. 151, 1–28; and Young,[3] pp. 2–10. For perhaps the most lucid recent discussion of the differences signified by and the ambiguities created by the terms "amateur," "professional," "generalist," and "specialist," see Cannon, pp. 137–165. I attempt to follow Cannon's usage throughout my analysis, but I am aware that the necessity to cite, for example, Huxley's usage of the term "professional" creates the possibility for confusion.

8. To cite just one example, see Edward Manier, *The Young Darwin and His Cultural Circle* (Boston: Reidel, 1978), pp. 40–47; and Sylvan S. Schweber, "The Origin of the *Origin* Revisited," *Journal of the History of Biology*, vol. 10 (1977), pp. 229–316, for a discussion of the impact that David Brewster's 1838 *Edinburgh Review* article on Comte had on Darwin's scientific thought.

9. Sidney Ross, "*Scientist*: The Story of a Word," *Annals of Science*, vol. 18 (1962), pp. 70–72, 75–78.

10. Young,[3] pp. 12–22.

11. MacLeod,[2] *Nature*, p. 426.

12. See George Foote, "Science and its Function in Early Nineteenth Century England," *Osiris*, vol. 11 (1954), p. 440, for Davy's statement and for a useful analysis of the popular understanding of science at the beginning of the Victorian period. For one early and one late instance of Huxley defining science as "nothing but trained and organized common sense," see his "On the Educational Value of the Natural History Sciences," *Collected Essays of Thomas Henry Huxley* (New York: D. Appleton, 1899), vol. III, p. 45, and *The Crayfish: An Introduction to the Study of Zoology* (London: C. Kegan Paul, 1880), p. 2.

13. MacLeod,[2] *Nature*, pp. 423–434; Meadows,[2] pp. 16–22; John F. Byrne, *"The Reader: A Review of Literature, Science and the Arts, 1863–1867,"* unpublished doctoral dissertation (Northwestern University, 1964), pp. 67–109, 305–333. Although MacLeod and Meadows claim to have made use of Byrne's dissertation, they both make certain mistakes of fact and emphasis, which Byrne's more extensive analysis avoids. For example, MacLeod claims that Huxley "wrote nothing for *The Reader*" (p. 434), although he agrees with Meadows that Huxley was the dominant scientific figure behind the editing of *The Reader*. Byrne lists Huxley's acknowledged contributions to *The Reader* (p. 320), and he corrects the false impression that Huxley exerted a dominant influence over the journal.

14. See Bishop Colenso's letter to Huxley in the Huxley Papers, 12. 274–5, housed in the archives of the Imperial College of Science and Technology, London.

15. See Turner,[2] pp. 364, 369, 371.

16. [Thomas H. Huxley], "Science and 'Church Policy'," *The Reader*, vol. 4 (1864), p. 821; and, Byrne,[13] pp. 157–159. When Huxley was unable to prevent *The Reader* from being sold to Thomas Bendyshe, "a violent atheist," he and other scientists who wished to maintain this broad vision of science and theology withdrew their support from the journal, thus hastening its demise; see Meadows,[2] pp. 21–22, and MacLeod,[2] *Nature*, p. 434.

17. Norman Lockyer, "Valedictory Memories," *Nature*, vol. 104 (1919), pp. 189–190. Most of this short article is reprinted in T.M. Lockyer and W.L. Lockyer, *Life and Work of Sir Norman Lockyer* (London: Macmillan, 1928), pp. 46–47. When Roy MacLeod asserts that this "prospectus was virtually identical with that in the 'revised' *Reader* of 1865" (*Nature*, p. 438), he purposely focuses only upon the natural sciences and ignores the literature, history, and theology that were always included in *The Reader*.

18. Young,[3] p. 33.

19. Turner,[2] p. 363.

20. See MacLeod,[2] *Nature*, pp. 438, 441; Diderick Roll-Hansen, "*The Academy*, 1869–1879: Victorian Intellectuals in Revolt," *Anglistica*, vol. 8 (1957), pp. 30–31, 120, 158; John Curtis Johnson, "*The Academy*, 1869–1896: Center of Informed Critical Opinion," unpublished doctoral dissertation (Northwestern University, 1958), pp. 55–58, 69–70, 290–291; J.H. Appleton and A.H. Sayce, *Dr. Appleton: His Life and Literary Relics* (London: Trubner, 1881), p. 27.

21. H.R. Fox Bourne revealed a limited, but clear sense of the economic and intellectual pressures that conditioned the intentions of *Nature* and *The Academy* when he wrote: "One of the functions proposed for itself at starting by *The Academy* was usurped and skilfully performed by a friendly rival, only a month later in the field. *Nature* gave a remarkable evidence of the demand that had arisen for sound yet popular information on scientific affairs." What I have attempted to outline in the preceding paragraphs is only the specific list of "friends" and the specific causes for "rivalry," which Bourne only suggestively identified; see his *English Newspapers* (London: Chatto and Windus, 1887), vol. II, p. 316.

22. Turner,[2] p. 363.

23. Meadows,[2] pp. 28–29. But see Charles Kingsley's letter to the editor, entitled "Cockroaches," before accepting Meadows' evaluation of Kingsley as "so able an amateur"; *Nature*, vol. 3 (1870), p. 148.

24. Thomas H. Huxley, *The Academy*, vol. 1 (1869), p. 13.

25. Thomas H. Huxley, *Nature*, vol.1 (1869), pp. 10–11. In case any readers were inclined to dismiss Huxley's lead article as mere celebratory ornament, Lockyer called specific attention to the "idealist," antireductionist, antimaterialist intentions of Huxley and other contemporary men of science in another article in the same issue (p. 18): "The Materialist

school of philosophy are just now getting very badly treated by men of science, much to the astonishment, it appears, of the general public. Mr. Huxley has startled the world by proclaiming himself in a way a disciple of Berkeley and Kant, and here is Rakitansky, the great master of modern pathological anatomy, walking in a similar path." It is important to note that although Lockyer is quite confident in his interpretation of Huxley's "idealist" intentions, he is also aware of the disbelief and "astonishment" of the "general public" about such an interpretation. Although neither scholar examines Huxley's intentions and beliefs in detail, some sense of the context of his "idealism" and his revival of teleology can be grasped by reading Peter J. Bowler, "Darwinism and the Argument from Design: Suggestions for a Reevaluation," *Journal of the History of Biology*, vol. 10 (1977), pp. 29–43, esp. p. 34, and Dov Ospovat, "Perfect Adaptation and Teleological Explanation: Approaches to the Problem of the History of Life in the Mid-nineteenth Century," *Studies in the History of Biology*, vol. 2 (1976), pp. 33–56, esp. p. 49.

26. Huxley insisted upon the irreducible "wonder and mystery of Nature" throughout his entire career and consistently criticized any "scientific" attempt to explain it away; see, for example, his vituperative review of Chamber's *Vestiges of the Natural History of Creation* in *The British and Foreign Medico-Chirurgical Review*, vol. 13 (1854), pp. 425–439.

27. [Norman Lockyer], "Government Aid to Science," Nature, vol. 1 (1870), p. 279.

28. See [Norman Lockyer], "The Education of Our Industrial Classes," *Nature*, vol. 27 (1883), p. 249; Roll-Hansen,[20] p. 79; and Roy M. MacLeod, "Resources of Science in Victorian England: The Endowment of Science Movement, 1868–1900." in *Science and Society, 1600–1900*, edited by Peter Mathias (Cambridge: Cambridge University Press, 1972), p. 132. In the latter essay, MacLeod once again chooses to focus almost entirely upon the "scientific" aspect of the "endowment of research" movement, although he is perfectly aware that Appleton and other Oxford-based reformers were just as much interested in humanistic as in scientific research.

29. [Norman Lockyer], "The Minister of Public Instruction," *Nature*, vol. 1 (1870), p. 423.

30. [Norman Lockyer], "The Inauguration of the Yorkshire College of Science," *Nature*, vol. 12 (1875), p. 509. See also, Haines,[1] pp. 60 ff., for a discussion of this and other articles in *Nature* expressing the same broad view of educational philosophy.

31. For Huxley's comment, see "Science and Culture," in *Collected Essays*,[12] vol. III, p. 148.

32. For a brief account of the contemporary context of this argument and of Lockyer's attempt to mediate the dispute, see [Norman Lockyer], "Science and Art," *Nature*, vol. 28 (1883), pp. 50–51; and D. Roos, "Matthew Arnold and Thomas Henry Huxley: Two Speeches at the Royal Academy, 1881 and 1883," *Modern Philology*, vol. 74 (1977), pp. 316–324.

33. John Brett, *Nature*, vol. 2 (1870), p. 158.

34. John Brett, *Nature*, vol. 1 (1869), p. 80.

35. *Nature*, vol. 1 (1870), pp. 286–287; 314; 384–385; 406–407; 430–431; 557–558; 651–653.

36. Norman Lockyer, "Physical Science for Artists, I–VI," *Nature*, vol. 18 (1878), pp. 29–31; 58–61; 87–89; 122–126; 154–157; 223–224. For Lockyer's two part review of the Royal Academy exhibition, see *Nature*, vol. 28 (1883), pp. 50–51; 73–76. For his reaction to the 1887 Royal Academy exhibition, see *Nature*, vol. 36 (1887), pp. 14; 145. Also see note 30 above for further indication of the context of the 1883 and 1887 reviews.

37. Lockyer and Lockyer,[17] *Life and Work of Sir Norman Lockyer*,p. 99

38. See *Nature*, vol. 18, (1878), pp. 66; 116; 249; 278; 329–331; 356–357.

39. For a more detailed discussion of this controversy, see Thomas A. Zaniello's essay elsewhere in this volume.

40. For Ley's criticism, see W. Clement Ley, "Mr. Ruskin's Bogies," *Nature*, vol. 29 (1884), pp. 353–354. Ruskin was by 1884 suffering from repeated bouts of mental distress, and part of the satire and venom of Ley's comments no doubt results from his knowledge of this condition. But the important point to note here is that Ley was quite serious when he recommended the readers of *Nature* to "study some of those little books which are beginning to be the delight of our children," since many eminent Victorian men of science, including Lockyer and Huxley, were involved at precisely this time in writing accessible and authoritative textbooks on various scientific topics. Ley's sarcasm may have been inappropriate and unkind, but his intention, finally, was to maintain and expand the public's understanding of modern science.

41. Turner,[2] pp. 363–365.

42. J.J. Sylvester, *Nature*, vol. 1 (1869), p. 238.

43. *Nature*, vol. 1 (1870), pp. 289; 314; 334–335; 360–361; 386–387; 407; *Nature*, vol. 2 (1870), pp. 296; 355; 375.

44. Whewell was already dead in 1869, and for some unknown reason Herbert Spencer chose not to respond to this attack on his philosophic ability. G.H. Lewes responded immediately and repeatedly, but eventually lost the argument despite his valiant and ingenious attempts to explain away the difficulty of his position. Lewes's empiricism was no match for a critical rationalism, and even Huxley eventually sided with the "intuitionists" against Lewes.

45. J.J. Sylvester, *Nature*, vol. 1 (1869), p. 238.

46. Young,[3] pp. 26–34.

47. MacLeod,[2] *Nature*, p. 440.

48. Thomas H. Huxley, *Nature*, vol. 51 (1894), pp.1–3. See Turner,[2] pp. 369–370, for a brief discussion of this article, which, I believe, seriously distorts the context and intentions of Huxley's remarks.

Mill, Arnold, and Scientific Humanism

ROBERT ALAN DONOVAN

Institute for Humanistic Studies
State University of New York at Albany
Albany, New York 12222

I N "Bentham" and "Coleridge," two early essays long considered classics,
John Stuart Mill defined the mental postures that, he thought, simultaneously polarized and rendered coherent the intellectual life of the century. Out of strikingly unlike minds and temperaments grew the basic
opposition in epistemological assumptions, in mental attitudes and
habits, that characterized the empiricist Bentham and the idealist Coleridge. It is clear enough that a similar opposition is to be observed
between Mill's own skeptical and nominalist views and the transcenden-
talism of Carlyle, but what may be even more far-reaching in its consequences is the less obvious antagonism that developed between Mill and
Matthew Arnold.

Like Bentham and Coleridge, Mill and Arnold belonged to different
spheres of life, separated by age, occupation, fortune, and circle of acquaintance, and there is no record of any meeting between them.[1] Indeed, during the early stages of their careers they had so little in common
that the likelihood of their entering the same intellectual arena must have
seemed remote, yet their responses to the implications of what was to be
the dominant movement of the midcentury, the emergence of political
democracy, were often strikingly parallel, but perhaps more often in
striking opposition to each other. Democracy, which they were both prepared to welcome, nevertheless raised certain insistent questions for both
of them, most notably the question of individual self-development in a
world where traditional authority and traditional values were being
superseded by cultural forces that could not yet be accurately understood
or measured. To deal with this absorbing question, Mill—in *On Liberty*
(1859)—and Arnold—in *Culture and Anarchy* (1869)—found it necessary
to make explicit their own epistemological assumptions, for they were

181

concerned not only with the social conditions and consequences of education, but also with the natural and inherent conditions of knowledge itself.

The words in which Mill and Arnold attempt to formulate the ends of education were curiously alike. In *On Liberty*, Mill writes, "The end of man. . . . is the highest and most harmonious development of his powers to a complete and consistent whole."[2] For Arnold, in *Culture and Anarchy*, perfection, as culture conceives it, "is a harmonious expansion of *all* the powers which make the beauty and worth of human nature, and is not consistent with the over-development of any one power at the expense of the rest."[3] The similarity is perhaps to be explained on the ground that Arnold is paraphrasing the same work that Mill quotes directly, *The Sphere and Duties of Government*, by Wilhelm von Humboldt, the first Prussian Minister of Public Instruction. At any rate, it is certain that Arnold knew and admired Humboldt's work, for in a later chapter of *Culture and Anarchy* he objects to an English reviewer's distortion of the book and refers to Humboldt himself as "one of the most beautiful souls that have ever existed." (vol. V, p. 161). Whatever the reason, it is clear that in stressing the harmonious development of diverse human powers to form a complete and balanced whole, Mill and Arnold are in entire agreement as to what constitutes the aim of individual self-development. Where they differ is in the role society or the state is to play in the process, a difference that grows out of a still more fundamental disagreement about the natural processes that govern human development, specifically the processes by which the mind attains to truth.

[I]

From the time of his earliest writings in the 1820s up to the publication in 1843 of his monumental *System of Logic Ratiocinative and Inductive*, Mill directed his studies towards discovering a way to apply the laws of Newtonian science to what he liked to call the "moral sciences," a term that reflects not so much a conviction as a hope, and perhaps an expectation, of what the study of man and his institutions might become. The first fruit of these studies was the precocious review of Whately's *Elements of Logic*, which appeared in the *Westminster Review* in 1828, when Mill was only 22. The following March, Macaulay's devastating attack in the *Edinburgh Review* on the *a priori* method of James Mill's *Essay of Government* forced the younger Mill to reexamine the question whether human society lent itself to scientific inquiry and, if so, what laws governed that inquiry. Macaulay's attack had shown Mill that to determine the natural laws that regulate the conduct of men toward one another more was needed than a set of assumptions about human nature from which the best practicable form of government could be deduced without reference to experience. On the other hand, the method that Macaulay seemed to be advancing, based entirely on experience without reference to principle, was equally deficient.

Macaulay's method of simple extrapolation from experience Mill calls the "chemical, or experimental, method," and he finds it wholly inadequate as a procedure in the social sciences since the complexity of human nature, and still more of human society, renders the causes of political phenomena impossible to isolate and bring under observation in the fashion of the more elementary researches in inorganic chemistry:

> In an age in which chemistry itself, when attempting to deal with the more complex chemical sequences, those of the animal or even the vegetable organism, has found it necessary to become, and has succeeded in becoming, a Deductive Science — it is not to be apprehended that any person of scientific habits, who has kept pace with the general progress of the knowledge of nature, can be in danger of applying the methods of elementary chemistry to explore the sequences of the most complex order of phenomena in existence. (vol. VIII, p. 886)

Since the Utilitarian position is at stake, Mill is somewhat less abrupt about dismissing the alternative fallacy, which he designates the "geometrical, or abstract, method," but here, too, he is led to the conclusion that social phenomena are too complex to be accounted for by one or a few assumptions about human nature:

> It is unphilosophical to construct a science out of a few of the agencies by which the phenomena are determined, and leave the rest to the routine of practice or the sagacity of conjecture. We either ought not to pretend to scientific forms, or we ought to study all the determining agencies equally, and endeavour, so far as it can be done, to include all of them within the pale of the science; else we shall infallibly bestow a disproportionate attention upon those which our theory takes into account, while we misestimate the rest, and probably underrate their importance. (vol. VIII, p. 893)

There remains a procedure that embraces both the experimental and abstract methods, or, more precisely, lies somewhere between them. Social science can never be truly experimental, since society does not submit to experiment, nor can any procedure remain truly abstract it if insists on referring the results of abstract speculation to experience. What is possible and fruitful, according to Mill, is to deduce the consequences of one's assumptions or hypotheses about states of society and then verify the deduced consequences by comparing them with the actual course of history. This is the "direct" or "concrete" deductive method, which Mill proposes as the true scientific method in the moral sciences. He is also, however, prepared to admit the validity of what he calls the "inverse deductive, or historical, method," which he derived from Comte.[4] Here the collection of historical evidence precedes the deductive process, which is then used to verify the results of historical generalization. That deduction forms an important part of all scientific method, and in fact gives its essential character to the highest and most mature form of scientific reasoning, is to be explained by the fact that empirical or experimental methods by themselves, though they can describe, or even predict the consequences of, the invariable processes of the phenomenal world, can never explain those processes or show why their consequences follow.

Curiously, however, although what he calls the deductive sciences are the most mature and least fallible, Mill has no high regard for deduction itself, rejecting the traditional concept of the syllogism as a valid form of inference. The conclusions of so-called syllogistic reasoning are already implicit in the premises, Mill insists, and since the conclusion can yield no new knowledge, there can be no inference. "All inference," he maintains, "is from particulars to particulars: General propositions are merely registers of such inferences already made, and short formulae for making more" (vol. VII, p. 193). The true form of the argument that we usually represent by the syllogism:

> All men are mortal.
> The Duke of Wellington is a man.
> The Duke of Wellington is mortal.

ought really to be: "If John, Peter, and Thomas are mortal, then the Duke of Wellington is mortal," and our so-called major premise, "All men are mortal," is merely a convenience for generalizing a procedure that actually depends on something else for its validity, namely that we expect nature to be uniform in its operations.

But why not take this last, most universal proposition and reason downward (that is, deductively) from it to less general and ultimately to particular conseqences? The answer is that we do not know the truth of this most general proposition until we can see it as the product of all the accumulated inferior inferences that can be subsumed under it (vol. VII,

p. 307). Ironically, our most universal law, the one logically antecedent to all the rest of our knowledge, is in fact dependent on that knowledge for whatever certainty it can possess. The consequence is that we can never *begin* to know anything, because what is necessary *a priori* is a truth that can be approached only *a posteriori*. Knowledge is possible only after knowledge already exists, or, to put it another way, any individual act of knowing presupposes a structure of knowledge. This is the structuralist premise, and it poses a dilemma that Lévi-Strauss and his followers call the problem of origins. How can one get back to the original structures of meaning or knowledge?

Induction itself, to which Mill looked for the knowledge that deduction could never supply, and for which he devised his four methods of experimental inquiry, is equally helpless to break out of the circle. Deduction, if it is constrained to proceed from particulars to particulars, is only a special case of induction; so induction, if, as Mill suggests (vol. VII, p. 308), it can be cast into the form of a syllogism with the major premise suppressed, is only a special case of deduction; and the two forms of inference are subject to the same inherent limitation. Unlike Whately, who assimilated induction to deduction, Mill preferred to think of induction as the universal type of all inference, but no induction that lays any claim to being scientific proceeds *ab initio*, for two reasons. In the first place, to assume with Descartes that nothing whatever is already known is not only impracticable, it is wholly at variance with our everyday experience, in which we constantly base our inferences and our actions on prior generalizations. And in the second place, an objection that cuts much deeper is that we have no means of testing the correctness of inductions except by reference to others already performed, for it is the accumulated result of all our inductive inferences, as we have already seen, that gives authority to the proposition that is itself the ground of all induction, the law of universal causation, so that the validity of any induction depends on that of all the rest. This is the reason behind Mill's assertion that "this mode of correcting one generalization by means of another, a narrower generalization by a wider, which common sense suggests and adopts in practice, is the real type of scientific Induction" (vol. VII, p. 319).

From the beginning Mill is careful to avoid metaphysical questions, such as the existence of matter or spirit, or the reality of time and space:

> This inquiry [he insists] has never been considered a portion of logic. Its place is in another and a perfectly distinct department of science, to which the name metaphysics more particularly belongs: that portion of mental

philosophy which attempts to determine what part of the furniture of the mind belongs to it originally, and what part is constructed out of materials furnished to it from without. (vol. VII, p. 8)

Logic, in fact, has little or nothing to do with the means by which we obtain knowledge or the conditions of our assent to the propositions that express that knowledge. Logic looks back at what the mind has done, not forward to what it may do. "Logic is not the science of Belief, but the science of Proof, or Evidence."[5] And what is true of logic is true of all science. The structure of knowledge in any department of science can never be predicted, but can only be known after the fact. Accordingly,

the definition of a science must necessarily be progressive and provisional. Any extension of knowledge or alteration in the current opinions respecting the subject matter, may lead to a change more or less extensive in the particulars included in the science; and its composition being thus altered, it may easily happen that a different set of characteristics will be found better adapted as differentiae for defining its name. (vol. VII, p. 140)

The progress of knowledge in any branch of science, besides being measurable in its direction and extent only by hindsight, is made more problematic still by the fact that it is determined not by natural law but by the unpredictable advent of new paradigms[6] for organizing the elements of our knowledge. In his review of Whately's *Elements of Logic*, Mill had stressed the fact that "classification is arbitrary" and that species or other logical categories are things that we make (vol. XI, p. 23). This view he was to modify in the *System of Logic*, recognizing that some species (for example, sulphur or phosphorus) are the work of nature (vol. VII, p. 122), but it remains true that the structure of knowledge depends less on natural distinctions than on the arbitrary distinctions of thought or nomenclature. Science, therefore, as the Cambridge cosmologist Herman Bondi once put it, has nothing to do with truth, for it has no means of perceiving or recognizing the truth of propositions about the natural world. What it can and perforce must do is either to *disprove* propositions that cannot be reconciled with a previously established paradigm, or to cast about for a new paradigm with which they *can* be reconciled. The apparatus of logic, like that of any science, can do no more than detect falsehood or, what amounts to the same thing, inconsistency with the results of previous inductions. Positive advances must always be the result of guesswork or inspiration or serendipity.

The formidable machinery of *A System of Logic*, by which Mill sought to settle the problem of induction and, more important, to provide a systematic methodology adequate to the requirements of the

human sciences, fails to do either. The underlying reason, as I believe
Mill would have readily acknowledged, is in the fatal discontinuity between the processes of the natural or the human world and man's cognitive apparatus. However profound or universal it may turn out to be,
truth has no inherent power by which it can command assent from men's
minds; nor are men's minds naturally equipped with any special faculty
for discerning truth. Such truth as mankind possesses owes more to exceptional insight than to any systematic process whatever, although the
laws of induction can provide a regular and systematic account of what
has already taken place without their aid. Nowhere is this made clearer
than in Mill's strictures against Macaulay for assuming that the natural
law of progress, the inevitable march of mind, has given mankind truths
that needed no exceptional insight to perceive. Macaulay's image is the
sun, which illuminates the hills while it is still below the horizon, but
eventually shines also on the plain:

> If this metaphor is to be carried out [Mill remarks], it follows that if there
> had been no Newton, the world would not only have had the Newtonian
> system, but would have had it equally soon; as the sun would have risen
> just as early to spectators in the plain if there had been no mountain at
> hand to catch still earlier rays. And so it would be, if truths, like the sun,
> rose by their own proper motion, without human effort; but not otherwise.
> I believe that if Newton had not lived, the world must have waited for the
> Newtonian philosophy until there had been another Newton, or his equiv-
> alent. No ordinary man, and no succession of ordinary men, could have
> achieved it.[7]

Mill's acute sense of the difficulties that remain standing in the way of
truth, in spite of his elaborate machinery of induction, accounts in large
part for the utility he assigns to the freedom of thought and discussion in
On Liberty. If, as he believes, truth is a precious commodity, and at the
same time so difficult of attainment as to threaten to elude all the snares
that care and foresight can set for it, mankind should not only refrain
from discouraging the free exchange of ideas, but should bend every effort towards fostering that exchange as the best, perhaps the only, means
of rectifying our opinions and guarding them against the intrusion of
error. Indeed, the assumption on which the argument rests is the idea put
forward many years earlier in the essay on Coleridge, that truth depends
on the "noisy conflict of half-truths." Englishmen, Mill believes, do not
yet sufficiently realize "the importance, in the present imperfect state of
mental and social science, of antagonist modes of thought: which, it will
one day be felt, are as necessary to one another in speculation, as

mutually checking powers are in a political constitution" (vol. X, p. 122). But what, in "Coleridge," had been a temperate plea for liberality and tolerance, becomes, in *On Liberty*, an impassioned warning against the dangers of allowing truth to shift for itself as somehow naturally recommending itself to mankind:

> The dictum that truth always triumphs over persecution is one of those pleasant falsehoods which men repeat after one another till they pass into commonplaces, but which all experience refutes. History teems with instances of truth put down by persecution. If not suppressed forever, if may be thrown back for centuries. To speak only of religious opinions: the Reformation broke out at least twenty times before Luther, and was put down. Arnold of Brescia was put down. Fra Dolcino was put down. Savonarola was put down. The Albigeois were put down. The Vaudois were put down. The Lollards were put down. The Hussites were put down. Even after the era of Luther, wherever persecution was persisted in, it was successful. In Spain, Italy, Flanders, the Austrian Empire, Protestantism was rooted out; and, most likely would have been so in England, had Queen Mary lived, or Queen Elizabeth died. Persecution has always succeeded, save where the heretics were too strong a party to be effectually persecuted. . . . It is a piece of idle sentimentality that truth, merely as truth, has any inherent power denied to error of prevailing against the dungeon and the stake. (vol. XVIII, p. 238)

What becomes for Mill by 1859 an extreme pessimism about the power of truth to prevail over error, notwithstanding the advantages that the science of induction affords, leads naturally to the value Mill places on individuality in the same work, for if the fallibility of opinion is such that truth cannot be clearly distinguished from falsehood, and no reliable calculus of the probabilities of happiness is possible, it is clear that all people need to be free to follow the promptings of their own reason and inclination, to attempt, in Mill's phrase, "different experiments of living," in order to realize their full human potential. That Mill does not attempt to specify what this full human potential is, beyond affirming, in the language of Humboldt already quoted, that it ought to be as many-sided as possible, has been made the focus of attack. R. P. Anschutz, for example, complains that if Mill's principle of individuality is to serve as a guide to life, "some content has to be found for it,"[8] but individuality, once we attempt to specify its "content," ceases to be individuality, in the sense of a course freely and spontaneously chosen. Anschutz's error, I believe, consists in confusing a necessary condition of human life with its end or goal. The fact is that Mill does not offer individuality as the fundamental principle of an ethical system, but merely as a practical necessity that presides over every man's quest for the meaning and purpose of his life.

[II]

For his part, Arnold never doubted the possibility of a scientific humanism, endorsing the opinion of the Homeric scholar F. A. Wolf: "There can be no doubt that Wolf is perfectly right; that all learning is scientific which is systematically laid out and followed up to its original sources, and that a genuine humanism is scientific."[9] But, although Arnold acknowledged the natural human need "to combine the pieces of our knowledge together, to bring them under general rules, to relate them to principles" (vol. X, p. 62), he had little concern with the method of scientific investigation beyond insisting that it be systematic and derive its light from the "original sources" of knowledge.

Arnold's preferred rhetorical stance is that of the amateur, and he never lays claim to any special fund of knowledge, scientific or otherwise. His self-characterization in the later chapters of *Culture and Anarchy* is typical:

> Knowing myself to be indeed sadly to seek, as one of my many critics says, in "a philosophy with coherent, interdependent, subordinate, and derivative principles," I continually have recourse to a plain man's expedient of trying to make what few simple notions I have, clearer and more intelligible to myself by means of example and illustration. (vol. V, p. 126)

The same rhetorical pose served him also in the later "Literature and Science," where he proclaims himself deficient in scientific learning (vol. X, p. 55). What Arnold intended, however, to be merely disarming candor proves to be a damaging admission indeed, for his understanding of science is extraordinarily naive. In "Literature and Science" he constantly stresses the importance of knowing the "results" of scientific study,[10] but he seems curiously insensitive to the claims of science as a mode of intellectual culture and thus as an important element in a liberal curriculum. Arnold acknowledges the value of what he calls "instrument-knowledges" as contributing to our store of interesting knowledge without themselves being, for the mass of mankind, interesting. "For the majority of mankind," he remarks, "a little of mathematics . . . goes a long way" (vol. X, p. 64). The natural sciences, although more intrinsically interesting than instrument-knowledges, are still primarily valuable for their "results," not for their procedures.

Arnold's naive conception of science is shown most clearly perhaps in the comparison he offers to display the superiority of literary to scientific knowledge. The passage must be quoted in full.

> I once mentioned in a school-report, how a young man in one of our English training colleges having to paraphrase the passage in *Macbeth* beginning,
>
> > "Can'st thou not minister to a mind diseased?"
>
> turned this line into, "Can you not wait upon the lunatic?" And I remarked what a curious state of things it would be, if every pupil of our national schools knew, let us say, that the moon is two thousand one hundred and sixty miles in diameter, and thought at the same time that a good paraphrase for
>
> > "Can'st thou not minister to a mind diseased?"
>
> was, "Can you not wait upon the lunatic?" If one is driven to choose, I think I would rather have a young person ignorant about the moon's diameter, but aware that "Can you not wait upon the lunatic?" is bad, than a young person whose education had been such as to manage things the other way. (vol. X, pp. 69–70)

Quite apart from the rhetorical effect of the iteration of the absurd "Can you not wait upon the lunatic?" or the unnecessary specificity with which the diameter of the moon is given, the comparison is a prejudicial one, for it presents as analogous knowledge of a bald, meaningless fact about the natural world and a complex judgment, involving both a determination of verbal meaning and a qualitative estimate, upon a line of poetry that can only be understood and evaluated in the light of the connotations it derives from its context, the play as a whole. It would be both fairer and more accurate to present the moon's diameter as comparable in educational importance with a fact that Arnold mentions earlier in the same essay: "If we are studying Greek accents, it is interesting to know that *pais* and *pas*, and some other monosyllables of the same form of declension, do not take the circumflex upon the last syllable of the genitive plural" (vol. X, p. 62). Or conversely, it might be appropriate to suggest that accurately paraphrasing a line of *Macbeth* is comparable in difficulty and importance to explaining how the moon's diameter is determined.

Science, for Arnold, is reducible to the facts that can be demonstrated by means of its ancillary "instrument-knowledges," and he is quite incapable of conceiving it as a structure of knowledge, humanly interesting in its own right and valuable, either as a form of intellectual training or discipline, or as a perspective from which to view humanistic thought.

To speak of humanistic learning as scientific, therefore, is not to lay claim to the coherence of what Mill would call the moral or human sciences, nor is it even to claim the authority of a philosophically grounded method common to the humanities and the natural sciences. It affirms no more than that the methods of humanistic scholarship are, or ought to be, in some unspecified way, "systematically laid out," and that their products therefore have a share in the authority we grant to the results of investigation in the natural sciences.

Ironically, however, neither Arnold's low opinion of the pretensions of science to an important place in liberal education nor his rather hazy notion of the claims of scientific methodology lead him to adopt a skeptical position about the status of knowledge, which he persists in regarding as not only possible, but, under certain conditions, easy of attainment. Not that Arnold subscribes to what he calls in *Culture and Anarchy* "a peculiarly British form of Quietism," the belief that we can attain to certitude without taking thought, "by the mercy of Providence, and by a kind of natural tendency of things" (vol. V, p. 156). On the contrary, Arnold sees formidable obstacles standing in the way of truth, but these do not, as they did for Mill, arise in the disparity between man's powers or faculties and the nature of things; they are not inherent in the cognitive act itself.

In *On Liberty* Mill had taken Humboldt's notion of a harmonious development of human powers as confirming his own view that, lacking any authoritative prescription, man's safest course was to experiment with as many modes of life as possible in an effort to realize his individual potential. Arnold also heeded Humboldt's words, although in his view they directed, not a restless search for one's individuality, but the realization of an ideal that was at once social and authoritative. Culture, he said, "is not satisfied till we *all* come to a perfect man" (vol. V, p. 112). The argument of *Culture and Anarchy* consists essentially of adapting the idea of culture — accessibility to ideas and the disposition to be moved by ideas — as an instrument to combat "anarchy," conceived in both political and intellectual terms. To counter political anarchy, culture proposes the instrumentality of the State, embodying the national and collective best self (vol. V, pp. 134–35); to counter intellectual anarchy, culture suggests a law of taste and the idea of right reason. Among the obstacles to the recognition of a law of taste Arnold reckons man's "natural taste for the bathos" (vol. V, p. 147); what chiefly militates against the recognition of right reason is a social structure that imprisons people within the attitudes and values peculiar to their class.

The effect of such a social arrangement, as Arnold was to sum it up in the later essay, "Equality," lies in "materialising our upper class, vulgarising our middle class, and brutalising our lower class. And this is to fail in civilisation" (vol. VIII, p. 299). The remedy proposed in *Culture and Anarchy* is to take the authority of the State out of the hands of those who are activated chiefly by class spirit or prejudice and place it in the hands of those who are so far able to transcend class spirit and prejudice as to be accessible to right reason and to be moved by a general humane spirit.

No one, surely, would quarrel with Arnold's notion that an indispensable condition of knowing the truth, or as Arnold was fond of putting it, knowing "the best that has been known and thought in the world,"[11] is to be freed from those impediments created and fostered by society itself in the form of class attitudes and prejudices. A more difficult problem arises when we begin to inquire into the method by which Arnold proposes to discover and recognize "the best that has been known and thought in the world," supposing those class attitudes and prejudices to be overcome. The phrase "right reason" begs the question, for how can we know when we are reasoning rightly? As Mill neatly demonstrates in his review of Whately's *Logic*, "Men may easily persuade themselves that they are able to reason although they are not; because the faculty which they want, is that by which alone they could detect the want of it" (vol. XI, p. 5). Has Arnold discovered some royal road to knowledge that Mill could not find in fifteen years of groping with the problem of induction?

Arnold confronts methodological questions rarely and reluctantly, and epistemological questions never, yet it is possible, by examining some of the conclusions he reaches, and by exploring the implications of his favorite terms and metaphors, to recover his tacit assumptions about the way the mind arrives at truth. Very early in *Culture and Anarchy* he speaks of "the universal order which seems to be intended and aimed at in the world" (vol. V, p. 93), a concept he apparently also had in mind as he rang the changes on the phrase, "the true, firm, intelligible law of things."[12] Now to speak of "the universal order" or "the firm intelligible law of things" clearly implies a faith in an inherent correspondence between things and the mind that comprehends them, a belief that the laws of nature, while possessing objective reality, are at the same time adapted to human intelligence, which is understood as specially empowered to grasp them. To Mill's nominalism, expressed in his belief that species are, for the most part, man-made, Arnold opposes a belief in the reality of universals; to Mill's skepticism, which can accord only a contingent probability to the results of induction, and looks in vain for a sign by which truth can be known, Arnold opposes the rationalist's faith in the intelligibility of a universal order, and thus also in man's natural capacity to seize truth.

It seems natural to suppose, furthermore, that man's innate capacity to grasp truth is the faculty to which Arnold had already given the name "right reason," and which needs only to be freed from external restraints to fullfill its function. Knowledge, accordingly, is not what we create but what we discover, and the law that both culture and education are to follow is to allow the mind's own light to shine upon its objects. The intrusion of abstract systems of thought turns out to be itself a major obstacle to the discernment of truth. "Here," Arnold wrote in "The Function of Criticism at the Present Time," "the great safeguard is never to let oneself become abstract, always to retain an intimate and lively consciousness of the truth of what one is saying, and, the moment this fails us, to be sure that something is wrong" (vol. III, p. 283). But when the mind frees itself from practical concerns, when its operations are not impeded by specious dogmas and claptrap, or even by too rigorous or slavish a dependence on intellectual method, then it can, by a free and inward action, seize and hold the truth. The nature of this mental act is suggested by Arnold's favorite metaphors for it: "giving our consciousness free play" and "letting the natural stream of our consciousness flow over [its objects] freely."[13] Here is no laborious seeking after truth, but a fluid, effortless, essentially *passive* operation, which we permit rather than actively will.

Arnold's insistence on freedom as a condition of the mind's most creative activity suggests, at least superficially, that like Mill he regards truth as the product of a dialectic conflict of half-truths and that human development is synonymous with free individual self-expression. But the resemblance is misleading. Of the two mutually supporting aspects of culture, Hebraism and Hellenism, it is unquestionably Hellenism that Arnold regards as most exigent (however he may shift the emphasis later), and to define Hellenism he coins the phrase, which he uses over and over again, "spontaneity of consciousness."[14] "Spontaneity" is a favorite word of Mill's, too,[15] but for Mill it refers to voluntary activity that is natural and unforced, while for Arnold it has the suggestion of disinterestedness and freedom from prejudice, conditions that, far from being natural, are achieved only by the transcendence of one's ordinary self. Thus Arnold's "free play of mind" means, not unconditioned, or even undisciplined, mental activity, but mental activity that is not subservient to circumstance. For Arnold a "free" play of mind is perfectly consistent with an attitude of deference toward the best that has been known and thought in the world, and with the willingness to suppress personal preferences in favor of objectively and even authoritatively determined values. In fact, Arnold objects strongly to Mill's more literal interpretation of freedom, attacking what he thinks of as the anarchy of "doing as one likes."

The practical and political consequences of this epistemological difference between Mill and Arnold emerge in their view of human rights. From one point of view, rights — and their correlative, duties — are nothing more than the defining characteristics of liberty, which may be regarded as directly proportional to the first and inversely proportional to the second, and therefore the relevant question for both Mill and Arnold — whether rights and duties are inherent in the human condition — is really the same as the question whether liberty is an essential condition of human life. Nevertheless, the former question is worth examining directly, for it shifts the emphasis from liberty as an end in itself, or as a means of perfecting knowledge, to the original sanctions of liberty. If, as Mill argues, liberty is indispensable to, or even identical with, human individuality, it follows that the rights that define liberty are also indispensable and belong to man even in the unaccommodated state of nature. Of course Mill is enough of a realist to know that even natural rights require the guarantees of law, and therefore of organized society, but he cannot enumerate the rights of man because they are infinite. Man has a right to do anything at all that does not interfere with the similar right of other men. Self-protection is the sole end that justifies society in abridging any man's liberty.

Arnold's view is diametrically opposite. "Now does any one, if he simply and naturally reads his own consciousness, discover that he has any rights at all? For my part, the deeper I go in my own consciousness, and the more simply I abandon myself to it, the more it seems to me that I have no rights at all, only duties" (vol. V, p. 201). Like Carlyle, Arnold holds that "renouncement," rather than liberty, "is the law of human life" (vol. V, p. 207), but this disparagement of liberty does not mean that Arnold has abandoned the ideal of spontaneity of consciousness; it means only that he rejects the notion of natural, a priori rights, discoverable by man's reason, substituting instead a notion of right and liberty that must be achieved through spiritual conquest. There is a kind of Hebraism in Arnold's Hellenism, which stresses not sensation, enjoyment, and untrammeled appreciation, but development, self-discipline, and self-improvement.

[III]

Modern structuralist thought has discovered in language the power that simultaneously constitutes human knowledge and renders it impossible. "The fundamental codes of a culture," Foucault argues persuasively, ". . . establish for every man, from the very first, the empirical orders with which he will be dealing and within which he will be at home."[16] But at the same time, language, which has become detached from the things it formerly signified, can now signify only itself, can tell us nothing about the world of things. It is a tissue of words, containing its own taxonomy, imposing its own order on what can be known or thought, and interposing itself between us and reality.

Obviously the radical gulf between words and things envisaged by Foucault amounts to a total repudiation of Arnold's comfortable assumption about "the firm intelligible law of things," knowable by a kind of introspection, or "spontaneity of consciousness." It is a curious paradox that Arnold, who was skeptical of the pretensions of natural science to an equal dignity with humanistic learning, yet believed firmly that humanistic learning could itself be scientific, and that real knowledge (science, *scientia*) was possible. It is perhaps a greater paradox that Mill, who did so much to make available for man's humane concerns the precision of method of Newtonian science, should remain so pessimistic about the possibility of any scientific knowledge at all, but perhaps both paradoxes reflect the immense changes that have overtaken both humanistic and scientific thought in the last hundred years. In his naive faith that the laws of nature were conformable to man's powers of cognition, Arnold remains a characteristic Victorian; in his fears that truth might elude man's most searching inquiries, Mill foreshadows the despairing skepticism of Foucault.

NOTES AND REFERENCES

1. Edward Alexander, *Matthew Arnold and John Stuart Mill* (New York: Columbia University Press, 1965), p. 28.
2. *Collected Works of John Stuart Mill* (Toronto: University of Toronto Press, 1963–   ), vol. XVIII, p. 261. All references are to this edition and will be identified parenthetically in the text, except where additional information is to be provided.
3. *The Complete Prose Works of Matthew Arnold*, ed. R. H. Super (Ann Arbor: University of Michigan Press, 1960–77), vol. V, p. 94. All references are to this edition and will be identified parenthetically in the text, except where additional information is to be provided.
4. John Stuart Mill, *Autobiography and Other Writings*, ed. Jack Stillinger (Boston: Houghton Mifflin, 1969), p. 126.
5. Mill,[2] vol. VII, p. 9. In his *Examination of Sir William Hamilton's Philosophy*, Mill was

strongly to endorse Hamilton's assertion that "logic considers Thought, not as the operation of thinking, but as its product." vol. IX, p. 361.

6. Mill does not use this term, which I have borrowed in this sense from Thomas Kuhn, *The Structure of Scientific Revolutions* (Chicago: University of Chicago Press, 1970), pp. 10–22.

7. Mill,[2] vol. VIII, p. 937. Macaulay's views are set forth in the Essay on Dryden.

8. R. P. Anschutz, *The Philosophy of John Stuart Mill* (Oxford: Clarendon Press, 1963), p. 26.

9. Arnold,[3] "Literature and Science", vol. X, p. 57.

10. Arnold,[2] vol. X, p. 59. Cf. also pp. 60, 64, 68.

11. Arnold,[3] "The Function of Criticism at the Present Time," vol. III, pp. 268, 270, 282, 283, 284.

12. Arnold,[3] vol. V, pp. 178, 184, 191, 201, 209, 213, 218, 219, 225.

13. Arnold,[3] "Giving our consciousness free play," vol. V, pp. 181, 186, 191, 192, 196, 201, 218, 221, 228, 229; "letting the natural stream of our consciousness flow" etc., vol. V, pp. 188, 209, 220, 221.

14. Super's index gives ten page references.

15. Mill,[2] vol. XVIII, pp. 261, 264, 277.

16. Michel Foucault, *The Order of Things* (New York: Vintage Books, 1973), p. xx.

Dickens and Victorian Dream Theory

CATHERINE A. BERNARD
Department of English
New York Institute of Technology
New York, New York 10023

WHEN a physician submitted an article on dreams to *Household Words*, its editor, Charles Dickens, diplomatically suggested that it could be "made a little more original and a little less recapitulative of the usual stories in the books."[1] He then proceeded to outline his own dream theories, claiming to have some scholarly and firsthand knowledge of the subject. Dickens' counsel on scientific rather than literary matters must have struck the doctor as a bit presumptuous, but his immodesty was not unwarranted. In his theories and fiction he reveals not only a considerable knowledge of contemporary dream theory, but also an apparently intuitive insight into the autobiographical meaning of dreams. Sensing that dreams were psychological in origin, rather than supernatural or physiological, Dickens offered a strikingly independent view of dreams that challenged many conventional beliefs of his day.

In the Victorian age, dreams belonged as much to the supernatural world as to science. Interpreters were sharply divided on the question of origins: spiritualists argued that dreams were miraculous events that permitted communication with the supernatural world, while scientists insisted they were natural phenomena that could be assigned governing laws.

Proponents of spiritualism, a movement that had attracted a fashionable following in Dickens' day, came in two varieties—secular and religious. The first embraced the occult and its supernatural trappings, arguing that dreams, along with such assorted paraphernalia as table rappings, seances, and house hauntings, permitted communication with the dead. Books in this vein provided little theory—they were mere catalogues of mysterious events, ghost stories, and dream narratives in which contact with the spiritual world had purportedly occurred. A title from Dickens' library should suffice to indicate the direction most of these took: *The Philosophy of Mystery (On Ghosts, Poetic Phantasy, Mysterious Forms and Signs, Mysterious Sounds, Fairy Mythology, Demonology, Dreams, Somnambulism, etc.)* (1841).[2]

Refuting any connection with the occult, other spiritualists supported a transcendental view of dreams, but one firmly grounded in Christian and moral concerns. Mrs. Blair, the editor of *Dreams and Dreaming* (1843), a Victorian collection of premonitory dreams, solemnly argued that dreams were divine revelations that God offered as moral directives "for the comforting of his tired people."[3] In *The Night Side of Nature* (1848), Catherine Crowe made similar claims. Taking her cue from the German transcendentalists, she argued that in special states such as dreams and mesmeric or magnetic sleep, the spirit could detach itself from the body and, through a "universal" sense, perceive spirits of the dead and messages from the Deity.[4]

While both women shunned the term "supernatural" and purported to have conducted serious and carefully documented investigations, their books remain ramshackle collections of popular ghost stories and dream narratives. Scientists viewed such books with disdain for, in the words of one, "they were more commonly read to amuse an idle hour than for any graver purpose."[5]

In Victorian England the tide in dream research was clearly shifting from the spiritual and speculative toward the scientific and mechanistic. Wanting nothing to do with miracles and wonders, philosophers and physicians sought to explain dreams according to natural principles and to place them within a genuine scientific framework. Their investigations focused upon three central points, the same ones that David Hartley had raised in *Observations of Man* a century earlier: (1) the law of association in determining the train of events in dreams, (2) the effects of recent impressions and ideas, and (3) the state of the body during sleep.

In *Elements of the Philosophy of the Human Mind* (1814), Dugald Stewart, a Scottish philosopher-psychologist, defined the theories of association that shaped many Victorian notions of dreams. He stated, simply, that our dreams followed the same laws of association as in waking life except that "the will loses its influence over those faculties of the mind, and those members of the body which during our waking hours are subjected to its authority."[6]

Once Stewart had accounted for the wildness and incongruity of dreams by this loss of will, he proceeded to argue that dreams reflected waking thoughts, particularly the prevailing habits and temper of the dreamer shortly before sleep. Thus, he observed, the poet will dream of literary endeavors, the miser of his money, and a cheerful person will experience happy dreams free of distressing events. While Stewart felt that dreams generally reflected recent impressions, he acknowledged that they sometimes recalled childhood memories because "the faculty of association is then much stronger than in advanced years."[7] Stewart, however, did not provide any principle for the associative process, as Freud later would, but simply viewed it as a succession of mechanical images which, once triggered, would recur in the same fashion.

Victorian medical science significantly modified and refined Stewart's theories. In *Inquiries Concerning the Intellectual Powers and Investigation of Truth* (1843), John Abercrombie, a Scottish physician, attempted to trace the patterns of association that emerged within the dreams themselves. In the process of categorizing dreams, he stumbled upon the insight that some dreams that at first seemed composed of unrelated events could be traced back to a common source, and that others revived "old associations respecting things that had passed out of the mind."[8] While Abercrombie comes close to defining the role of the unconscious, his evidence was not convincing. For example, he traced those forgotten memories not to important emotional events in the dreamer's life, but to forgotten bits of information such as lost documents and forgotten parcels.

The effect of sensory impressions upon dreams formed a second popular source of inquiry. The first to conduct experiments of this phenomenon was a Frenchman, L.A. Maury, whose classical book, *Sleep and Dreams*, appeared in 1861. Maury, who experimented upon himself, observed that during sleep certain external stimuli would elicit complementary images in dreams. When his lips and nose were tickled by a feather, he dreamed he was subject to horrible tortures; a scent of perfume led him to dream of being in a perfume shop in Cairo.

While Victorian dream studies did, on occasion, allude to Maury's experiments, most physicians cited Abercrombie's evidence of the sensory impressions of dreams. Abercrombie, however, did not conduct experiments in this vein as Maury did, but relied upon the accounts of friends. For example, he tells of Dr. Gregory who went to bed with a vessel of hot water at his feet and dreamed of walking up the crater of Mount Etna and of Dr. Reid who, when a poultice on his head became ruffled, dreamed of being scalped by savages.[9] Abercrombie's inquiries into this phenomenon, though, were somewhat shortsighted. However complex the dream, he reduced all of its meaning to the single stimulus, failing to recognize, as Maury did, that stimuli could trigger a dream but did not necessarily account for all of its meaning.

Physiological disorders were also believed to produce dreams. Indigestion was the most common culprit, but other "morbid" conditions of the body such as "sanguineous derangement," liver disorders, and cerebral diseases were cited in medical literature as the causes of disturbed dream visions.[10] Nightmares were attributed to certain bodily positions during sleep that produced pressure on the heart.[11]

Some noted physicians of Dickens' day, including John Elliotson in *Human Physiology* (1840) and Robert MacNish in *The Philosophy of Sleep* (1838), established a phrenological basis for interpreting dreams. Citing Francis J. Gall's theory that the brain was composed of several mental faculties, they asserted that the nature of a dream was determined by the type and number of those faculties that remained active during sleep. Thus, if color were in excess, it was believed that dreams would be brilliant, if cautiousness, that the dreamer would be filled with terror.[12]

While physicians did not always agree upon the causes of dreams, their research reflected two tendencies basic to the Victorian age. First, reacting against the speculative approach of the supernaturalists, physicians and scientists insisted upon basing new laws upon facts and documented evidence. Their commitment to facts was frequently expressed in somber, cautious terms. Dr. Abercrombie's introductory comments in *Inquiries* suggest the tone and approach adopted in most dream studies of the period: "Our duty is to keep steadily in view, that the objects of true science are facts alone, and the relations of these facts to each other. . . . It is entirely out of the reach of our faculties to advance a single step beyond the facts which are before us."[13]

Second, like so much else in mid-nineteenth-century England, dreams were firmly entrenched in Victorian morality. It was commonly assumed that dreams not only reflected a man's habits and temper, but his moral character as well: "A person whose habits are virtuous does not in his dreams plunge into a series of crimes; nor are the vicious reformed when they pass into the imaginary world."[14]

Physicians did note some exceptions to this rule. Thomas Laycock, a neurophysiologist, reported having "been consulted more than once by persons of a highly religous turn of mind" who were horrified by what appeared to them to be "hideously immoral dreams."[15] But while physicians recorded such aberrations, they attributed them to physiological, not psychological causes. It was believed that during sleep the higher or moral instincts remained passive, giving rein to the lower or sensual ones and resulting in morally objectionable and incongruous dream images.[16]

Nevertheless, many individuals believed that dreams not only reflected moral character, but served as a means of moral improvement. While a man might not be held culpable for dreams that contradicted his waking moral character, physicians still felt that all such aberrations in sleep should be carefully regarded. As Dr. Beattie advised, if we "attended to what passes in sleep, we may sometimes discover what passions are predominant and so receive good hints for the regulation of them."[17]

Inherent in the assumptions of Dr. Beattie and others was the confident belief that if the dreamer followed a proper regimen, he could control his dreams. It was commonly assumed that if the dreamer remained temperate and morally sound, particularly in pre-bedtime hours, the odds for restful sleep were high. And the benefits were far-reaching. As one physician promised, if we avoid excess, "we ensure refreshing sleep, undisturbed by dreaming visions of any kind, and our lives would then be useful to others, agreeable to ourselves, and ensure a lasting satisfaction."[18]

However naïve such moral precepts may appear to the modern reader, Victorian physicians such as Abercrombie, MacNish, and Beattie should be credited with one major advancement in the field of dream research. By focusing on natural causes rather than supernatural or divine agencies, they finally removed dreams from the realm of superstition and placed them firmly within that of science. Their intentions, unfortunately, were often more promising than their achievements. For example, while they insisted upon verified evidence, their supply of dream narratives remained paltry and "documentation" often simply meant that they relied on a friend's dream account rather than on historical sources or clinical research. Furthermore, while they insisted upon basing their new laws upon facts and observable phenomena, they often stressed the trivial, the outward physiological signs such as snoring patterns, bodily positions, and facial expressions during sleep, and disregarded altogether any inquiries into the dream's content and meaning. Confined by their rigid moral concerns, Victorian researchers too often denied or ignored the dream's seemingly immoral and irrational nature. Focusing on the way dreams reflected the immediate thoughts and experience of daily life, they failed to speculate on the dream's deeper, psychological origins.

While theorists on the Continent were making much better headway in this direction, English dream theorists were generally unaware of such trends. In Germany, figures like C.G. Carus, K.A. Scherner and G. Von Schubert were recognizing the dream's dynamic relationship with the unconscious, setting the stage for Freud's dramatic breakthrough at the turn of the century. In their studies of the brain and nervous system, English physicians such as Thomas Laycock and W.B. Carpenter did define the unconscious, but they viewed it as a separate pocket of activity where forgotten memories and impressions could be stored and recalled by reflex action.[19] While they, like the associationists, pointed out that such memories returned, they did not speculate on what motivated their return. English theorists were scattering plenty of clues, but stopped short of recognizing the important link between the conscious and unconscious realm.

It is not surprising that Dickens speculated more freely about dreams than did most scientists of his day. In those pre-Freudian days, literary figures openly discussed their dreams and frequently arrived at important insights into their psychic function. Dickens had a rich dream life and, while he derived most of his theories from his own dreams, he also, as he put it, had "read something about the subject."[20] It is difficult to say just how scholarly Dickens was in this matter. We do know that his Gad's Hill library contained many of the major studies on dreams, including Abercrombie's *Inquiries Concerning the Intellectual Powers and the Investigation of Truth* (1838), Robert MacNish's *The Philosophy of Sleep* (1838) and Dugald Stewart's *Elements of the Philosophy of the Human Mind* (1814).[21] It is true, as Leonard Manheim has pointed out, that some of these books were presentation copies that Dickens may not have read.[22] But since so many of these books recapitulated scientific and popular theories of the day, we can assume that Dickens' reading of a limited number of them would have familiarized him with the major currents of Victorian dream theory.

In his correspondence and essays, Dickens made frequent reference to dreams, a subject that he had "long observed with the greatest attention and interest."[23] At times he seemed to side with the scientists on this matter. For example, in a review of Catherine Crowe's book, *The Night Side of Nature* (1848), he shared their skepticism about spiritualism. Although Dickens was always fond of ghost stories, he found most of Crowe's testimony unconvincing because, as he put it, "there is the common fault of seeking to prove too much. . . . She stands by her weakest ghost at least as manfully as her strongest."[24]

Like physicians of his day, Dickens also sensed the need for collecting and analyzing dream data. In *Doctor Marigold's Prescriptions* (1865), he described the hushed embarrassment that had long surrounded the subject and urged his readers to impart "their knowledge of their own psychological experiences when they have been of a strange sort." He warned his readers that if they did not disclose their "subjective creations" such as dreams and visions, "the general stock of experience in this vein would remain miserably imperfect."[25]

Dickens' own lack of reticence in such matters is evident in a letter he wrote, dated February 2, 1851, to Dr. Stone, the physician who had submitted an article on dreams to *Household Words*.[26] Dickens was clearly unimpressed by Stone's material, which he found too recapitulative of popular beliefs. As Warrington Winters has pointed out, Dr. Stone eventually did get his paper published, but only after he included most of the points Dickens outlined in his letter.[27] If we compare the contents of this letter with the final article, it is easy enough to detect Dr. Stone's original contributions. He covered the standard fare: the dreamer's expression during sleep, the effect of bodily sensations and recent impressions, and the effect of waking habits and temperament upon dreams. Predictably, he took a firm stand on the moral character of dreams, insisting that "the identity of moral goodness" could never be "perverted in the dream state."[28]

In his letter to Dr. Stone, Dickens objected to most of these points and offered his own theories in their place. First, he questioned the "conventional philosophy" that dreams merely echoed the previous day's occurrences. He conceded that dreams sometimes reflect waking thoughts and recent events, but added that these were usually of an insignificant nature and were depicted allegorically.

Dickens made two further points in relation to this claim. He questioned the belief that dreams could be controlled by the dreamer: "Did you ever hear of any person who, by trying and resolving the mind on any subject, could dream of it, or who did not, under any circumstances, dream preposterously wide of the mark?" He then cited the notion of recurrent dreams, "unhealthy and morbid species of those visions," which, he claimed, were perpetuated in secrecy on the part of the dreamer.

Dickens went on to note the similarities among dreams, stating that if we take into "consideration our vast differences in point of mental and physical constitution, there is a remarkable sameness" to our dreams. Finally, he observed that, even while we sleep, there is a waking and reasoning part of the brain which knows that we are dreaming and which sometimes corrects certain delusions manifested by the dream state.

This last point Dickens may have derived from medical texts, for it was a point often cited by physicians.[29] But his other claims he derived from firsthand observation and these frequently clashed with popularly held beliefs. In his own dreams, which he informed Dr. Stone were of

events from "twenty years ago," Dickens began to recognize the retrospective nature of dreams. He was more specific about the content of his
dreams in his autobiographical fragment, which he wrote and permitted
his friend, John Forster, to read in 1847. This document had curious
origins. Dickens had never divulged the events of his childhood to
anyone, but in the spring of that year, Forster happened to mention that
an acquaintance recalled seeing Dickens as a child working in a blacking
factory. At last Dickens had no choice but to confront his past and to
disclose some long-buried childhood memories.

When Dickens was twelve years old his parents had been imprisoned
for debt and had abandoned him to a job at Warren's blacking factory in
London. To Dickens no fate could have been more devastating or cruel.
His memories of the experience are filled with a sense of betrayal, injustice, and self-pity: "The deep remembrance of the sense I had of being
utterly neglected and hopeless; of the shame I felt in my position; of the
misery that it was to my young heart to believe that, day by day, what I
learned, and thought, and delighted in, and raised my fancy and my
emulation up by, was passing away from me, never to be brought back
any more; cannot be written."[30] Dickens, a man rarely at a loss for
words, could barely articulate the pain: "That I suffered in secret, and
that I suffered exquisitely, no one knew but I. How much I suffered it is
. . . beyond my power to tell."

Dickens would never be able to forget this experience nor to forgive
his parents for "very willingly" abandoning him to such circumstances
even when financial matters improved. While his sister was permitted to
continue her music lessons at the Royal Academy of Music, he, "a child
of singular abilities, quick, eager, delicate," saw no prospects for relief.
But Dickens could never confront this issue either. Even as an adult he
could not muster the courage to walk on the same side of the street that
housed the warehouse for fear of being reminded of "what I once was."
The entire episode was a subject to which his parents had never alluded:

> From that hour until this, my father and mother had been stricken dumb
> upon it. I have never heard the least allusion to it, however far off and
> remote, from either one of them. I have never until I now impart it to this
> paper, in any burst of confidence with anyone, my wife not excepted, raised
> the curtain then dropped, thank God.

But these unmentionable childhood memories returned to haunt Dickens
in his dreams, for as he confessed to Forster, "even now, famous and
caressed and happy, I often forget in my dreams that I have a dear wife

and children; even that I am a man; and wander desolately back to that time of my life."

Dickens had derived another of his theories—his notion about recurrent dreams—from his own experience. As he explained in his letter to Dr. Stone, he dreamed repeatedly of his sister-in-law, Mary Hogarth, who died in Dickens' home after a sudden illness. Shortly after her death, Dickens quietly admitted, "I could have spared a much nearer relation."[31] While Dickens' wife became a source of increased disappointment, Mary, who had moved into the household, had become his feminine ideal. Overwhelmed with grief, he begged to be buried next to her when he died. Dickens' ardent worship of Mary during her life and his obsessive idealization of her after death are well known biographical matters, and surely Edgar Johnson's appraisal that Dickens' "devotion to Mary was an emotion unique in his entire life" does not overstate the case.[32] It is curious that in his letter to Dr. Stone, Dickens referred to his dreams of Mary as an "unhealthy and morbid species," which, he suggests, he had kept secret only out of reverence for his wife's sense of bereavement for Mary. Some years earlier, however, he had described these dreams as happy visions which he longed for "always with a kind of quiet happiness, which became so pleasant to me that I never lay down at night without a hope of the vision coming back in one shape or another."[33] These dreams, he informed Dr. Stone, ceased abruptly when he told his wife of them.

Although Dickens was surely unaware of the sort of psychological ramifications his experiences would suggest to modern audiences, he comes close in his theories to recognizing certain aspects of dream psychology that Freud and his followers would later formulate. By recognizing that dreams often reflect the past, particularly crucial early memories that the adult could not confront, Dickens points to the role of memory in dreams and to the formative nature of our early experiences. While some of Dickens' contemporaries acknowledged that dreams could recall childhood scenes, they failed to recognize, as Dickens did, the volatile, emotionally sensitive nature of these memories. By recognizing that recent events could figure in our dreams, but that they are generally of an insignificant nature, Dickens anticipated Freud's idea of day residues and the distinction between latent and manifest content.[34] Most importantly, because his own dreams derived from two of the deepest experiences of his life—one that he could not confront and one that he could not resolve—Dickens had intuitively perceived the autobiographical origins of dreams.

In light of Dickens' own dreams, it is easy enough to see why he questioned certain theories of his day. His own experiences had proved to him that dreams could not be regulated and controlled at will and that they derived from more than sensory stimuli or physiological disorders. While, as cherished moral guardian of his age, he never explicitly challenged the Victorian belief in the moral character of dreams, he implies such a criticism in his proposal that we all share similar symbolic experiences in dreams. As he put it, "we all fall off the same tower . . . we all take unheard of trouble to go to the theater and never get in; we all go to public places in our nightdress and are horribly disconcerted lest the company observe it." In "Lying Awake," he even makes the more shocking claim that in their sleep he, Her Majesty Queen Victoria, and the vagrant Winking Charlie "have all fallen many thousands of times from the same tower . . . and have all three committed murders and hidden bodies."[35] Surely Dickens chose these particular images from a random sampling of his own dreams, not because he had any sense, as Freud did, that such "typical" patterns could be traced to an infantile, sexual source, but because he believed that dreams could not be explained simply in terms of their moral qualities. For if, regardless of circumstances, habits, and moral character, we are all subjected to the same nightly patterns, then the Victorian precept that dreams were neat moral packages would no longer hold.

Dickens was a novelist, not a theorist, and so many of the ideas that he merely hints at in his letter to Stone find much fuller expression in his novels. His invented dreams differed just as sharply from conventional literary usage as did his theories from popular and scientific beliefs of his day.

With the exception of the use of dreams made by such literary figures as Bulwer-Lytton and Le Fanu, who were boldly exploring the supernatural realms of trance and vision, dreams were not the standard ingredient of the Victorian novel.[36] Novelists writing in a realistic vein such as Trollope, Thackeray, and Eliot, shunned them altogether in their fiction. Intent upon accurately recording social realism, not the interior landscapes of distorted imagination, they viewed dreams as suitable material for Gothic romances, but not for the domestic novel. When John Blackwood, keeping in mind the family nature of his magazine, urged his client, George Eliot, to substitute her heroine's carrying a dagger with the mere dream of it (for as he put it, "I am pretty sure that his dear little heroine would be more sure of universal sympathy if she only dreamed . . . she could stab the cur in the heart"), Eliot quickly dismissed the idea.

Her response neatly sums up the stance taken by the realistic writers: "Dreams usually play an important part in fiction, but rarely, I think, in real life."[37]

The sensationalist writers of Dickens' day used dreams frequently in their novels, but these were Gothic props, serving the needs of plot, not character. Novelists like Wilkie Collins and Mary Braddon assigned ominous dream warnings to their heroes and heroines to heighten the atmosphere of fear and to develop suspense by anticipating the story's final outcome. But they were more intent upon relaying the events of the story than in recreating a character's state of mind and, as a result, their dreams appear contrived and artificial. Colored by the theatrical sets of melodrama, these dreams mirrored neatly formed patterns of good and evil.

Dickens sometimes used dreams in a similar fashion. His early novels contain numerous instances of contrived dream usage that conformed to Victorian literary conventions and moral codes. Wrongdoers such as Ralph Nickleby and Barnaby Rudge are plagued by punitive dreams and visions that force their guilty deeds upon them in predictably stagey terms. But eventually Dickens began to explore dreams in a more subtle psychological fashion. Forfeiting the popular Victorian tenet that dreams serve as a type of moral index to character, he permitted his loathsome villains like Headstone and Orlick to sleep away in peaceful slumber, while he assigned long, troubled dreams to his pure innocents, those who seem totally shut off from dark imaginings of any sort.[38]

These later dreams function in a peculiarly modern way. Unlike comparable dream passages in Gothic and sensational novels, they rarely serve to promote plot or to heighten atmosphere, but instead reveal underlying conflicts and strivings that contradicted the dreamer's statements and actions in waking life. In many instances, they reveal egotistic impulses and wishes that permit the dreamer to satisfy in hallucinatory fashion what he or she cannot state consciously. I would not suggest that Dickens, like Freud, perceived these wishes as sexual ones, although any post-Freudian reader will quickly recognize the sexual nature of his symbols. Dickens, I am sure, had no conscious understanding of this. He was merely using the kinds of symbols and images that he knew occurred in his and everyone else's dreams. But by assigning such dreams to his heroines, all of whom remain curiously devoid of the slightest hint of sexuality, Dickens, intuitively and unconsciously, was channeling into a literary dream what he could not bring himself to confront elsewhere.

On one level, Dickens' use of the wish-fulfilling dream served a very practical purpose. It provided him with a convenient means for expressing thoughts that, as George Eliot's editor well knew, Victorian audiences would have deemed unsuitable for their heroes and heroines. It permitted Dickens to attribute some egoistic strivings to characters whose natures are so self-sacrificing and totally innocent that their waking thoughts turn only to martyrdom, never to self-preservation. Thus, in his waking life, a figure like Stephen Blackpool in *Hard Times* (1854) never reproaches his wife, although she is a drunken wretch who abandons him periodically and prevents him from marrying his beloved, Rachael. When he sights a bottle of poison on his wife's nightstand, his thoughts turn to suicide, not to murder. But shortly afterward, Stephen falls asleep and experiences a long, troubled dream in which he is clearly the murderer of his wife and is about to take a new bride[39] Stephen's dream thus permitted him to express some threatening wishes that he could never consciously imagine; it also managed to keep his goodness intact.

On a deeper level, Dickens' invented dreams derived from sources closer to his own psyche and this permitted him to exorcise some of those memories of the blacking factory and all of the resentment associated with it. The connection between Dickens' life and fiction becomes apparent if we look at the kinds of wishes manifested in his fictional dreams.

One, that of patricide, permitted him to project in fantasy the anger that he felt, but could only express obliquely, towards his own father. Throughout his fiction Dickens unconsciously deflected this anger in an array of negligent, abusive father figures. He was clearly obsessed with the thought of a son killing his father, but was so horrified by the deed, or so afraid that his readers would be, that even in fiction he could not complete the act. While he never spares his readers the gory details of a villain's death, when it came time for a son to kill his father, courage fails the character at the last moment or his efforts are deemed unnecessary by the father's timely, miraculous death.

Even that sunlit realm of Dickens' first novel, *Pickwick Papers* (1837), is darkened by a dream of patricide. The teller of the interpolated tale, "The Old Man's Tale about the Queer Client," relates the story of a man named Heyling whose family has been imprisoned for debt. While his wife and son die in prison, Heyling vengefully seeks out his father-in-law, who has mercilessly refused to spare their plight. In his long bouts of delirium, he fantasizes various ways of killing the father. First he drags him to the bottom of the sea and drowns him. Then, discovering him in a scorching desert, he denies him a drop of water and greedily watches as the old man dies in his grip. Within the tale itself, Heyling's father-in-law dies conveniently of a stroke, but in the dream the son has had his wishes graphically satisfied.

A second subject to which Dickens was obsessively drawn in his fiction was the plight of the rejected child. Many of his pure innocents are subject to some form of parental rejection. They also experience strange dreams that reveal resentments and more fervent desires for self-preservation than critics have generally credited them with having. By assigning such dreams to his "angels," Dickens was not attempting to flesh out their one-dimensional natures. We know enought about Victorian literary tastes and Dickens' bondage to audience demands to know that these figures were cherished simply because of their unblemished purity. He was unconsciously exploring in their dreams some of the conflicts and resentments he had privately nurtured, but could never openly confront about his own childhood abandonment.

While Dickens' earliest angels in such works as *Oliver Twist* (1837–39) and *The Old Curiosity Shop* (1840–41) are abandoned orphans who must endure endless suffering, it wasn't until *Dombey and Son* (1846–48) that he rendered the subject of childhood rejection in a psychological light. Unlike Oliver and Little Nell, Florence Dombey has not been denied the material comforts of a home. She is victim to a decidedly more subtle form of rejection—her father's chilling reminders that she was a daughter when a son was called for, "a piece of coin that couldn't be invested—a bad boy—nothing more."[40] Having absorbed numerous rebukes on this account, Florence continues to offer her love, unreservedly and selflessly.

But martyrdom does not come as easily to Florence as it had to her predecessor, Little Nell, and one senses that she has had to work hard at coping with the cruel circumstances of her birth. She has learned, we are told, to acquire "the habit, unusual to a child . . . of being quiet, and repressing what she felt" (chap. 6). She has learned to live with her father's hatred by fantasizing enchanted visions of "what her life would have been if her father could have loved her and she had been a favorite child" (chap. 23). Finally, though, Florence must confront the fact that she is not, as she diplomatically puts it, "a favorite child." After admitting this sad fact to her stepmother Edith, Florence experiences her dream:

> She dreamed of seeking her father in wilderness, of following his track up
> fearful heights, and down into deep mines and caverns; of being charged

with something that would release him from extraordinary suffering — she knew not what, or why — yet never being able to attain the goal and set him free. Then she saw him dead, upon that very bed, and in that very room, and knew that he had never loved her to the last, and fell upon his cold breast, passionately weeping. Then a prospect opened, and a river flowed, and a plaintive voice she knew, cried, "It is running on, Floy! It has never stopped! You are moving with it!" And she saw him at a distance stretching out his arms towards her, while a figure such as Walter's used to be, stood near him, awfully serene and still. In every vision, Edith came and went, sometimes to her joy, sometimes to her sorrow, until they were alone upon the brink of a dark grave, and Edith pointing down, she looked and saw — what! — another Edith lying at the bottom. (chap. 35)

It is not surprising that Florence seeks her father's love in this dream, nor that she fails to find it — this has been the pattern of her mission throughout. What is curious, though, is the arrangement of figures in the dream. Her father and Edith are left for dead, while Florence joins her beloved, Walter, in the realm of immortal love. In her waking life Florence used to fantasize about joining her mother and brother in this realm. But she soon denied herself even this respite for fear that "she might stir the spirits of the dead against her father" (chap. 23). In this dream Florence's fear becomes a wish, and she can finally maneuver, according to her own designs, all those whom she has frantically sought to love.

Dombey's transport from his earthly deathbed to the immortal realm foreshadows the novel's sentimental ending, where, in the tearful embrace of Florence, he succumbs to her redemptive love. It also reinforces the strange power of this love, which, in her waking life, Florence has repeatedly been denied the chance to exercise.[41]

Edith's relegation to the pit of a grave might merely reflect her mention of the word "grave" in a conversation with Florence shortly before the dream occurs. But Edith's exclusion from the final love embrace might also suggest some unconscious resentments Florence has harbored against her stepmother. The surrounding narrative would supply ample motivation for such feelings. Twice, on the eve of this dream, Edith has halted Florence's attempts at gaining the love she desperately seeks. The first instance of this occurs when Florence and her father sit alone in a room. Dombey, pretending to sleep, but secretly observing his daughter, began to soften toward her, for "as he looked, she became blended with the child he loved" (chap. 35). He was just about to call out to her when, hearing Edith's footsteps, the moment is abruptly shattered. Edith is responsible for one more setback before Florence experiences her dream.

Drawn instinctively to the child's sad plight, when they are alone in a room, Edith attempts to comfort her. Florence eagerly responds, clinging to this new source of hope "with some fervent words of gratitude and endearment," and then, in a strangely confessional moment, acknowledges her suffering. Suddenly, remembering her own role in the Dombey household, Edith warns the child: "Never seek to find in me what is not here. . . . There should be, so far, a division and a silence between the two, like the grave itself" (chap. 35). The words jolt Florence, but it is only in her dream that she can retaliate for this last, and most unexpected rejection.

In some ways, this dream functions as a conventional plot device. Drenched in Victorian sentiment, it merely foreshadows the eventual reunion of Florence and Dombey and the less fortunate downfall of Edith. But, in a limited way, it also reveals Florence's true feelings, which she has been unable to express throughout. It reveals some of the resentments that we believe such a character as Florence should have felt at many instances and it provides her with a compensatory way of dealing with a father who hates her.

Florence's fate, that of being the "less favorite" child, was one Dickens must have felt keenly. His memories of his experiences at the blacking factory were heightened by his seemingly total abandonment by a family who managed to house and educate his sister Fanny. Dickens' description of sitting through an awards ceremony in which Fanny was honored reveals the resentments he felt but could not admit to: "I could not bear to think of myself—beyond the reach of all such honorable emulation and success. The tears ran down my face. I felt as if my heart were rent . . . I never had suffered so much before. There was no envy in this."[42] That last quickly added refrain sounds very much like the narrative voice in *Dombey and Son*, which throughout the novel has leveled harsh reproaches against Dombey, but which insisted that Florence "had no such thoughts" (chap. 18). Florence's dream permits her to compensate for the power and love denied her in waking life; the wish-fulfilling element of it and of the novel's ending also permitted her creator to fantasize what, in real life, had been denied.

In *Bleak House* (1852–53), Esther Summerson's dreams are psychologically more impressive, for they are convincingly consistent with her character, providing a dramatic climax to a carefully evolved personality straining to express itself throughout the novel. Esther has had a cruel start in life. An illegitimate child, she is raised by an unloving godmother who glumly informs her of her tainted birth and warns her that the only prospect for a life like hers, "begun with a shadow upon it," was for "submission, self denial and diligent work."[43] Esther meekly accepts this sentence and devotes her life to serving others, gratefully accepting whatever love is thrown her away. In her daily behavior, she is the standard version of a self-effacing, duty-bound heroine, which is the way critics viewed her for many years. But as recent, more sympathetic assessments would attest, Esther's character is more complex and more subtle than that of the standard variety of Victorian angels.[44] There is more suppression than meekness in Esther.

Throughout the narrative, Esther has kept many of her feelings pent up, methodically immersing her real feelings in carefully contrived defense tactics. Whenever she is threatened emotionally, she busies herself in a frenzy of household chores. Whenever she suffers a rebuke or disappointment, she frantically offers gratitude, even when outrage would have been the more appropriate response. After her godmother's scorching indictment, she states, "I was fervently grateful to her in my heart" (chap. 3). When she leaves a school where she has at last found some affection, Esther sobs uncontrollably but repeats to herself the refrain, "O, I am so thankful, I am so thankful" many times over.

It is only in dreams that Esther's true feelings find an outlet. In her waking life she has cheerfully insisted that she was "the happiest creature of all" (chap. 13), that she had "everything to be grateful for, nothing in the world to desire" (chap. 13). But when Esther falls sick, her dreams and delirium reveal quite another thing. As Esther describes it, "in falling ill, I seemed to have crossed a dark lake, and to have left all my experiences, mingled together by the great distance on the healthy shore." (chap. 35). Her mind first travels to the past — "to those summer afternoons when I went home from school with my portfolio under my arm, and my childish shadow at my side, to my godmother's house." At this point in the dream her roles in life become jumbled, and the dream images become more painful:

> At once a child, an elder girl, and the little woman I had once been so happy as, I was not only oppressed by care and difficulties adapted to each station, but by the greatest perplexity of trying to reconcile them. . . . Dare I hint at that worse time when, strung together somewhere in great black space, there was a flaming necklace, or ring, or starry circle of some kind, of which I was one of the beads! And when my only prayer was to be taken off from the rest, and when it was such inexplicable agony and misery to be part of the dreadful thing. (chap. 35)

This dream reveals several important elements of Esther's repressed self. In it Esther finally discards her guise of modesty and coyness and confronts at last her unhappy past and all the compromises in life it has forced her to make. While she had flippantly dismissed the injustice of her godmother's scorching rejection with a mere sigh, "It was not for me to muse over bygones" (chap. 6), her dream recalls the scars it left behind. No one would have understood this better than Dickens, who, in the course of tracing Esther's stifled responses to rejection, was projecting the trauma of his own early abandonment. Esther, like her creator, is drawn in her dreams to the early scenes of her rejection. As she tells it, "I am not sufficiently acquainted with such subjects to know whether it is at all remarkable that I almost always dreamed of that period of my life." (chap. 9).

Esther's dream also reveals some resentments that she has secretly harbored about her present position. In her waking life she assumed all her duties with a cheerful smile. As housekeeper, duenna, and moral guardian to an assorted array of outcasts, she has excelled in all her roles. But her dream reveals what careful readers have sensed all along—she may have been happy in none of them. Selflessly devoting her energies to healing the emotional wounds of others, she has ignored her own.

The worst part of the dream, that flaming necklace and Esther's struggle to remove herself from it, may, as some critics have suggested, symbolize broad social conflicts—society's corruption, Chancery's legal labyrinth, or her parents' sexual guilt.[45] But the roots of this dream are more private than public. What Esther finally struggles to release herself from is a stifling servitude in which her personal identity has been lost in a series of cloying nicknames—Dame Durden, Little Old Woman, Mother Hubbard. Having lived under her godmother's stinging indictment that she has been "set apart" from birth, in her waking life Esther strove to be "useful, amiable, and serviceable" (chap. 74). But this dream suggests striving of a different sort—both for an identity other than the old-maidish one assigned to her and for a form of love and attention other than the patronizing one she has come to receive.

The symbolism of this dream also suggests fears and strivings of a sexual nature. Amidst her suffering, Esther tells us, "I labored up colossal staircases, ever striving to reach the top, and ever turned, as I have seen a worm in a garden path, by some obstruction and laboring again. . . . Oh, more of those never-ending stairs . . . piled high to the sky, I think!" (chap. 35). Given the nature of Esther's romantic associations up to this point, such fears would be understandable. From the unsavory Guppy, she has received one marriage proposal, which first elicits her laughter, then her tears. And soon, as she must suspect, she will receive another proposal from her benefactor, Jarndyce, a kind, generous offer for a woman without means, but hardly a romantic match. Her response to Jarndyce's offer suggests the ambivalence Esther feels at many moments: "I was very happy, very grateful, but I cried very much." (chap. 44). Jarndyce's relationship to her has been a puzzling one. At one point she suspects he might be her father, at other times, she cares for him as a daughter, and finally, she is designated to be his wife. It is the tension of the many roles that Esther is expected to fill that finally erupts in fiery images of suffering and despair. If her shriek of despair seems too dramatic an outcry for the coy and modest Esther we have come to know, it is only because she has succeeded so well at repressing what she has felt all along.

Dickens could not sustain this kind of psychological probing, and in predictable fashion, he resolved Esther's conflicts by granting her a suitable husband and a new Bleak House safely away from all problems of the old one. Just as predictably, Esther briskly dismisses her dreams, announcing that she "was not the least unhappy in remembering them," and blandly observes: "It may be that if we knew more of such strange afflictions we might be better able to alleviate their intensity." (chap. 35).

Dickens, by observing his own dreams rather than relying upon the standard stories in the books, had come to know a great deal more about such "afflictions" than he would ever let Esther admit to. Years before dreams had been accorded the status of psychic acts, he had perceived that they were rooted in our deepest experiences and could express conflicts and wishes routinely suppressed in waking life.

No one would have been less surprised at Dickens' achievements in dream psychology than Freud. Understanding full well how truly creative writers could intuit meanings well before scientists had formulated them, he considered writers to be "valuable allies." As Freud put it: "They are apt to know a whole host of things between heaven and earth of which our philosophy has not yet let us dream. In their knowledge of the mind they are far in advance of us everyday people, for they draw upon sources which have not yet opened up for science."[46]

NOTES AND REFERENCES

1. Charles Dickens, *The Nonesuch Edition of the Letters of Charles Dickens*, ed. Walter Dexter (Bloomsbury: The Nonesuch Press, 1938), vol. II, p. 267, 1/2/51; hereafter cited as *Nonesuch*.

2. See *Catalogues of the Libraries of Charles Dickens and W.M. Thackeray*, ed. J.H. Stonehouse (London: Picadilly Fountain Press, 1935).

3. Mrs. Blair, *Dreams and Dreaming* (London: G. Groombridge, 1843), p. i.

4. Catherine Crowe, *The Night Side of Nature* (1848; rpt. Philadelphia: Henry T. Coates, 1901).

5. Robert Dale Owen, *Footfalls on the Boundary of Another World* (Philadelphia: J.B. Lippincott, 1860), p. 22.

6. Dugald Stewart, *Elements of the Philosophy of the Human Mind* (New York: Harper & Row, 1814), vol. I, p. 277.

7. Stewart,[6] vol. I, p. 284.

8. John Abercrombie, *Inquiries Concerning the Intellectual Powers and the Investigation of Truth* (Boston: Otis, Broaders, 1837), p. 203. In *Principles of Mental Physiology* (1884), W.B. Carpenter would define this process as unconscious activity, although he relied upon the same stories cited by Abercrombie to support this insight.

9. Abercrombie,[8] pp. 199, 201.

10. Most physicians of Dickens' day acknowledged that some dreams were caused by such physiological disorders. In *Sketches of the Philosophy of Apparitions* (Edinburgh: Oliver & Boyd, 1825), Samuel Hibbert proposed the theory of sanguineous derangement.

11. "On the Physiological and Psychological Phenomena of Dreams and Apparitions," *The Journal of Psychological Medicine and Mental Pathology*, vol. 9 (1851), p. 548.

12. John Elliotson, *Human Physiology* (London: Longman, Orme, Brown, Green, and Longmans, 1840); John MacNish, *The Philosophy of Sleep* (Glasgow: W.B. McPhun Publisher, 1838). Dickens frequently poked fun at the phrenologists, but became an avid practitioner of mesmerism, which Elliotson had introduced him to in 1838. For a study of Dickens' involvement with mesmerism, see Fred Kaplan's, *Dickens and Mesmerism: The Hidden Springs of Fiction* (Princeton: Princeton University Press, 1975).

13. Abercrombie,[8] p. 27, p. 33.

14. "Dreams," *Encyclopaedia Britannica*, 7th ed. (1842).

15. Thomas Laycock, "A Chapter on Some Organic Laws of Personal and Ancestral Memory" (George P. Bacon, Steam Printing Offices), p. 25.

16. Laycock,[15] p. 25.

17. Quoted in Frank Seafield, *The Literature and Curiosities of Dreams* (London: Chapman & Hall), p. 16.

18. *The Journal of Psychological Medicine and Mental Pathology*, vol. 9, p. 298.

19. See Thomas Laycock, *Mind and Brain* (London: Simpken Marshall, 1859) and W.B. Carpenter, *Principles of Mental Physiology* (New York: D. Appleton, 1884).

20. *Nonesuch,*[1] vol. II, 2/2/51.

21. See *Catalogues of the Libraries of Charles Dickens and W.M. Thackeray*. Among other titles in Dickens' library were: Catherine Crowe's *The Night Side of Nature* (1848); Samuel Hibbert's *Sketches of the Philosophy of Apparitions* (1825); R.D. Owen's *Footfalls on the Boundary of Another World* (1860); and Frank Seafield's *Literature and Curiosities of Dreams* (1865).

22. Leonard Manheim, "Dickens' Fools and Madmen," *Dickens Studies Annual*, ed. Robert Partlow, Jr. (Carbondale, Ill.: Southern Illinois University Press, 1972), vol. 2, p. 71.

23. *Nonesuch,*[1] vol. II, p. 267, 2/2/51.

24. Reprinted in "Dickens on Ghosts: An Uncollected Article," Philip Collins, *The Dickensian*, vol. 59 (1963), pp. 6–7.

25. "Doctor Marigold's Prescriptions," *All the Year Round*, December 7, 1865, p. 33.

26. *Nonesuch*,[1] vol. II, pp. 267–1270, 2/2/51. Any further references to Dickens' dream theories, unless otherwise indicated, can be assumed to be from this source.
27. Warrington Winters, "Dickens and the Psychology of Dreams," *PMLA*, vol. 62 (1946), pp. 984–1006.
28. "Dreams," *Household Words*, March 8, 1851, p. 568.
29. See, for example, Abercrombie,[8] p. 216.
30. John Forster, *The Life of Charles Dickens* (Philadelphia: J.B. Lippincott, 1872), vol I, p. 69. All further references to Dickens' autobiographical fragment are taken from this volume, pp. 47–70.
31. *The Pilgrim Edition of the Letters of Charles Dickens*, ed. Madeline House and Graham Storey (Oxford: The Clarendon Press, 1965–74), vol. I, p. 257, 5/8/1837; hereafter cited as *Pilgrim*.
32. Edgar Johnson, *Charles Dickens: His Tragedy and Triumph* (New York: Simon and Schuster, 1952), vol. I, p. 196.
33. *Pilgrim*,[31] vol. III, p. 484, 5/8/43.
34. By asserting that recent, not past events assumed symbolic forms in dreams, Dickens, however, reversed Freud's principle.
35. "Lying Awake," *Household Words*, October 30, 1852.
36. The literary dreams of Emily and Charlotte Brontë are notable exceptions. The dream passages in *Wuthering Heights*, *Jane Eyre*, and *Villette*, some of which have received critical attention, are impressive psychological portraits.
37. *The George Eliot Letters*, ed. Gordon S. Haight (New Haven: Yale University Press, 1954), vol. II, p. 308, 3/11/57. The work to which Blackwood was referring here was Eliot's "Mr. Gilfil's Love Story" in *Scenes of Clerical Life*.
38. The dreams of Montague Tigg and Jonas Chuzzlewit in *Martin Chuzzlewit* are two exceptions to this rule. For an analysis of these dreams, see Joseph Brongunier, "The Dreams of Montague Tigg and Jonas Chuzzlewit," *The Dickensian*, vol. 58 (1962), pp. 164–170.
39. For an illuminating analysis of this dream, see Warrington Winters, "Dickens's Hard Times: The Lost Childhood," *Dickens Studies Annual*, ed. Robert Partlow, Jr. (Carbondale, Ill.: Southern Illinois University Press, 1972), vol. 2, pp. 217–36.
40. Charles Dickens, *Dombey and Son* (London: Oxford University Press, 1950), p. 3, chap. 3 (*The Oxford Illustrated Dickens*). Subsequent citations are to this edition.
41. For a critical discussion of Florence's redemptive love in this novel, see Julian Moynihan, "Dealings with the Firm of Dombey and Son: Firmness versus Wetness," in *Dickens and the Twentieth Century*, ed. John Gross and Gabriel Pearson (Toronto: University of Toronto Press, 1962), pp. 121–311.
42. Forster,[30] vol. I, p. 66.
43. Charles Dickens, *Bleak House* (New York: Oxford University Press, 1948), pp. 17–18, chap. 3 (*The Oxford Illustrated Dickens*). Subsequent citations are to this edition.
44. For a sampling of such criticism, see Alex Zwerdling, "Esther Summerson Rehabilitated, *PMLA*, vol. 88 (1973), pp. 429–1439; Crawford Killian, "In Defense of Esther Summerson," *Dalhousie Review*, vol. 54 (1974), pp. 318–328; William Axton, "The Trouble with Esther," *Modern Language Quarterly*, vol. 26 (1965), pp. 545–557; Martha Russo, "Dickens and Esther," *The Dickensian*, vol. 65 (1969), pp. 90–94.
45. See, for example, Richard Dunn, "Esther's Role in *Bleak House*," *The Dickensian*, vol. 62 (1966), pp. 163–966; Constance McCoy, "Another Interpretation of Esther's Dream," *The Dickensian*, vol. 62 (1967), p. 181.
46. Sigmund Freud, *The Standard Edition of the Complete Psychological Works of Sigmund Freud*, vol. 9, trans. James Strachey (London: The Hogarth Press, 1959), p. 8.

Ruskin and Tyndall: The Poetry of Matter and The Poetry of Spirit

PAUL L. SAWYER

Department of English
Cornell University
Ithaca, New York 14853

> We can generate, in air, artificial skies.
> — JOHN TYNDALL
> Ah, masters of modern science, give me back my Athena out of
> your vials. . . .
> — JOHN RUSKIN

I N 1873, John Ruskin entered a long-standing scientific controversy by writing a violent attack on John Tyndall's theory of glacial movement. This encounter, one of the last between a literary figure and a scientist in the Victorian period, has never been closely examined, no doubt because of Ruskin's quirkiness; the assault seems hysterical and unmotivated, and Tyndall never deigned to reply. Yet I would like to suggest that few incidents in Victorian literature are so rich in implication. In the first place, Ruskin's treatment of the glacier question belongs to his general critique of scientific materialism — a phase in what one commentator has called the "most comprehensive, if not the most orderly treatment by a Victorian man of letters of the place of science in modern culture."[1] In the second place, both men were alike in having attempted to mediate between science and poetry: Tyndall, for example, wrote popular evocations of the Alpine sublime, while Ruskin published specialized papers in geology. Ruskin's attack is rich in irony, therefore, since his true relationship to Tyndall is not one of simple antagonism, but (with important exceptions) an antagonism of resemblances. These resemblances are important because they illuminate a crucial intersection in Victorian culture: the intersection of Romantic tradition with the triumph of scientific naturalism.

217

In the following pages, I have used the glacier controversy as the starting point for a comparison of two important Victorian philosophers of nature. After briefly discussing the substance of Ruskin's attack, I will move to those elements of Tyndall's character and career on which the attack bears, and then to a consideration of how both men sought to bring science into harmony with religion and poetry. Throughout the essay, I have tried to keep two broad themes in view: first, the struggle for a language that can reconcile scientific detachment with the direct, sensuous apprehension — in a word, the livingness — of nature; and second, the debate over the moral implications of scientific investigation.

[I]

In the middle years of the nineteenth century, the movement of glaciers was a question of great interest to both geologists and physicists. How could a brittle substance like ice appear to flow like a river? Since the freezing point of ice is lowered under pressure, might not the new science of thermodynamics contribute to the understanding of glacial movement? In 1841 James David Forbes of Edinburgh visited the Swiss scientist Louis Agassiz (later the well-known opponent of Darwin) in a year-round hut built by Agassiz on the Aletsch glacier. Both men were intrigued by the transverse fissures in glacial ice, which Forbes decided were caused by stresses caused by different velocities of flow. In his influential book *Travels in the Alps* (1843), he presented detailed measurements suggesting that glaciers, like rivers, move at different rates — more sluggishly at the edges and rapidly in the center. A glacier, then, according to this thesis, is "an imperfect fluid, or a viscous body, which is urged down slopes of a certain inclination by the mutual pressure of its parts."[2] Forbes's critics accepted his data, but charged that he had given no real explanation; even his use of the crucial word "viscous" was vague.[3] Thirteen years later, Tyndall, who had experimented with substances that develop fissures under pressure, read Forbes on the suggestion of his friend T.H. Huxley. "For some reason," according to his biographers, Tyndall "disapproved of the viscous theory of Forbes and fought it tooth and nail."[4] After an Alpine excursion with Huxley in 1856, Tyndall presented his alternative theory in several memoes and then in his first book, *The Glaciers of the Alps* (1860), a blend of travel memoir and popularized exposition. Among other opinions, he held that glaciers do not conform to their beds, as rivers do, but melt or fracture under pressure and then freeze again at the point of contact — a phenomenon called "regelation" by his great mentor, Faraday.

Tyndall's theory ignited a bitter dispute over the question of Forbes's personal honor. In 1853, the year he was elected to the Royal Society, Tyndall delivered a detailed refutation of a paper by Forbes; afterwards, he noted in his journal that "they all seemed amused at the manner in which I have 'demolished Forbes' as they express it."[5] In 1855 he criticized a second paper by Forbes. Whether or not his interest in glaciers was heightened by the wish to demolish Forbes again, Tyndall nevertheless cited the work of an earlier researcher, Canon Rendu (later Bishop of Annecy), to suggest that Forbes had suppressed or ignored evidence provided by Rendu. Rendu had indeed compared ice to a soft paste, as Forbes had acknowledged, but in fact he provided no rigorous argument or demonstration; most of his information came from Alpine guides (Life, p. 85). Forbes gave a point-by-point response to Tyndall, and the argument raged throughout the late 1850s. In the midst of this, Forbes was considered for the Copley Medal of the Royal Institutition, but then passed by (it is now known that Huxley wrote a confidential letter to one of the judges questioning Forbes's treatment of Rendu), which encouraged him to think the Londoners were persecuting him. And, indeed, the glacier controversy was carried out entirely by Londoners and Scotsmen. When Tyndall restated his charges in The Forms of Water (1873), the warfare flared up again, with increased virulence. Forbes was now dead, but his associates answered Tyndall in a biography of Forbes; Tyndall answered the biography with another pamphlet, then carried on a hostile exchange of letters in Nature that was finally canceled by a weary editor; and so on. When he learned that Ruskin's "diatribe" was being circulated at Belfast, Tyndall asked Huxley if he shouldn't publish a psychological portrait of Forbes. Huxley advised him to refrain.[6]

Ruskin's "diatribe" in letter 34 of Fors Clavigera (1873) added little to the debate except overstatements: Forbes, he claims, "solved the glacier problem forever," while Tyndall, trying to usurp Forbes's place, is guilty of "absurdity," "iniquity," "willful avoidance," and so forth. Ruskin's aim in exposing him is to show his readers "the degree in which general science is corrupted and retarded by these jealousies of the schools; and how important it is to the cause of all true education, that the criminal indulgence of them should be chastised."[7] Tyndall deeply resented the attack, but was advised by Huxley and Carlyle to ignore it; all the same, Ruskin returned to the battle again, most importantly in Deucalion, a serial publication begun in 1875, in which he attempted to set in order all his contributions, suggestions, and memoirs concerning geological questions. This remarkable and almost totally unread book contains a mass of fragments stitched together without obvious connections, consisting of catalogues, queries, lectures, polemical digressions, and landscape meditations of great beauty and force. Into these chaotic jottings the glacier question intrudes like an obsession that cannot be exorcized.

Part of the extended attack in *Deucalion* is scientific — after a fashion. For example, Ruskin's evidence in favor of Forbes comes from experiments in compressing pie crust and pouring treacle, an apparent attempt at overthrowing Tyndall's own experiments with wax, dough, and bismuth. (Ruskin rightly claimed that Tyndall's experiments do not contradict Forbes.) Of these "contributions" John D. Rosenberg has said they are "not science but play,"[8] and indeed their effect seems to mock the very experimental method they imitate, as though Ruskin were impudently reintroducing sensuous description into a field of inquiry that had become increasingly quantitative and abstract. But the greater part of the attack on Tyndall is once again moral, consisting most simply in a near deification of Forbes and a total vilification of Tyndall. Forbes, Ruskin claims, had been his "master" in geology ever since he read *Travels in the Alps*, and he treats readers of *Deucalion* to a hazily nostalgic reminiscence of a chance meeting with Forbes — the only one in his life — during an Alpine tour in 1844. Ruskin entitles this reminiscence "Thirty Years Past," an adaptation of the subtitle of Scott's *Waverley;* thus, in his mind, Forbes is ranged with all that is noble and beautiful about the past as evoked in landscape — the Highlands of Scott, the Alps of the old guides and naturalist-explorers, and the summers of Ruskin's youth. By contrast, Tyndall is a monument to the follies of an impious age: his sloppy drawings, imprecise language, and dishonesty in debate arise together from a "scum of vanity"; he is "unsteadied by conceit and paralyzed by envy"; he and his associates have published "such a quantity of accumulated rubbish of past dejection, and moraine of finely triturated mistake, clogging together gigantic heaped blocks of far-traveled blunder, — as it takes away one's breath to approach the shadow of" (XXVI, p. 228).[9]

These geological errors, however, are only particular instances of the moral condition that Ruskin associates with "materialist science," for which Tyndall was a leading spokesman. Ruskin's earlier references to him set Tyndall in opposition to the livingness of nature, but they are not directly hostile. Thus, in *The Ethics of the Dust* (1866), the children learn that Professor Tyndall has defined life as heat, a mere mode of motion (XVIII, p. 238). (The reference is to Tyndall's popular exposition of conservation of energy, called *Heat, a Mode of Motion*.) In the Preface to

The Queen of the Air (1869), Ruskin relates Tyndall's discoveries about firmamental blue to the mythical "meaning" of Athena's blue eyes, although there is irony in his citing of Tyndall's claim that he can create "a bit of sky more perfect than the sky itself" — an irony that turns into rebuke a few pages later (XIX, p. 228; see epigraph). But in the later works, materialism becomes essentially the cold rational power feared and hated by the Romantics. In *Deucalion*, "materialist" philosophers in general seem interested in horrific discoveries — the organs of vivisected animals, prehistoric beasts, gigantic primordial conflagrations of heat and gas. Ruskin's most considered critique, in a paper of 1883, claims that the

> lower conditions of intellect which are concerned in the pursuit of natural science . . . are . . . dependent for their perfection on the lower feelings of admiration and affection which can be attached to material things. . . but they differ from the imaginative powers in that they are concerned with things seen — not with the evidences of things unseen — and it would be well for them if the understanding of this restriction prevented them in the present day as severely from speculation as it does from devotion. (XXVI, p. 338)

By "speculation" Ruskin means enquiries like those into the age of the earth or the descent of man which, along with a drive toward minute complexities of research and an unconcern for public enlightenment, have made of science not the human endeavor it once was, but a thing reserved for the few and the proud (XXVI, pp. 340–342).[10] This description contrasts with the statement in *Modern Painters IV* that in the old days De Saussure, the great Alpine explorer, had gone to the Alps "as I desired to go myself, only to *look* at them, and describe them as they were, loving them heartily — loving them, the positive Alps, more than himself, or than science, or than any theories of science" (VI, p. 476). Although he did not always hold this view, Ruskin claims in *Deucalion* that true science, like art and social theory, must have social and spiritual value. For the older naturalists like Linnaeus or von Humboldt, "natural religion was always a part of natural science," and in the fruits of their discoveries one knew them. This is the crucial distinction between the great ones and the "wild theories or foul curiosities" of the moderns, whose discoveries have "almost without exception, provoked new furies of avarice, and new tyrannies of individual interest; or else have directly contributed to the means of violent and sudden destruction, already incalculably too potent in the hands of the idle and the wicked" (XXVI, p. 339). The analogy with Ruskin's earlier attack on economic science will be apparent.

However blind, even desperate, these sallies may seem, they are nevertheless part of a consistent interpretation of his own times that dominates Ruskin's work almost from the beginning and receives its fullest expression in *The Stones of Venice* (1851, 1853) and the later volumes of *Modern Painters* (1856, 1860). That interpretation rests on the central thesis that modern men have willfully alienated themselves from the sources of being through a kind of voracious asceticism, substituting in place of natural piety the specters of their own insatiable egos. In this critique science, by becoming "materialist," severs rather than reaffirms the human connection with nature, leaving only the strength to pollute, control, and destroy. But what does all this have to do with glaciers? We must still ask why Ruskin found the Forbes-Tyndall debate so crucially important to his larger aims.

In the first place, and most obviously, the debate revealed a weak spot in the opposition, the opposition being all the theories Ruskin lumped together as "materialism." The scientists' violent rhetoric and tireless squabbling over matters of priority and reputation (of which Ruskin's side was as guilty as Tyndall's) must have proved to Ruskin that the theories he hated were products of a spiritual defect: Tyndall, the self-proclaimed materialist, was clearly and openly exploiting the Alps for his own professional vanity. However much anguish evolutionism may have caused him, Ruskin never made the ignorant error of Wilberforce (Ruskin respected Darwin both as a person and as a scientist); instead, he proposed in *Proserpina* and parts of *Deucalion* complementary ways of thinking about botany and geology without directly refuting the new theories. But Tyndall on glaciers was eminently refutable (to Ruskin's mind); the attack on him, therefore, may be read in part as an ultimate, indirect response to Lyell's "dreadful hammers," with the rage directed at a lesser scientist and a weaker theory. A second reason, which I have touched on already, is the personal significance for Ruskin of the challenge to Forbes. In order to take up the glacier question in *Fors Clavigera*, Ruskin had to break off writing a series of reflections on the life of Scott, but the connection between Scott and Forbes is one Ruskin never got around to explaining to his readers: in fact, Forbes was the son of Scott's first love (XXVII, p. 614n). By defending Forbes, Ruskin puts himself, as it were, in apostolic succession to his favorite boyhood author and, by extension, to the land of his fathers; the defense of Forbes is a defense of the fathers against impious sons, a situation recalling the *Blackwood's* attack on Turner and Ruskin's famous response (*Travels in the Alps* and *Modern Painters I* were published in the same year, 1843).

Moreover, by juxtaposing Forbes and the past with Tyndall and the present, Ruskin creates an historical allegory of the decay of science, doing for the Alps what he had done for Venice (*Deucalion* might have been subtitled *The Stones of Switzerland*).

In this historical myth Tyndall and the glaciers become emblems of things larger than themselves. Tyndall and his like are ranged with the forces of evil, but they are less like the sinister wizards of Romantic mythology than the stock pedants of Alexander Pope's *Dunciad* and the satirical tradition in general. Feverish and babbling, these men of science quarrel among themselves for priority, climb the Alps for conquest, dissimulate in their pamphlets, and remain blind to all the essential facts about nature, particularly its sanctity. They are failures, finally, even as scientists. Shrunk by the lens of Ruskin's irony, they seem climbers, tourists, or mere motes of desecration:

> little jumping black things, who appear, under the photographic microscope, active on the ice-waves, or even inside of them; — giving to most of the great view of the Alps . . . a more or less animatedly punctuate and pulicarious character. (XXVI, p. 227)

And so, in his characteristic fashion, Ruskin converts a private response into a dramatic situation emblematic of the spiritual life of his time. Deucalion, the Greek Noah, threw over his back the "bones of his mother" (that is, the stones of the earth), which landed and created a new race of men (XXVI, pp. xlvi–xlvii); but in Ruskin's text, blasphemous pedants parody Deucalion by turning the living world lifeless, in anticipation of the Great Day foreseen by "modern scientific prophets"—a day, Ruskin says in his introduction, "when even these lower voices [of nature] shall be also silent; and leaf cease to wave, and stream to murmur, in the grasp of an eternal cold" (XXVI, p. 99). Thus, although the glacier question fails in any rational sense to bear the significance Ruskin tries to give it, he is nevertheless able by metaphorical suggestion to present Tyndall's attack on Forbes as an attack on nature herself, and to turn the materialist vision into a kind of universal glacier—the condition of the lowest circle in Dante's hell and of the apocalypse hinted at by Robert Frost: "I think I know enough of hate/To say that for destruction ice/Is also great".

Having viewed Ruskin's case in its broadest terms, both as statement and as symbol, we must return to the focal point of the original controversy. Tyndall is for Ruskin both the stock pedant and the archetype of the new atheist philosopher, but he was also for Ruskin a distinct personality, the object, first and last, of a personal attack. Yet the two men

had not been on bad terms before 1873. Both were members of the
Metaphysical Society and attended many of the same meetings — even,
remarkably enough, after the attacks in *Deucalion*.[11] In 1869 Ruskin sent
Tyndall a copy of *The Queen of the Air*, along with a nervously con-
ciliatory letter (*Life*, p. 138). But by at least the 1880s, he had settled into
a fixed hostility, as is suggested by a friend's published memoir of a conversation in which Ruskin commented on Tyndall:

> I admire his splendid courage (I am a dreadful physical coward myself . . .),
> and his schoolboy love of adventure . . . naive to a degree — incurably so;
> but he has never felt himself to be a sinner against science in the least be-
> cause of his all-overwhelming vanity. His conduct to James Forbes . . . was
> the outcome of the schoolboy feeling when he sees the Alps for the first
> time: "Good gracious! no one ever saw this before; and I can tell the world
> all about it as no one ever did before!". . . . But before long people will find
> that this theory was all decided before this conceited, careless schoolboy
> was born. And that is why I always attack him, and shall continue to do so
> until I die. (XXVI, pp. xxxix–xl n)

Clearly, Ruskin's attitude to Tyndall (which contains, one suspects, a
strong element of projection) is too complex to speculate about here; but
the subject is important since the whole argument of *Deucalion* presupposes a psychology of the scientist. In turning now to Tyndall's career, I
will try to describe him both as a thinker and as a man.

[II]

John Tyndall (1820–1893) is remembered by science for a number of
contributions, of which perhaps two stand out: first, he demonstrated
the power of vapor to radiate heat and scatter particles of light (he suggested that the "firmamental blue" is actually the scattering of light by
droplets of water vapor); second, he disproved the hypothesis of spontaneous generation by showing experimentally that living particles in the
air do not generate themselves *ex nihilo*. These along with researches into
the magnetic properties of crystalline planes and the behavior of sound
waves show that many of his interests duplicate many of Ruskin's: the
structure of crystals, the origins of life, the purity of air, the blueness of
the sky, the whiteness of clouds, the radiance of the sun. Ruskin insisted
that, like the work of all good scientists, his own geology had a social
aim — the conservation of Alpine streams and farmlands; Tyndall's work
led to the beginnings of modern epidemic control and, as a result of his
measurements of sound waves, to the perfection of fog signals. Like
Ruskin, his love of nature was too many-sided to be satisfied by

research; he traveled, hiked and wrote books recording his experiences for a wide audience. Lecturer, popularizer, benefactor, researcher, word-painter — in more ways than anyone but Huxley, Tyndall was the "compleat" Victorian man of science.

He was born into a Protestant family in county Carlow, the son of a sergeant of police whom he venerated and who, in Mrs. Tyndall's words, taught his son "to love a life of manly independence" (*Life*, p. 2). The few existing anecdotes from his boyhood show an extraordinary physical drive that flourished both in discipline and in play; the schoolboy who often stopped on the way home "just for the love of fighting" could also determine, when he was at the National School, to rise before dawn and work "with a coffee bean in his mouth to keep him from feeling sleepy" (*Life*, pp. 3, 4). Even grammar, as he recalled, could be "a discipline of the highest value, and a source of unflagging delight. How I rejoiced when I found a great author tripping. . ." (*Life*, p. 4). Here "discipline" and "delight" come together tellingly, and the combat with great authors (the example he uses, interestingly, is Biblical) seems less irreverence than a young man's testing-out his mental muscles (his favorite activity, one might add, for the rest of his life). But Tyndall's thinking was "physical" in another way, since for him knowledge passed readily into action — a fact nicely caught in the anecdote that as a student he worked out the solutions of geometry problems upon the snow with his teacher (*Life*, p. 4). Later, at the same time that young Ruskin was moving by carriage through Switzerland, young Tyndall put his mathematics to use in the Irish land survey, literally by working in the mud of the shoals. In 1842, when unemployed laborers were starving and demonstrating throughout the industrial Midlands, Tyndall was transferred to Preston ("I could pass very quietly through 10,000 Chartists and be taken for a brother," he wrote his father), and was soon afterwards deeply impressed by the compassion and "manly radicalism" of Carlyle's *Past and Present*. At almost the same time, T.H. Huxley was avidly reading Carlyle's *Sartor Resartus* on board the *Rattlesnake*. James G. Paradis has shown that for Huxley, "the idea of the scientist had strong proletarian roots".[12] The same is obviously true for Tyndall, which helps to explain not only why Tyndall became a devoted follower of Carlyle for the rest of his life, but also why reading Carlyle helped him to decide on a scientific career. We will return to this point shortly.

After several years in England, Tyndall saved enough money to attend the University of Marburg to study physics and chemistry, where, through grueling labor and force of will, he managed to finish the Ph.D.

in two years instead of the usual three. The months after his return to England, when he wrote his first scientific papers in considerable poverty, were also very difficult; yet by 1853, a decade after beginning the land survey in Ireland, Tyndall had become a Fellow of the Royal Society and Professor of Natural Philosophy at the Royal Institution. "My health during the last half year was exceedingly feeble," he wrote at this time: "a perpetual headache and a certain indolent helplessness as if I were going to crumble away altogether" (*Life*, p. 41). Whether or not this feebleness was Tyndall's version of the Victorian malaise — his price for following the anti-self-consciousness philosophy of Carlyle — the pattern of enormous expenditures of energy accompanied by insomnia, headaches, and dyspepsia continued the rest of his life.

Tyndall earned his professorship essentially with a single lecture, the circumstances of which give another important insight into his motivations. "I consented," he wrote later, "not without fear and trembling. For the Royal Institution was to me a kind of dragon's den, where tact and strength would be necessary to save me from destruction." The lecture proposed a view of the magnetic force opposed to that of Faraday, yet it did not produce in the great man "either anger or enmity. At the conclusion of the lecture, he acquitted his accustomed seat, crossed the theatre to the corner into which I had shrunk, shook me by the hand, and brought me back to the table" (*Life*, p. 40). It was not the last time Tyndall opposed Faraday's views, yet he remained a devoted disciple and friend of the older man, writing in later years a memoir, *Faraday as a Discoverer* (in which the preceding account appears). The incident interestingly parallels Tyndall's first meeting with Carlyle. At a dinner, Carlyle mistook him for a homeopathist and began arguing vigorously. In a letter Tyndall wrote, "What I have seen thus far of Carlyle makes me revere the old brute more than ever. Nobody seems to agree with him, but he pushes his way through all dealing out doom and praise with unhesitating faith." At the end of their meeting he concluded that, "I know I am nearer to the man than I was this morning. I told you I would reach him, and I am doing so, though in a crooked way" (*Life*, pp. 74–75). This also became a powerful friendship, ending only at Carlyle's grave in Ecclefechan; for like Ruskin, Tyndall, as should be clear by now, was a hero-worshipper. Yet two different modes can hardly be imagined: one has only to think of Ruskin's kneeling at Carlyle's coach and hovering as the "old lion" reclined and held forth. And unlike Ruskin, who suppressed his dissent yet remained a radical in all things, Tyndall seemed to spar with his heroes in order to reaffirm his devotion. One suspects that for him the masculine world of science was a place where one could lock horns with the great in a kind of continual rite of passage and so prove oneself worthy of their love; but more important, of course, was the disciplined expenditure of energy that was scientific research itself—a discipline not unlike that of language for the schoolboy who sparred with Milton's verbal constructions.

In 1856 Tyndall's first Alpine tour with Huxley opened up a new locus of activity. Eventually he became as famous for his mountaineering as for his science: he was the first to climb the Weisshorn and one of the first to scale the Matterhorn and other difficult peaks. (The monument erected above the Aletsch glacier, which Ruskin fortunately never saw, honors this part of his career.) Once again, it is hard to imagine two stronger contrasts than Ruskin and Tyndall as Alpine explorers. Ruskin, whose model was the older naturalists, contented himself with short hikes with a guide, during which he collected rocks and crystals, ferns and flowers; his eye was keen and sensuous, his emotional response essentially contemplative, his point of "contact" the skilled hand reproducing images in the sketchbook. But for Tyndall the discipline was athletic. It may be fanciful to see his kind of science as essentially assaultive—a kind of combative questioning or loving conquest by measurement of the great primary forces—but it is undeniable that his zest for physical excitement often blended with his investigations, as when he stood waist-deep as a youth in the mud of the Irish Sea. On one occasion, he took advantage of eleven days between terms at London to measure the movement of the Mer de Glace in winter; two rows of stakes were driven into the ice 150 feet apart, the second in a blinding snowstorm (*Life*, p. 83). Eventually, Tyndall became notorious for taking risks: he crossed glaciers alone, for example, where many had fallen and perished in crevasses, and on one occasion he made an *escalier* up the side of a glacier despite the warnings of his guide:

> The guide's attention had been divided between his work and his safety, and he had to retreat more than a dozen times from the falling boulders and debris. I, on the other hand . . . I took my axe, placed a stake and an auger against my heart, buttoned my coat upon them, and cut an oblique staircase up the wall of ice, until I reached a height of forty feet from the bottom. Here . . . I pierced the ice with the auger, drove in the stake, and descended without injury. During the whole operation, however, my guide growled audibly.[13]

This passage appears in Tyndall's first book, where Ruskin later singled it out for his contempt (XXVI, pp. 142–143). In fact, *Glaciers of the Alps*,

which combines a popular account of Tyndall's theories with landscape
description and many anecdotes about climbing, is full of the first person
in an almost naïve vanity that irritated other readers besides Ruskin. Yet
although Tyndall obviously climbed for glory, it was the Alps that provided solace and renewed strengths drained by his exhausting labors in
London. As Huxley recalled, "When things got bad with him, his one
remedy was to rush off to the nearest hills and walk himself into
quietude."[14]

Thus, almost everywhere one touches Tyndall, one finds a love of
energy almost for energy's sake and a closely related instinct for
showmanship. In this regard his character developed with singular consistency: the student showing off in Marburg is not far from the eminent
scientist braving danger on the glaciers. Hard work did not make him a
drudge, and success did not make him idle. This is perhaps because his
need for self-punishing labors shaded over into a love of sportive combat
that gives more strength than it takes; the Calvinist, that is, shades over
into the Alpinist, spiritual discipline becomes physical skill, and the mor-
tification of the flesh becomes the exuberance of animal spirit. In so
many ways Tyndall is a Victorian, often to a platitudinous degree. To a
friend who thought that peace and blessedness lay in "equilibrium," Tyndall replied that life consists in the passage *toward* equilibrium: "The
passage often involves a fight. Every natural growth is more or less of a
struggle with other growths, in which the fittest survive. In times of strife
and commotion we may long for peace; but knowledge and progress are
the fruits of action."[15]

Tyndall's love of energy and of the dramatic is inseparable from his
scientific interests. As a popularizer of science he offered his own version
of what James Paradis has called the "Huxley theater," lecturing with skill
and flair to audiences of the educated and the laboring classes; but if
anything, his subject was more intrinsically dramatic than Huxley's. In
one lecture-demonstration, he placed an opaque object in front of a light-
source, then put a cigar in his mouth and passed in front of the object;
the cigar burst spontaneously into flame—showing that non-visible
radiation can pass through a barrier impervious to visible light. In
another lecture he created an "artificial sky" by passing a beam of light
through a vessel containing sulphurous acid: the gas was at first invisible, then turned an intense sky-blue, then white, then an imperfect,
cloudy white. These are spectacular examples, yet they illustrate a concern that unifies nearly all Tyndall's experimental work, which is his interest in the bridge between the sensible and the insensible—the

manifestations in sound, light, and motion of forces acting on molecules. He describes this bridge in the lecture, "The Scientific Use of the Imagination": "that composite and creative power, in which imagination and reason are united," leads us "into a world not less real than that of the senses, and of which the world of sense itself is the suggestion and, to a great extent, the outcome."[16] In his public efforts at returning wonder to the province of science, Tyndall becomes the reverse of the stereotypical Newtonian who analyzes the rainbow into dry formulations; he becomes, that is, a wonder-worker, an alchemist of light who recombines the colors of the original Creation. "We can generate, in air, artificial skies," he tells the same audience. "By a continuous process of growth, moreover, we are able to connect sky-matter, if I may use the term, with molecular matter on the one side, and with molar matter, or matter in sensible masses, on the other."[17] His showmanship is finally inseparable from his love of natural beauty, his desire to manifest it to others, and even his very conception of the physical world — and one might say the same of Ruskin.

Tyndall's basic interest in bridging the sensible and the insensible reminds us that the science of his time still owed much to Idealism. In an address to students at University College, London, Tyndall said that theologians often describe the world as a carefully-built clock; but, he continues, if the Lord called His creation "good," He surely meant to include "the wheels and pinions, the springs and jewelled pivots of the works within — . . . those qualities and powers, in short, which enable the watch to perform its work as a keeper of time." Some religious persons know only the surface; yet it is "the inner works of the universe which science reverently uncovers. . ." The strategy is clever. The Romantic critique held that science is irreverent and superficial; but Tyndall connects the ultimate reduction of materialist physics with a Romantic depth metaphor. Physics, then, is profound and the final solution to which it tends will be "more one of spiritual insight than of actual observation."[18] "Spiritual" here is sleight-of-hand, since the primary meaning in this context is not "pertaining to spirit or soul" but "intellectual." Yet Tyndall's interest in religion and idealism is more than rhetorical — so much so that one cannot understand his career or his thinking without some account of it.

The most valuable part of his early schooling, Tyndall once said, was learning the arguments for "Popery and Protestantism" (*Life*, p. 4). The mark of Calvinism on Huxley's thought is well known;[19] with Tyndall, however, the mark is not so much a habit of thought or a doctrinal residue as a habit of feeling — the combination, once again, of a love of discipline with a love of energy. Huxley put his finger on the matter in a witty comparison in a review of *The Glaciers of the Alps:*

> A popular and influential school of modern theologians requires muscularity as well as meekness in candidates for the Kingdom of Heaven; and in science the same manly and vigorous spirit has evoked that sect of muscular philosophers whose best-known church is the Alpine Club, and whose mightiest evangel up to the present time is assuredly the work before us.[20]

Once, after watching some Methodists passing out tracts, Tyndall wrote to a friend that "there was something at the heart of those Methodist fanatics which I lacked, and which I longed for. . . . My power over science is, I think, increasing greatly, but the power and vigour of what I call my soul is not commensurate." By dwelling on this subject, he hoped to "clear away the moss which chokes those inner fountains, and allow the abundant spirit to gush forth once more" (*Life*, p. 70). That "abundant spirit" is clearly what he felt in Carlyle as well as in his Protestant ancestor, Tyndale the martyr, with whom he identified strongly. Earlier, in his journal, he contrasted the demands of intellect — his "scientific haste" — with the religious devotion of still another strong figure, Faraday: "He drinks from a fount on Sunday which refreshes his soul for a week. I think I will try and do the same according to my own methods; for I believe the same source of power is substantially open to me and to Faraday, although we approach it by different routes. I think that what in the New Testament is called 'faith' merits more attention from me than it has received" (*Life*, p. 36).

That source, as we know, he partly recovered through "Alpine" Christianity; but elsewhere he speaks of science not in opposition to faith but in analogy with it. "I often think," he wrote to Faraday, "that the qualities which go to constitute a good christian [sic] are those essential to a man of science, and that above all things it is necessary to become 'as a little child.' But how apt is a man to forget this docile spirit, how apt to rise disaffected and unhappy from his tasks . . ." (*Life*, p. 53). His friend Hirst recorded that Tyndall told him "how, after immense and arduous toil, he succeeded in making himself a fit recipient of the truths which dawned on him without effort as reward . . . . Self-chastening is the condition of inspiration in religious matters, as well as in science; and thus there is a discipline in science of immense value . . ." (*Life*, p. 86). Seven years later, in a lecture on miracles, Tyndall spoke of the discipline of discerning natural laws: "The material universe is the complement of intellect; and without the study of its laws, reason could never have awakened to the higher forms of self-consciousness at all. It is the Non-Ego through and by which the Ego is endowed with self-discernment." Then, by a reversal that would have enraged Ruskin, he suggests that a belief in miracles has encouraged the worst forms of religious fanaticism: witch-hunters, for example, could burn their victims with impunity because "they could not put Nature into the witness-box, and question her — of her voiceless 'testimony' they knew nothing."[21]

This last image is, of course, Carlylean, except for the implication that it is the man of science who can best secure nature's "voiceless 'testimony.' " Recent scholarship has shown that despite his antiscientific tone, Carlyle's writings profoundly influenced a generation of scientific naturalists in his conceptions of the unity of nature, of a Calvinistic fixity of natural law, of the primacy of "Fact" and effort, of an "aristocracy of talent" capable of using knowledge for social purposes.[22] Huxley is the best known of these younger men, but on Tyndall the influence was deepest. In *Past and Present* Carlyle wrote, "Equitable Nature herself, who carries her mathematics and architectonics not on the face of her, but deep in the hidden heart of her, — Nature herself is but partially for him; will be wholly against him, if he constrains her not!"[23] This image of salutary struggle describes not only Tyndall's conception of science but also his code of conduct. In a lecture to students, he recalled that "through three long cold German winters Carlyle placed me in my tub, even when ice was on its surface, at five o'clock every morning — not slavishly, but cheerfully. . . ." Carlyle, Emerson, and Fichte, he continues, "told me what I ought to do in a way that caused me to do it, and all my consequent intellectual action is to be traced to this purely moral source. . . . These three unscientific men made me a practical scientific worker."[24] That is, he was himself both an idealist and a practical worker, like the Carlylean hero. Huxley was also enamoured with Carlylean heroism, but unlike Huxley, who was a more complicated man in every way, Tyndall remained a loyal Carlylean to the end (the two scientists ended up on opposite sides in the Governor Eyre controversy).[25] There is, of course, a paradox in this loyalty since mountaineering would seem to consort ill with the stern "deep-heartedness" of Carlyle's heroes. Tyndall is the case, rather, of a man who keeps open both his Carlyle and his Byron; as Calvinism shades over into muscular Christianity, the idea of heroism shades over into heroics, and the Alpine sublime takes the place of moral grandeur. Perhaps Tyndall's philosophy of life can best be summed up as a grafting of Carlyle's gospel onto Burke's theory of the sublime.

Tyndall's cheerful Byronism brings us back to Ruskin and his caricature of Tyndall as an impious schoolboy. What, if any, of Ruskin's charges survive after this brief look at Tyndall's thought and career? In many of them there is a grain of truth — one thinks of Tyndall's combativeness, his breakneck capers, his love of fame, and especially the naïve vanity that made possible comments like the following, describing one of his own lectures: "It is a comfort to me to feel that I did not disappoint my friends. Some, however, in shaking hands with me, said I was two centuries in advance of my time, and I replied that Galileo was the same" (*Life*, p. 51).[26] But on the whole, Ruskin was not only unfair, but profoundly mistaken. To balance, and to summarize, we may compare the best description of Tyndall, written by one who knew him far better than Ruskin did:

> Impulsive vehemence was associated with a singular power of self-control and a deep-seated reserve, not easily penetrated. Free-handed generosity lay side by side with much tenacity of insistence on any right, small or great; intense self-respect and a somewhat stern independence, with a sympathetic geniality of manner, especially towards children . . . a singularly clear and hard-headed reasoner, over-scrupulous, if that may be, about keeping within the strictest limits of logical demonstration; and sincere to the core.

Throughout his eulogy, Huxley stresses above all Tyndall's sincerity, veracity, and unfailing loyalty.[27] Of course, if Tyndall did not exist, Ruskin would have had to invent him; and in large part he did.

We will return later to the question of science and ethics, pausing here for a last look at the glacier controversy in relation to Ruskin's and Tyndall's general approaches to science. In reviewing Forbes's collected papers, Huxley noted that phenomena "form a portion of a continuous chain of causation, and the business of the natural philosopher is to trace out the successive connexion of cause and effect, link by link. . . ."[28] In a lecture to students, Tyndall said that "the whole body of phenomena is instinct with law; the facts are hung on principles, and the value of physical science as a means of discipline consists in the motion of the intellect, both inductively and deductively, along the lines of law marked out by phenomena."[29] Both of these link or chain metaphors characterize Tyndall's thinking about glaciers; the laws linking the behavior of glaciers would be the known laws of force acting on crystalline bodies and Faraday's principle of regelation. But Ruskin's whole approach is to "read" glaciers like a language, with little concern for laws or "inner workings"; his "science of aspects" (to use the term he applied to art) dwells rather on the sensuous surface of things. So abundant is the play of language in *Deucalion*, that verbal fastidiousness seems to take the place of empirical proof, as though words were in absolute correspondence with states and things: "You can *stretch* a piece of India-rubber, but you can only *diffuse* treacle, or oil, or water . . . let [honey] be candied, and you can't pull it into a thin string . . . You can't stretch mortar either. It cracks even in the hod, as it is heaped" (XXVI, p. 141). In passages like these, Ruskin tried to overwhelm the argument that Forbes's terminology was vague, while failing to understand what Forbes's critics meant by "explanation." Yet, in fact, Forbes's essential perception has won out: later experiments have shown that ice, though brittle, can indeed be deformed under pressure, like paste or dough.[30]

[III]

Between Ruskin's attack in *Fors Clavigera* and the publication of *Deucalion*, Tyndall delivered his most famous lecture, the Belfast Address, on August 19, 1874. He had just completed the experimental work that all but disproved spontaneous generation, which to his mind confirmed the unity of nature; as he had told an audience the year before, "all our philosophy, all our poetry, all our science, and all our art — Plato, Shakespeare, Newton and Raphael — are potential in the fires of the sun."[31] The Belfast Address works toward the same conclusion by means of a sketchy history of Western science, beginning with the bold insights of Democritus, Epicurus and Lucretius, then moving through the Christian era, when the twin authorities of Aristotle and Church dogma fettered the scientific spirit, and concluding with the rise of evolutionism in physics and biology. According to Tyndall, the followers of Darwin have once again hypothesized a universe built, like that of Lucretius, upon the elementary forces and aggregations of matter; there is, in other words, no mystical line dividing the living from the unliving:

> By a necessity engendered and justified by science I cross the boundary of the experimental evidence, and discern in that Matter which we, in ignorance of its latent powers, and notwithstanding our professed reverence for its Creator, have hitherto covered with opprobrium, the promise and potency of terrestrial life.[32]

It is not immediately clear why this statement aroused such an instant and violent storm of response; the idea, after all, was not new (one thinks of Huxley's lecture of 1868, "On the Physical Basis of Life"). To explain this one must keep in mind not only Tyndall's authority as a researcher but also the occasion: the annual presidential address of the

British Association for the Advancement of Science, delivered in this case to an audience of 1800 people. To that audience Tyndall presented the broad thesis (as a modern commentator has paraphrased it) that "science not merely may ignore theology, but must replace her in any sphere in which the two may come in conflict. This had been said many times before . . . but rarely had it been presented with such confidence and such seeming authority to the entire intellectual world of a nation."[33] Frank M. Turner has argued that by associating scientific naturalism with an ancient Greek cosmology, Tyndall muddled the term "materialism" and committed a rhetorical blunder; a number of opponents were now free to claim that there was nothing new in the atomistic theory.[34] However that may be, the example of Lucretius is essential to Tyndall's first aim in the Address, which is not to bait the religious, but to reconcile the claims of religion and science in the only way he thought possible. This is also the aim of a series of addresses and essays.

The first and most obvious step in Tyndall's reconciliation of the data of science and the emotions of religion is to assert the scientist's ignorance of the real bond between mind and matter, a bond he calls "intellectually impassable."[35] In the Belfast Address he agrees with the argument he puts into the mouth of Bishop Butler: "But can you see, or dream, on in any way imagine, how . . . from these individually dead atoms, sensation, thought, and emotion, are to arise? Are you likely to extract Homer out of the rattling of dice, or the Differential Calculus out of the clash of billiard-balls?"[36] This dualism leaves the separate realms of religion and science, the emotions and the intellect, to flourish without conflict; but then Tyndall draws the two realms together by refashioning the old metaphors for organicism and "materialism." For him the unity of nature under law, stretching from the simple to the complex and from the inorganic through the organic, is a conception of great emotional power, verifiable by the new principle of the conservative of energy. James Prescott Joule had shown that a 1-pound weight falling 772 feet produces enough heat to raise a pound of water 1 degree Fahrenheit; in other words, forms of energy are convertible in a strict balance of "payment." Slightly earlier, J.R. Mayer, a physician, had demonstrated conservation of energy in muscular heat, showing that the animal body and the steam engine work according to the same principles. Findings like these eventually prevailed against vitalistic theories, as Tyndall explained to his general audiences. (To do so, he used colorful simplifications such as: "the vegetable is produced while a weight rises, the animal is produced while a weight falls.") By implication, all energy comes literally, not poetically, from the sun, but molecular forces "condition" the form that energy assumes; this interaction of force and matter Tyndall calls "structural energy," or, using Fichte's phrase, "formative power," and to describe it he invokes a familiar metaphor: "the molecular machinery through which the combining energy acts may, in one case, weave the texture of a frog, while in another it may weave the texture of a man." "In an amorphous drop of water," he continues, "lie latent all the marvels of crystalline force; and who will set limits to the possible play of molecules in a cooling planet? If these statements startle, it is because matter has been defined and maligned by philosophers and theologians, who were equally unaware that it is, at bottom, essentially mystical and transcendental."[37] This, of course, is the natural supernaturalism of *Sartor Resartus*, just as the weaving metaphor derives from the Earth Spirit in *Faust*.

In a lecture-demonstration to the working men of Dundee in 1867, Tyndall attempted to make literal and visible the Goethean weaving. A beam of light was passed through a slab of ice, causing six-petalled ice-flowers to break out all over. A solution containing dissolved crystals evaporated under heat, producing "Crystalline spears, feathered right and left by other spears." Still further demonstrations showed that "the common matter of our earth" could arrange itself "into forms which rival in beauty those of the vegetable world."[38] Thus, by a combination of metaphor and evidence, he reduces the distinction between organic and inorganic almost to a figure of speech: in a system of continual evolution toward higher and higher forms, it is no more scientific — if anything, it is less so — to claim that matter is "dead" than to claim that it is alive. Tyndall even refers to molecular forms as "incipient life," and in the Belfast Address he criticises the mechanical philosophers as "but partial students because they were not biologists, but mathematicians . . . . Their science was mechanical science. not the science of life. With matter in its wholeness they never dealt; and, denuded by their imperfect definitions, 'the gentle mother of all' became the object of her children's dread."[50] But the fullest statement of Tyndall's organicism came in an address of 1877, in which he compared the Strasbourg clock with Carlyle's "umbrageous Igdrasil":

A machine may be defined as an organism with life and direction outside; a tree may be defined as an organism with life and direction within. In the light of these definitions, I close with the conception of Carlyle . . . . But the two conceptions are not so much opposed to each other after all . . . . They equally imply the interdependence and harmonious interaction of parts,

and the subordination of the individual powers of the universal organism
to the working of the whole.[39]

In extending organic imagery beyond biology to include all the sciences,
Tyndall once again is Carlylean, for, as he correctly pointed out, Carlyle
had "ground" from all the sciences paint for the accurate metaphors he
drew. Carlyle's images are as often as not nonbiological: fire and light,
storms and combustions, the roaring of waters, the trembling of mountains; and in the passage from *Sartor Resartus* in which Teufelsdröckh
makes of a Schwarzwald smithy the emblem of cosmic unity, Tyndall
saw a poetic foreshadowing of the conservation of energy.

Thus, at the very least, Tyndall was able to undermine some of the
false oppositions between nature philosophy and skeptical naturalism.
He showed that the Romantic faith in a living universe need not stand or
fall on the evidence for vitalism; and he did this by demonstrating that
the mechanical model of eighteenth-century philosophy, against which
the nature philosophers had rebelled, was less suited than the organic
model to describing the new physics. Tyndall's testimony is particularly
important because unlike Spencer, he was a practicing scientist, and unlike Huxley, he was a physicist rather than a biologist. His universe is an
evolving pattern of complex, brilliantly colored forms continually emerging from an invisible region, of which the visible is but a partial "suggestion"; and instead of the endless and aimless collison, he substitutes a
"formative power" acting in a system of interdependencies more complex
than any the previous century had known. Carlyle had said,

> Detached, separated! I say there is no such separation: nothing hitherto
> was ever stranded, cast aside; but all, were it only a withered leaf, works
> together with all; is borne forward on the bottomless, shoreless flood of
> Action, and lives through perpetual metamorphoses. The withered leaf is
> not dead and lost, there are Forces in it and around it, through working in
> inverse order; else how could it *rot?* . . . all objects are as windows,
> through which the philosophic eye looks into Infinitude itself.[40]

Thus, reconciliation of poetry and science culminates in a union of the
two metaphors that had long kept them in opposition. This union allows
Tyndall two separate descriptions of nature that complement each other
point by point in a kind of physicopoetical parallelism; indeed, he may
be one of the last thinkers to have assimilated an advanced scientific
discovery to traditional poetic metaphor. But if Tyndall's description is
inviting, it is so because of what it conspicuously omits. Although he was
an evolutionist, his universe is oddly atemporal: we see a weaving of

forms, but not the horrific forms of the evolutionary past—the giant bones fossilized in Lyell's lava beds, or the subhuman faces that haunt Huxley's *Man's Place in Nature*. By replacing Young's "brute matter" with Carlyle's "living garment" of God, Tyndall turns his face from the brute altogether; in this paradoxical way, his organicism is almost an escape from biology.

[IV]

Having viewed Tyndall's "materialism" in its most philosophical form, one may wonder whether he and Ruskin (had Ruskin wished to know it) were not in essential agreement. *The Queen of the Air* begins, as I have mentioned, with a sneer at Tyndall, yet certain passages in that book are so close to Tyndall's lectures of the same years that one has to wonder if Ruskin knew them. In the second part, for example, he paints a prose portrait of Athena touching the soil and creating life, which is an allegory of oxygen uniting under the direction of a "formative power." Like Tyndall, he uses that phrase (without attributing it to Fichte or to anyone else) to assert a teleology of matter and force, although unlike Tyndall he associates that power with a vitalistic notion of "life." His Athena is essentially a weaver, like Goethe's Earth Spirit, as comes clear in his peroration to the "essential" creature of the air, the bird:

> Also, upon the plumes of the bird are put the colours of the air . . . the vermilion of the cloud-bar, and the flame of the cloud-crest, and the snow of the cloud, and its shadow, and the melted blue of the deep wells of the sky—all these, seized by the creating spirit, and woven by Athena herself into films and threads of plume; with wave on wave following and fading along breast, and throat, and opened wings, infinite as the dividing of the foam and the shifting of the sea-sand . . . . (XIX, pp. 360–361)

How does Ruskin's Athena differ, then, from the transcendent "weaving" of Tyndall's molecules? For Ruskin the difference is crucial and leads in turn to the two central distinctions dividing the two men. The first distinction concerns the theory of natural myth; the second concerns the morality of science.

For Ruskin all religion grows from natural myths, which draw together factual observations of nature and a passionate apprehension of personal divinity. This means that the creation is a book of scriptures or of divine hieroglyphs; personality and moral truth, then, are both inherent in the structure of things. According to Tyndall, molecular machinery, acting through the formative or structural power, "may, in one case, weave the texture of a frog, while in another it may weave the texture of a man." According to Ruskin, "the calcareous earth, soft, may beget crocodiles, and dry and hard, sheep . . . representing to [man] states of moral evil and good, and becoming myths to him of destruction and redemption, and, in the most literal sense, 'Words' of God" (XIX, p. 359). In *The Eagle's Nest* (1872), a series of lectures on science and art, Ruskin quotes a rhapsodic (and in some ways Ruskinian) passage from Tyndall personifying the sun as the source of all heat, then follows the quotation with the remark that the serpent's motion also comes from the sun. "But where did its *device* come from? There is no wisdom, no device in the dust, any more than there is warmth in the dust. The springing of the serpent is from the sun: — the wisdom of the serpent, — whence that?" (XXII, p. 197). The complexity, and indeed the passion, of Ruskin's position come out of his private religious doubts, as we can see in the hedging in certain passages in *The Queen of the Air:* about "first heat" or "first cause," for example, "we can show no scientific proof of its not being personal"; and of "a creative wisdom, proceeding from the Supreme Deity," he says that "every formative art hitherto, and the best states of human happiness and order, have depended on the apprehension of its mystery (which is certain), and of its personality (which is probable)" (XIX, p. 378). By the time of *Deucalion,* Ruskin had converted back to Christianity (it is at least plausible that the fury with which he turned on Tyndall reflected his continuing dread of unbelief), but despite his religious shifts, the important distinction in that book between myth and "theory" is consistent with a long-held position: noble myths, that is, are "truer" than any "materialistic" theory, including the Darwinian, because they are "instinctive products of the natural human mind, conscious of certain facts relating to its fate and peace" (XXVI, p. 336).

If, therefore, all our "best and happiest moments" — in effect the very salvation of our souls — depend on seeing nature as alive and on seeing moral truths as words inscribed in her brow, profound consequences follow for any serious philosopher of science. Tyndall, having reconciled the metaphors of organism and machine, had also been able to reconcile imagination and analysis; for if the source of value in the universe is traceable to the infinitesimal and the insensible — the ultimate particles and laws through which the All manifests itself and which contain "the promise and potency of terrestrial life" — why should any observer fear analysis? Ruskin, on the contrary, preserved the Romantic opposition between analysis and moral awareness, and developed it in his later years into an almost fearful dialectic.

In *Modern Painters III* (1856), Ruskin rebuked Wordsworth for not understanding that "to dissect a flower may sometimes be as proper as to dream over it"; but a more typical statement is the warning, at the end of the same chapter, that scientific pursuits may check "the impulses towards higher contemplation" and in that case are "to be feared or blamed. They may in certain minds be consistent with such contemplation; but only by an effort: in their nature they are always adverse to it, having a tendency to chill and subdue the feelings, and to resolve all things into atoms and numbers. For most men, an ignorant enjoyment is better than an informed one; it is better to conceive the sky as a blue dome than a dark cavity, and the cloud as a golden throne than a sleety mist" (V, pp. 386–387). This conventionally Romantic position may be compared with the extreme statements of a late lecture like "Yewdale and Its Streamlets" (1877), reprinted in *Deucalion*, in which Ruskin outdoes Wordsworth by describing the wrong kind of science as literally diabolical. He warns his audience that they should keep their children away from the Mylodon in the British Museum: "The devils always will exhibit to you what is loathsome, ugly, and, above all, dead; and the angels, what is pure, beautiful, and, above all, living." Science, that is, ought to blinker itself from certain pursuits and adopt, with the rest of us, an almost childlike form of natural piety. "All true science," he concludes, "begins in the love, not the dissection, of your fellow-creatures; and it ends in the love, not the analysis of God. Your alphabet of science is in the nearest knowledge, as your alphabet of morals is in the nearest duty" (XXVI, pp. 263, 265–266). Here "analysis," linked with "dissection," contains two specific references: the study of prehistoric monsters and the vivisection of household pets (it was ostensibly over the issue of vivisection that Ruskin resigned the Slade Professorship at Oxford). Perhaps the best way of summing up Ruskin's shifting views on science and nature in the late 1870s is to say first that he believed that nature is partly diabolic and that those who would ghoulishly explore this part of her are, in a possibly literal sense, in league with the devils; and second, that he believed that nature is also alive and beautiful, and that those who would desecrate and analyze her are also in league with the devils. In either case, bad science is a form of violence or pollution, a position that follows directly from the apprehension of nature as a personal presence. Thus, to the pure eyes of myth, the air is blue, not with Tyndall's gases, but with the cloak of Athena; yet Ruskin's Tyndall, as we have seen, figuratively encapsulates Athena in the test tube, leaving a stench of smoke in the Alps (XIX, p. 294).

The connection emerges most clearly, perhaps, in the letter in *Fors Clavigera* containing Ruskin's original attack on Tyndall, which is entitled "La Douce Dame." The argument is not explicit, but instead takes the form of implied links. The attack on Tyndall follows directly upon a commentary on the *Roman de la Rose* and a long excerpt from a Swiss children's tale. The hero of this tale is Hansli, the broom-maker, who knows his willows personally and gives many of them names ("Lizzie, Little Mary-Anne, Rosie, and so on"). Occasionally the trees are ravaged by thieves: "But when he arrived thus, all joyous, at his willows, and found his Lizzie or his Rosie all cut and torn from top to bottom, his heart was so strained that the tears ran down his cheeks, and his blood became so hot that one could have lighted matches at it." Eventually, he fights off the thieves until they leave the willows alone — "as happens always when a thing is defended with valour and perseverance" (XXVII, pp. 633–634). Clearly, Ruskin saw his ensuing attack on Tyndall as a similarly valorous defense — a defense not only of Forbes, the son of Scott's first love, but of Ruskin's first love as well. We recall, finally, that in the Deucalion myth, stones are the bones of our mother.

At their most troubled moments, Ruskin's late essays on science come close to nightmares — raging struggles of sanctity against blasphemy, purity against the devil, life against death; mythopoeia draws close to hallucination, and one senses the darkness of the morning in 1878 when Ruskin awoke to find the Evil One in his room. By contrast the Lucretian universe of Tyndall, purged of troublesome projections, seems blandly salutary; for Tyndall feared neither analysis nor Mother Earth, nor life, nor death. When the Reverend James Martineau objected to the Belfast Address, Tyndall replied,

> Nature, according to his picturing, is base and cruel: what is the inference to be drawn from its Author? If Nature be "red in tooth and claw", who is responsible? On a Mindless nature Mr. Martineau pours the full torrent of his gorgeous invective; but could the "assumption" of "an Eternal Mind" — even of a Beneficent Eternal Mind — render the world objectively a whit less mean and ugly than it is? . . . [H]e rashly . . . kicks away the only philosophic foundation on which it is possible for him to build his religion.[41]

But did Tyndall, who exulted in the impersonal, believe his Lucretian "faith" could satisfy the emotional needs that Christianity had fulfilled? In W.H. Mallock's *The New Republic* (1877), the character representing Tyndall is made to tell a young Catholic that when the "awe-struck eye," guided by science, "sees that all that is has unfolded itself . . . from a brainless, senseless, lifeless gas," our feelings before "this stupendous truth" must be religious.[42] The caricature is of course inaccurate, yet it was a common response to Tyndall and it points to a real difficulty and a real ambiguity. In his speech at the unveiling of Carlyle's statue, he said, "Out of pure Unintelligence [Carlyle] held that Intelligence never could have sprung, and so, at the heart of things, he placed an Intelligence — an Energy which, 'to avoid circuitous periphrasis, we call God.' " He adds, "For the operation of Force — the scientific agent — his deep and yearning soul substituted the operation of the Energy before referred to . . . ."[43] But where did Tyndall himself place the Intelligence that Carlyle had said he could not live without? In his readiness to exchange Carlyle's metaphors for scientific terms — Energy for Force, or the philosophy of Might for the survival of the fittest, or the unity of the All for the Conservation of Energy — does Tyndall sacrifice the emotional power of the original vision and reduce both science and religion in the process?

There are persuasive reasons for answering *yes*. As regards science, Tyndall's literal belief in minute material entities was outmoded in his time. Huxley had pointed out in 1868 that "matter and spirit are but names for the imaginary substrate of groups of natural phenomena"; and although, according to him, the progress of science will more and more depend on the use of "materialistic formulae and symbols," the scientist who mistakes these for "what is commonly understood by materialism" places himself "on a level with the mathematician who should mistake the x's and y's with which he works his problems, for real entities . . . ."[44] By contrast Tyndall wrote (in a letter to a relative) that he believed William Tyndale's molecules "are now working in each of our brains . . . . His temper is our family temper, and the cast of his intellect the prototype of ours" (*Life*, p. 87). As Huxley wrote in the eulogy to his friend, "I really think that he, in a manner, saw the atoms and molecules, and felt their pushes and pulls."[45] And as regards his metaphysics, some of Tyndall's dabblings seem similarly naïve. In a scrap of blank verse composed during retirement in the Alps, he compares the "vast brain" that created the mountain passes with the mind and power of the human conquerors, like Hannibal and Napoleon, who traversed them and whose names are now without value. The poem evokes Carlyle ("But well you knew/Might, to be Might, must base itself on Right") and gives warning to those whose work may lack "that true core which gives to Right and Might/ One meaning in the end."[46] The speculativeness here sounds dilettantish; is the assumption of a "vast brain" a poetic convention rather than a belief? The connection between Divine and human purposes remains vague, and "right" is all but drained of meaning.

Undoubtedly, Tyndall lacked the power of language and the moral imagination to create a genuinely imaginative vision; but our final appraisal cannot rest on his weakest moments. What he did achieve is perhaps best seen by recalling a famous Coleridgean distinction, one that lies at the heart of Romantic nature philosphy and therefore at the heart of both Ruskin and Tyndall's experience of the created world. Every act of perception, according to Coleridge, includes an identification of the self with the object and a contradistinction of the self from the object; thus, Coleridge distinguished between

> that intuition of things which arises when we possess ourselves, as one with the whole, which is substantial knowledge, and that which presents itself when . . . we think of ourselves as separated beings, and place nature in antithesis to the mind, as object to subject, thing to thought, death to life. This is abstract knowledge, or the science of the mere understanding. By the former, we know that existence is its own predicate, self-affirmation, the one attribute in which all others are contained, not as parts, but as manifestations.[47]

Ruskin and Tyndall would have agreed that both modes of perception must exist and that in their proper relations (however differently our writers understood them) lies the proper relations of science and poetry. And both men, as we have seen, attempted to "speak" in the two modes. As for Tyndall, the completeness with which he could possess himself "as one with the whole" is evident not only in his habitual metaphors, but in all the ways, traced throughout this essay, that he conceived of life and energy as an ultimate unity. The effect on him of the great men he emulated is less the absorption of their ideas than the absorption, if one can put it this way, of their moral force; and this transmission, both reverent and combative, is close to Tyndall's experience of nature, the ultimate source of all energy. For him that energy inspired both the man of science and the man of religion. Perhaps because of this conviction, he was able to prophesy at the close of the Belfast Address a time when the human mind might "still turn to the Mystery from which it has emerged, seeking so to fashion it as to give unity to thought and faith." This unity would be performed by "what, in contrast with the *knowing* faculties, may be called the *creative* faculties of man. Here, however, I touch a theme too great for me to handle, but which will assuredly be handled by the loftiest minds, when you and I, like streaks of morning cloud, shall have melted into the infinite azure of the past."[48]

These sentiments seem far removed from the villain of *Deucalion* — the impious schoolboy scrambling over Ruskin's sacred mountains — but they seem so not only because of Ruskin's bigotry but also because of Tyndall's many-sidedness. A practical man with a speculative bent; an athlete and a metaphysician; a disciplined experimentalist and an exuberant performer; an iconoclast and a believer — these are all sides of Tyndall and, one might add, of the Victorian age in general. To the degree that we find it difficult to reconcile all the aspects of Tyndall — the mystic, for example, with the muscular Christian carrying his pickaxe and black bulb thermometer — we have inherited Ruskin's tradition. By depicting bad science as a rape or dissection of the earth's body, Ruskin anticipates, in all but the extremeness of his metaphor, those modern critics of scientific rationality for whom science is a drive for mastery over nature that excludes the contemplative union of the mystic.

But if in this regard one feels closer to Ruskin than to Tyndall, Tyndall becomes all the more interesting for the ambiguity of his relationship to Romanticism, an ambiguity that is nowhere more marked than in his descriptions of Alpine scenery, where he is both scientist and prose-poet at once. The Romantic sublime is itself profoundly ambiguous. Is it, for example, an experience of humility or pride? Is the ego annihilated or exalted? Is the reverent poet or painter, transfixing natural phenomena with the power of his art (as Ruskin did) so different from the Faustian scientist, who would submit those phenomena to his control? Coleridge's woodman sees the Specter of the Brocken "gliding without tread,/An image with a glory round its head," and "worships its fair hues,/nor knows he makes the shadow, he pursues!" These lines echo in the mind as one reads Tyndall's late essay, "The Rainbow and Its Congeners," in which Tyndall combines reverence and the drive for mastery like a hunter or an explorer, taking from the poets the idiom of the sublime and capturing it, so to speak, for science. The essay begins with a discussion of Newton's analysis of the rainbow into measurable angles, and then goes on to describe the varieties of rainbows in the Alps. Near the end, Tyndall quotes from the record of a Himalayan explorer in search of the "Glory of Buddha": "It was described to me as a circle of brilliant and many-coloured radiance, broken on the outside with quick flashes, and surrounding a central disc as bright as the sun, but more beautiful. Devout Buddhists assert that it is an emanation from the aureole of Buddha, and a visible sign of the holiness of Mount O." "The shadow of the head,"

Tyndall comments, "must have always occupied the centre of the 'Glory,' " a fact he knows because, as he has explained to his readers, the Glory is a circular rainbow with colors similar to those produced by mixing turpentine with water. The conclusion gives Tyndall's characteristic twist to natural supernaturalism:

> Thus, starting from the first faint circle seen in the thick darkness at Alp Lusgen, we have steadily followed and developed our phenomenon, and ended by rendering the "Glory of Buddha" a captive of the laboratory. The result might be taken as typical of larger things.[49]

NOTES AND REFERENCES

1. Edward Alexander, "Ruskin and Science," *Modern Language Review*, vol. 64 (1969), p. 508.
2. James David Forbes, *Travels in the Alps*, (Edinburgh: A. & C. Black, 1843), p. 365.
3. See J.S. Rowlinson, "The Theory of Glaciers," *Notes Received by the Royal Society of London*, vol. 26 (1971), pp. 189–204.
4. Arthur S. Eve and C. H. Creasey, *The Life and Work of John Tyndall* (London: Macmillan, 1945), p. 85. Hereafter cited in the text as *Life*.
5. John Tyndall, Journal for 26 January 1854; quoted by Rowlinson,[3] p. 192.
6. Rowlinson,[3] pp. 194–199. This essay argues that Tyndall's advocacy of Rendu "uncannily" anticipates his most famous controversy, the championing of Julius Mayer in 1862. "In both cases the man championed by Tyndall had had some undeniably correct views of the problem, but was not the one who formulated them most clearly nor who undertook their experimental verification" (p. 198).
7. *Works of John Ruskin*, ed. E. T. Cook and Alexander Wedderburn, 39 vols. (London: George Allen, 1903–1912), vol XXVII, pp. 639, 642. Hereafter cited in the text by volume and page number.
8. John D. Rosenberg, *The Darkening Glass* (New York: Columbia University Press, 1961), p. 180.
9. The following is an example of Ruskin's rigor in treating his opponents' views: "You may best conceive the gist of the Regelation theory by considering the parallel statement. . . that if [you] put a large piece of barley-sugar on the staircase landing, it will walk downstairs by alternately cracking and mending itself" (vol. XXVI, p. 230).
10. On these views see also Alexander,[1] *passim*.
11. Alan W. Brown, *The Metaphysical Society: Victorian Minds in Conflict* (New York: Columbia University Press, 1947), p. 333.
12. James G. Paradis, *T.H. Huxley: Man's Place in Nature* (Lincoln: University of Nebraska Press, 1978), pp. 30 ff.
13. John Tyndall, *The Glaciers of the Alps* (London: J. Murray, 1860), pp. 289–290.
14. T.H. Huxley, "Professor Tyndall," *Nineteenth Century*, vol. 35 (1894), p. 7.
15. John Tyndall, *New Fragments of Science*, 3rd ed. (New York: Appleton, 1915), p. 10.
16. John Tyndall, *Fragments of Science for Unscientific People* (New York: Appleton, 1872), vol. II, p. 107.
17. Tyndall,[16] p. 119.
18. Tyndall,[16] p. 94.
19. Paradis,[12] pp. 50ff.

20. T.H. Huxley, "The Glaciers of the Alps," *Saturday Review*, vol. 10 (July 21, 1860), p. 81.

21. Tyndall,[16] pp. 31–32.

22. Frank M. Turner, "Victorian Scientific Naturalism and Thomas Carlyle," *Victorian Studies*, vol. 17 (1975), pp. 325–343.

23. Thomas Carlyle, *Past and Present*, ed. R. D. Altick (Boston: Houghton Mifflin, 1965), p. 198.

24. Tyndall,[16] p. 96. Interestingly, Tyndall first introduced himself to Carlyle by writing a letter on behalf of his friend Thomas Archer Hirst. His own advice to Hirst is pure Carlyle in imagery and doctrine: "There is no stability on earth, Tom, life is a flowing river; dam it up, and it will create festering sedge-marshes and unwholesome fens; let it flow out in an honourable effort, and it is beautiful as the figure I have used to express it. In a word, Tom, it will never do to sit for hours looking disconsolately inwards, for by this manner of looking there is nothing found. . . ." This letter and the letter to Carlyle are quoted in *Life*,[4] pp. 26–27.

25. For the importance of the Carlylean hero for Huxley and Tyndall and the significance of the Governor Eyre controversy, see Paradis,[12] pp. 63–71.

26. Hirst, Tyndall's best friend, perhaps provides the most candid glimpse into this aspect of his personality. In two journal entries of 1855, a year before Tyndall became interested in glaciers, he wrote: "John is changing a little; I think he is less accommodating than he was, his conversation has unconsciously become more egotistical. . . . In his opinions he is perhaps less generous to others, and more peremptory, abrupt, and dogmatic. . ." ". . . [H]e is in debate a terribly rough and unconquerable antagonist. . . he enjoys an intellectual fence for its own sake, and I am not sure that his own dexterity in inflicting sharp lashes is not a source of amusement to him." To a letter Hirst wrote on these matters, Tyndall replied, "I have read your pretty note to me with a great deal of interest; for it held up a glass before me, which revealed myself to myself in a manner in which I had not appeared to myself previously. . . . never has mortal father, I think, less to complain of regarding his son that I have in this respect regarding you" (*Life*, p. 71). Hirst was ten years younger than Tyndall; the friendship, which lasted until death, seems analogous to the friendship with Carlyle, except that Tyndall's role was reversed.

27. Huxley,[14] p. 2.

28. T.H. Huxley, "The Structure of Glaciers," *Saturday Review*, vol. 8 (July 16, 1859), p. 80.

29. Tyndall,[16] p. 95.

30. See, for example, Rowlinson,[3] pp. 203–204. Regelation does occur, particularly near the source of glaciers, but not to the extent Tyndall thought.

31. Tyndall,[16] p. 131.

32. Tyndall,[16] p. 191.

33. Brown,[11] p. 237.

34. Frank M. Turner, "Lucretius Among the Victorians," *Victorian Studies*, vol. 16 (1973), pp. 327–348.

35. Tyndall,[16] p. 87.

36. Tyndall,[16] p. 167.

37. Tyndall,[16] pp. 49, 51.

38. Tyndall,[16] pp. 67–68.

39. Tyndall,[16] p. 337.

40. Thomas Carlyle, *Sartor Resartus* (New York: Dutton, 1967), p. 53.

41. Tyndall,[16] p. 230–231.
42. W.H. Mallock, *The New Republic* (London: Chatto & Windus, 1877).
43. Tyndall,[15] p. 396.
44. T. H. Huxley, *Collected Essays* (London: Macmillan, 1893–94), vol. I, p. 160.
45. Huxley,[14] p. 7.
46. Tyndall,[15] p. 500.
47. S.T. Coleridge, *Works*, ed. W. G. T. Shedd (New York: Hayes, 1884), vol. II, pp. 469–470.
48. Tyndall,[16] p. 201.
49. Tyndall,[15] pp. 222–223.
50. Tyndall,[16] p. 190.

The Spectacular English Sunsets of the 1880s*

THOMAS A. ZANIELLO
Department of English
Northern Kentucky University
Highland Heights, Kentucky 41706

THE SUNSETS AND VICTORIAN CULTURE

ON THE 26th and 27th of August 1883, the volcano on the island of Krakatoa in the Straits of Java erupted violently and continuously, bringing death and destruction to a vast area of the South Pacific and changing an island into a rocky wasteland in a few days. The eruptions and subsequent atmospheric fallout of volcanic debris also brought a relatively sudden and dramatic change to the appearance of sunrises and sunsets all over the world: in Madras the sun "turned green," near Hawaii blue sunlight woke the startled passengers of a steamer, and in England the sun set in such a blaze of color that fire brigades were called out in a number of cities. Although these unusual solar phenomena were associated in the popular mind with Krakatoa, there was some controversy in scientific circles about a facile indentification of the two, since a number of scientists stated publicly that it was impossible for volcanic dust to be ejected to such a great height and to be carried such great distances. But the majority of Victorian scientists believed that an inordinate concentration of Krakatoa dust particles in the atmosphere had refracted in the sun's light in novel ways that needed scientific investigation.

* This work has been supported by the office of the Provost and the Faculty Benefits Commitee of Northern Kentucky University, Highland Heights, Kentucky.

Although newspapers in virtually every country in the world were flooded with verbal descriptions of the solar phenomena, the visual record of the sunsets is limited primarily to the work of two London artists, John Sanford Dyason, a Fellow of the Royal Meteorological Society, and William Ascroft, a Royal Academy of Arts exhibitor with scientific interests. Both Ascroft and Dyason recorded numerous sunsets in "crayon" or pastel chalk; it is Ascroft's work that illustrates the present essay. In FIGURES 1 and 2, some of the unusual tones of green and red may be appreciated, but what is not evident from the reproductions and hard for us to comprehend today is that these two drawings were made only forty-five minutes apart. In fact these two form part of a series of six drawings made shortly after sunset on the Thames River in Chelsea when the "afterglow" of the Krakatoa effects lasted about sixty-five minutes on November 26, 1883; this series was printed by the Royal Society in its official report, *The Eruption of Krakatoa and Subsequent Phenomena* (1888). Dyason had also exhibited his drawings to the Krakatoa Committee, which wrote the Royal Society report, but his work was not printed in the end. Both artists' drawings, however, were exhibited at a number of Royal Society *conversazioni* or public meetings held at Carleton House, London, from 1884–1889. At one particular *conversazione* on June 8, 1887, the usual crowd of notables were present—the Dean of Westminster Abbey, Sir Frederick Leighton of the Royal Academy, and Robert Browning, for example—but the evening also brought together Thomas Huxley and Matthew Arnold, the two Victorians who have come to represent opposing sides of the "two cultures" debate. And although both men had much to say about the role of science in Victorian culture, they never (as far as I know) commented on the quite startling marriage of art and science represented by Ascroft's and Dyason's drawings. They had exchanged toasts and barbs about science and culture at the Royal Academy's banquets in 1881 and 1883,[1] but this night, at least, they were publicly silent.

We must turn elsewhere then for a consideration of the spectacular sunsets and their impact on Victorian culture. Although J.M.W. Turner's controversial paintings had certainly demonstrated the chromatic range of the traditional English sunset in his depiction of light and color in swirling patterns of paint with the sun as cynosure (a feature of such paintings as "Ulysses" and "Slavers," exhibited in the 1840s), by midcentury both ridicule in the press and new trends in realistic art had tended to drive Turner's special style out of public favor.[2] Ascroft's work (and Dyason's to a lesser extent) may be called Turneresque without the sting

Figure 1. William Ascroft, *Sunset on the Thames, November 26, 1883* (pastel).

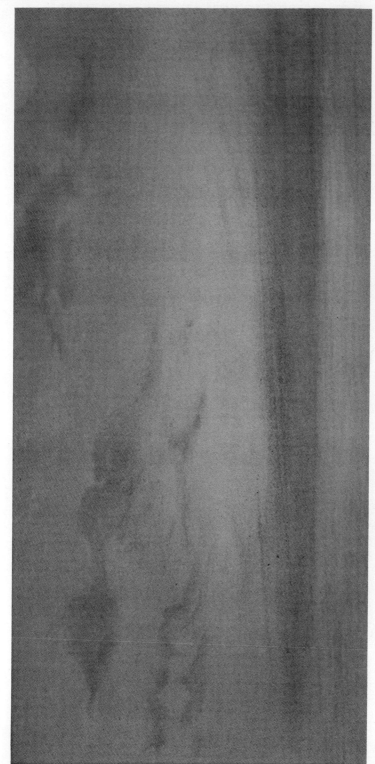

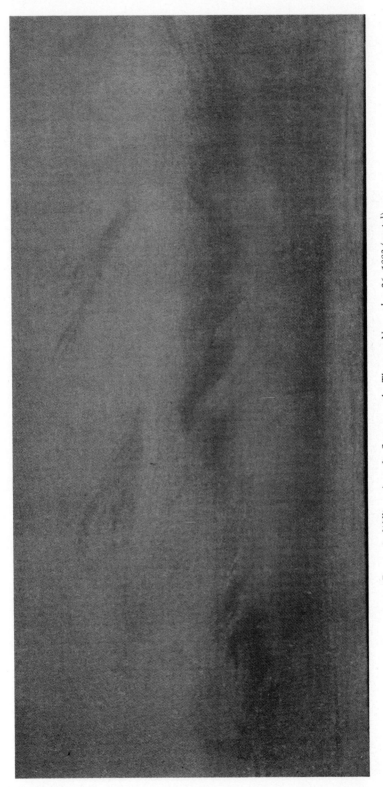

FIGURE 2. William Ascroft, *Sunset on the Thames, November 26, 1883* (pastel).

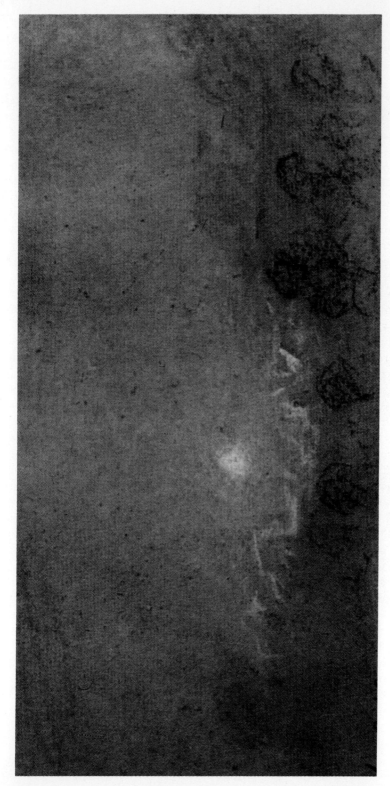

FIGURE 3. William Ascroft, *Bishop's Ring, September 2, 1884* (pastel).

of controversy, for Ascroft sought a rendering of light and color that was first, scientifically sound, and second, aesthetically pleasing; the second consideration was not far behind the first, as his letters to the Royal Society make clear. When the Royal Society was preparing his drawings for the Krakatoa volume, he was quite insistent that they avoid the errors other publishers had made in printing Turner's work by ordering the printer to work only in "bright weather" so that the colors would be true.[3] Dyason's work was more strictly scientific in intention: he called his drawings "photographic experiments" and a "faithful transcript of Nature."[4] Dyason's drawings were exhibited under the label, "The Chromatics of the Sky" and they included notations that indicated their function as means for weather forecasts. The London *Times'* reviewer of the *conversazione* of June 8th wrote that Ascroft's and Dyason's drawings (as well as another exhibitor's photographs of "beautiful cloud form") demonstrated that "evidently science as well as art can find a use for cloud effects and the evanescent glories of sunsets" (June 9, 1887).

How far this reviewer's optimism about the relationship of science and art represented a genuine moment in Victorian culture is the subject of the present essay. And since two of the most likely candidates for this study — Huxley and Arnold — are not available in this instance, let us look instead at the reactions of two other Victorians whose interests were both artistic and scientific: John Ruskin (1819–1900) and Gerard Manley Hopkins (1844–1889). Neither would have been invited to a Royal Society *conversazione*, although their interest in the Krakatoa sunset drawings would have been great: Ruskin, the art and social critic whose long career both shaped and harassed British taste, had spent the 1880s bitterly attacking British scientists for ignoring a meteorological sign — an omnipresent storm-cloud — which he believed to portend the moral collapse of English society; and Hopkins, today regarded as the key inspiration for many of the great changes in modern poetry, was in his lifetime a virtually unknown priest, poet, and amateur scientist.

Although in a number of ways Hopkins' and Ruskin's interests coincided — they were both avidly concerned about architecture and painting — we do not usually think of their having strictly scientific leanings. But Ruskin, as we will see, really began his career as a geologist, and Hopkins, in the 1880s, contributed four letters to *Nature* (then as now the leading general scientific forum) on his observations of various solar phenomena, including two extended analyses of the Krakatoa sunsets.[5]

In fact it is the Krakatoa eruptions and their atmospheric aftereffects that provide a unique connection between the two men and reveal key aspects of Victorian science and culture. While the debate about Krakatoa and atmospheric conditions was joined by Ruskin in polemical terms, one of Hopkins' letters was being considered for its report by the Krakatoa Committee of the Royal Society: every one of Ruskin's pronouncements would generate extensive publicity in English newspapers and even such journals as *Nature* and *Knowledge* (*Nature's* only real competitor) would at least be compelled to reply to his charges; Hopkins' poetry and perceptive prose were known only to a small circle of friends, although his letters to *Nature* presumably had some impact on interested readers. And although we have no evidence of Hopkins' direct knowledge of Ruskin's ideas during this period, we have a middleman in all these matters: Robert C. Leslie, a painter and Ruskin's friend. (Leslie's father, Charles R. Leslie, was a major painter and public figure in the early Victorian art world, counting not only Turner but also John Constable as his friends: his son, therefore, grew up among the canvases of Turner's sunsets and Constable's cloud studies.) Leslie's letters on Krakatoa and associated phenomena also appeared in *Nature* throughout the 1880s; furthermore, Ruskin relied, in both his lectures on the "storm-cloud" and in his autobiography, *Praeterita* (1885–1889), on Leslie's letters to buttress his prophetic message of doom about an industrializing society that no longer cared about the individual or the individual's workmanship.

By examining Ruskin's and Hopkins' reactions and investigations into the Krakatoa events, a number of important aspects of Victorian culture are revealed. First, we have the critical interaction of the two men who may be said to represent, respectively, the outstanding artistic and social critic and one of the outstanding poets. Second, although Ruskin was trained as a scientist, he takes the position of the inadequacy of science, while Hopkins, poet and priest, argues for the dispassionate precision of scientific objectivity. Third, the "debate" between the two is thrown literally against the backdrop of Victorian painting, because — as one correspondent to *Knowledge* wrote (December 14, 1883) — if the Krakatoa sunsets had been painted by Turner, it would be assumed that he was "either delirious or drunk"; in a lighter vein, but still characteristic of the times, a poetaster in *The St. James Gazette* (January 14, 1884) suggested facetiously that Nature herself had arranged the sunsets to outdo her nearest competitor, Turner. And finally, we have here the use of art to *document* the scientific observation of phenomena: in this instance color drawings became a necessity since black and white photographs (the only kind then available) would not provide the accurate documentation needed. The debate over the spectacular English sunsets of the 1880s illustrates some of the tensions in Victorian culture and some of their tentative resolutions: if some, like Ruskin and Leslie, thought that science could not provide the interpretation of the world that art and morality demanded, others, like Hopkins, Ascroft, Dyason, and numerous scientists, tried to see nature with a unified artistic and scientific eye.

Let us first examine the separate paths Ruskin and Hopkins took before Krakatoa brought them together in the 1880s.

In a sense Ruskin may be said to have begun and ended an astonishing career as an artistic and social critic by publishing articles not on art or society, but on meteorology. He had intended a career as a geologist, modelling himself at Oxford after William Buckland, Canon of Christ Church, Oxford, and a well-known geologist and mineralogist. Ruskin's first published prose was "Enquiries on the Causes of the Color of the Water of the Rhine," an analysis differentiating the color of the sea as an optical effect from the intrinsic color of fresh water. Ruskin published this early effort in J.C. Loudon's *Magazine of Natural History* in 1834, and continued to publish in the magazine essays of mostly geological interest. His debut at the London Meteorological Society was the presentation of a paper, "On the Formation and Color of Such Clouds as are Caused by the Agency of Mountains," which he read in 1837 but did not publish. But his major statement on the beauty and utility of his field, "Remarks on the Present State of Meteorological Science" (1839), was published in the *Transactions* of this society and reprinted as late as 1870 in *Symons's Monthly Meteorological Magazine*, the standard journal of weather reporting. Even when Ruskin's first piece of artistic analysis appeared in 1837 (*The Poetry of Architecture*), he was still contributing scientific pieces, with geometric diagrams, to Loudon's *Architecture Magazine* on problems of perspective. By the late 1840s Ruskin had turned to the analysis and defense of Turner's paintings, but even in *Modern Painters* (1843–1860), he illustrated such topics as Turner's clouds with diagrams of cloud structures and formations.

The publication of *Modern Painters* established Ruskin as an expert in the artistic rendering of sunsets and the other dramatic subjects of light and color that Turner loved—fires at sea, for example, or the conflict between steamships and the elements. By 1880s, just twenty years after the publication of the last volume of *Modern Painters*, Ruskin became convinced that traditional English sunsets were no longer to be seen; in 1884 he delivered two public lectures in London on what he called, alternately, "the storm-cloud of the nineteenth century" and the "plague-cloud." In fact there was an increase of smog in many areas of England during this period of heavy industrialization and, perhaps not coincidentally, a series of bad-weather years in which sunny days were at a premium.[6] Ruskin's association of bad weather with a plague-like visitation of God's wrath was his way of predicting the collapse of English civilization.

Hopkins was the younger man, and had entered Oxford after Ruskin had established a national reputation with *Modern Painters*. Although he never met Ruskin or heard him lecture, Hopkins became in architectural and other aesthetic matters Ruskin's disciple and to a marked degree even sketched from nature, to use his own words from a letter in 1863, in "a Ruskinese point of view." As his undergraduate studies in classics drew to a close in 1867, he recorded with meticulous care his observations of the forms or what he called the "inscapes" (perceived unique structures) of flowers, trees, clouds, and the varying color patterns of the sky. The following passage from Hopkins' *Journals* (published long after his death in 1889) attempts to describe the essential form or "inscape" of the bluebell:

> I do not think I have ever seen anything more beautiful than the bluebell I have been looking at. I know the beauty of our Lord by it. Its inscape is mixed of strength and grace, like an ash tree. The head is strongly drawn over backwards and arched down like a cutwater drawing itself back from the line of the keel. The lines of the bells strike and overlie this, rayed but not symmetrically, some lie parallel. They look steely against the paper, the shades lying between the bells and behind the coiled petal-ends and nursing up the precision of their distinctness, the petal-ends themselves being delicately lit. Then there is the straightness of the trumpets in the bells softened by the slight entasis and by the square spray of the mouth.
>
> (May, 1870)

The moral of the passage (the perception of divinity in nature), although characteristic of Hopkins' romantic and religious views, is only lightly drawn out; his concentration is on the observed naturalistic detail, simply and precisely described, with only one technical word (*entasis*, the almost imperceptible swelling of a column) which is not drawn from science, but from architecture. It is not the somewhat more strictly scientific writing of *Nature* in the 1870s but it is quite close to it, and Hopkins would only need to delete the religious references to make his prose ready for that journal. The prose is actually closer to the popular scientific writing of the Victorian naturalists, such as Philip Henry Gosse, whose bestselling books on the Devonshire seacoasts and the ocean were a mixture of precisely observed details of nature and brief reminders of the "endless variety of structure and form...manifested in God's marvellous works."[7]

If we look at Ruskin's and Hopkins' responses to the rainbow, a *locus classicus* of the interaction of science and art in the nineteenth century, the different qualities of their sensibilities will become apparent. The rainbow is an appropriate subject to consider, for during this period scientists and mathematicians were still codifying its laws, moralists were always concerned about it as a symbol of God's mercy, and painters were forever including it in their landscapes.[8] Although Hopkins wrote a number of lines describing rainbows, the following untitled poem (1864) concerns the epistemological status of the phenomenon, that is, the fact that no one sees the same rainbow; furthermore, Hopkins explains that the rainbow cannot exist independently of a perceiving consciousness and that it cannot exist without very specific limiting conditions (although Hopkins only insists on the water droplets in the air, the rainbow also requires that the sun be behind the observer and not higher than 42° above the horizon):

> It was a hard thing to undo this knot.
> The rainbow shines, but only in the thought
> Of him that looks. Yet not in that alone,
> For who makes the rainbows by invention?
> And many standing round a waterfall
> See one bow each, yet not the same to all,
> But each hand's breadth further than the next.
> The sun on falling waters writes the text
> Which yet is in the eye or in the thought.
> It was a hard thing to undo this knot.

Hopkins characteristically is interested in epistemology, and so locates his poem in the act of perceiving the rainbow. Ruskin, on the other hand, wants art to verify independently the truth of science, as in a controversy he participated in over a painting, "Shiplake, on the Thames," shown by Edward Duncan at the Exhibition of the Society of Water-Colors in 1861. The artist had rendered a rainbow *reflected* in water, a phenomenon various critics had argued was impossible. Ruskin's reaction to the controversy, published in *The London Review* (May 16, 1861), presents a geometric demonstration that a reflected rainbow is theoretically possible, but "the thing can hardly ever be seen in nature," for the reflected bow requires rain close to the water's surface and hence that surface would be "ruffled by the drops and incapable of reflection." It is perhaps symptomatic of Ruskin's relationship with science that when the "water-reflection" rainbow came back "in the news" in scientific circles in the 1870s, not only was Ruskin's contribution to the debate ignored, but also the main point of the argument was the second or third rainbow visible *in the sky*, caused by the rays' being reflected from the water surface. In fact, Ruskin's 1861 diagram is almost identical with that published in Symons' article, "Extraordinary or Water-Reflection Rainbow" (*Symons's Monthly Meteorological Magazine*, December 1875), except that Ruskin's "second" rainbow lies on the water, while Symons' reflected rainbow (actually, in this case, seen by a number of people) rises above the horizon and is contained within the arc of the primary (or usual) rainbow.[9]

We have in Ruskin and Hopkins, then, two related but somewhat opposed sensibilities: a critic warily expecting science to support art, and a poet with a keen and rigorous scientific eye.

HOPKINS, RUSKIN, AND THE SCIENTIFIC RECORD OF THE SUNSETS IN *NATURE*

In the summer of 1882, when Hopkins was teaching classics at Stonyhurst College in Lancashire and Ruskin was writing at his home in Brantwood in Cumbria, just north of Lancashire, the columns in *Nature* were filling up with letters reporting unusual and striking phenomena. J.P. O'Reilly of the Royal College of Science in Dublin reported seeing "a sort of halo" of dark beams in the "east-south-east" sky late at night. "All those beams were *dark*," he emphasized, and he added that he was writing about them because "the weather has been singularly cold and rainy for the season" (July 20, 1882). Within a week, another scientist, Silvanus P. Thompson, then at Glasgow, reported that O'Reilly's "curious halo" was an example of *rayons du crépuscule*, "frequently visible near sundown in the eastern sky." Thompson noted that he had described "similar rays" in the 1870s in scientific journals (July 27, 1882). An astronomer from Shanghai, the Jesuit Marc Dechevrens, also wrote, giving examples of similar phenomena, not only from the 1870s, but also from the seventeenth century. Dechevrens wrote that he was "at a loss to give a more satisfactory explanation" than "that the phenomenon is due to the atmospheric vapor" (November 9, 1882). We should note at this point — because these reports will soon lead these scientists and naturalists into debate and then into controversy — that these observers are writing a full year before the first Krakatoa eruption on August 26, 1883, and that at least one observer, an astronomer, had attributed the phenomenon to "atmospheric vapor," an explanation that will recur in other contexts and especially in Ruskin's "camp."

Two of the four letters Hopkins wrote to *Nature* refer directly to the *rayons du crépuscule* phenomenon, although they appear a year apart (November 16, 1882, and November 12, 1883). Hopkins wrote from the Jesuits' Stonyhurst College, where he had spent three years as a student from 1870 to 1873; at the beginning of his second residency there, from 1882 to 1884 as a classics instructor, he boasted to his friend, Robert Bridges, about the scientific furnishings: Stonyhurst had an observatory, laboratories, anemometer, a sunshine gauge, and magnetic instruments. Readers of *Nature* would certainly have recognized the name of the college appended to Hopkins' letters, because one of the leading Victorian astronomers and solar scientists, Stephen Perry, was the Director of the Stonyhurst Observatory, a Fellow of the Royal Society, and a close colleague of Norman Lockyer (founding editor of *Nature*).

Hopkins' first letter agreed with Dechevrens' observations from China, but Hopkins was less concerned with details than about the process of observation itself:

> There seems to be no reason why the phenomenon should not be common, and perhaps if looked out for would be found to be. But who looks east at sunset? Something in the same way everybody has seen the rainbow; but the solar halo, which is really commoner, few people, not readers of scientific works, have ever seen at all....I may remark that things common at home have sometimes first been remarked abroad. The stars in snow were first observed in the polar regions; it was thought that they only arose there, but now everyone sees them with the naked eye in his coatsleeve.   (November 16, 1882)

"But who looks east at sunset?" Indeed this is the key question for a scientist or naturalist or, one might add, a poet, since the ability to "look" where often no one else bothers to look is characteristic of both scientist and artist. As it turned out, numerous Victorians did turn eastward as the letters from scientists, painters, and poets in *Nature* testify.

Hopkins' second letter, a year later, was on the same topic, "shadow-beams in the east at sunset." Hopkins called the phenomenon "merely an effect of perspective, but a strange and beautiful one," that is, the phenomenon depends not simply upon the object, but involves the position and perception of the observer: this careful analysis of subjectivity, characteristic of Hopkins' prose, cannot be too strongly emphasized, for it will be of importance in both Ruskin's critique of scientists and Hopkins' indirect replies.

Hopkins' phrasing in this second letter also had the detail and beauty of his prose and poetry: "Yesterday the sky was striped with cirrus clouds like the swaths of a hayfield; only in the east there was a bay or reach of clear blue sky, and in this the shadow-beams appeared, slender, colorless, and radiating every way like a fan open" (November 15, 1883). This second letter appeared *after* the Krakatoa eruptions, although Hopkins does not mention them; furthermore, a number of other letters appeared in this month (November), discussing a "green sun" observed all over the world. In fact one of the "green sun" letters, from "W." of Colombo [Ceylon], was the first in *Nature* to suggest the connection between the "green sun" and the recent eruptions (November 1, 1883). A typical letter on the "green sun" was from C. Michie Smith, an English meteorologist then in Madras, India, who noted that the sun's appearance at setting was both green and blue, and that "most people ascribe the phenomena to the great eruption in Java, but there are difficulties in the way of accepting this view" (November 8, 1883).

The situation, then, by November 1883, had a number of aspects: there were Hopkins' (and others') observations of *rayons du crépuscule* in the eastern sky at sunset, observations of a green or blue sun at sunset, and suggestions of atmospheric disturbance possibly caused by the Krakatoa eruptions. A number of writers pointed out, however, that both *rayons du crépuscule* and a "green sun" had been observed in years past (C. Michie Smith in the letter cited above, and William Swan, a scientist from Scotland, in a letter on November 22, 1883, to mention just two).

For the next two months, through January 1884, *Nature* published extensive analyses by its own editorial staff and from its correspondents on the Krakatoa eruptions and their possible atmospheric ramifications. Krakatoa became a minor obsession with both scientists and nonscientists. Both Robert Bridges, Hopkins' friend, and the poet, A.C. Swinburne, used descriptions of the dramatic sunsets in their poetry, and Tennyson placed the disturbance retroactively in the fifth century in his poem "St. Telemachus" (1812):

> Had the fierce ashes of some fiery peak
> Been hurled so high they ranged about the globe?
> For day by day, through many a blood-red eve,
> In that four-hundredth summer after Christ,
> The wrathful sunset glared....[10]

If aesthetic reactions are expected of poets, then what are we to make of C. Piazzi Smyth, Astronomer Royal for Scotland, who, in his letter to

Nature (December 13, 1883), mentioned that he had spent months (in the past) making "quick colored drawings of . . . any exceptionally fine sunset" and (more recently) making records of the light at sunsets with various instruments, including the "meteorological spectroscope" (for measuring the spectrum and intensity of light from the sun)? Simply this: that Victorian culture had room for and publicly encouraged scientists who knew art in a practical way and artists who had scientific leanings. In fact, Smyth concluded his letter in this way:

> . . . it is to be earnestly hoped, as an outcome of the late remarkable sunsets, and the great numbers of the public by whom they have been witnessed, that our painters will no longer be content to give us so generally mere afternoon pictures slightly yellow ochred and 'light red'–ed near the horizon before the sun goes down, as sunsets, but will more frequently paint the deep red afterglows at their richest.

Critics certainly lectured artists, but do scientists? The answer is yes, at least in the Victorian period: although Smyth's comments seem casual, *Nature* ran a series of columns, written by Lockyer, its editor, assessing the scientific accuracy of landscape paintings in the Royal Academy exhibitions.

Smyth's letter, on the purely scientific level, suggested that the unusual sunsets may easily be the result of the dust from various meteor masses "rather than from a supposed continual accent of one particular charge of volcanic dust from Java, full three months after the cessation of all violent disturbance there." Smyth's letter raised the first of the two alternative hypotheses to account for the origin of the sunsets: meteoric dust; the second, the high concentration of water vapor in the atmosphere, was offered by Ruskin's friend, Leslie, and gained a few supporters. Both were, in the end, rejected by the Royal Society report on Krakatoa.

Considering Smyth's position — Astronomer Royal of Scotland — Hopkins' third letter was certainly bold if not downright aggressive: he wrote that "the body of evidence now brought in from all parts of the world must, I think, by this time have convinced Mr. Piazzi Smyth that the late sunrises and sunsets do need some explanation, more particular than he was willing to give them" (January 3, 1884). Hopkins then proceeded, in a long and rigorous argument (of about 2,000 words), to outline, using his own observations and those of others, a rather dizzying array of phenomena. His letter apparently attracted some attention in scientific circles, for it earned him the title of "observer" in the Royal Society report, which reprinted parts of his letter.

I cannot quote this letter in full, but a number of passages reveal Hopkins' facility with both scientific detail and the painter's palette. In discussing the green phase of the phenomena, he wrote that "the green is between apple-green or pea-green (which are pure greens) and an olive (which is a tertiary color). One of Ascroft's drawings (FIGURE 1), shows a green afterglow on November 26, 1883, just one week before Hopkins' observations were made, and the unusual tones of green are apparent even in the reproduction. Hopkins' description of the curious nature of the red phase of the afterglow of the setting sun is also noteworthy, since the details concerning the "lustreless" quality of the light are echoed in the word "blanched", which was first used by Ruskin and later by Leslie. These sunsets differ from others, he wrote,

> ...in the nature of the glow, which is both intense and lustreless, and that both in the sky and on the earth. The glow is intense, this is what strikes everyone; it has prolonged the daylight, and optically changed the season; it bathes the whole sky, it is mistaken for the reflection of a great fire; at the sundown itself and southwards from that on December 4, I took a note of it as more like inflamed flesh than the lucid reds of ordinary sunsets. On the same evening the fields facing west glowed as if overlaid with yellow wax.

Ascroft's drawing (FIGURE 2) gives us this sense of "inflamed flesh" nicely. Hopkins continued with a comparison with Rembrandt's paintings, surely an unusual citation for a scientific journal:

> The two things together, that is intensity of light and want of lustre, give to objects on earth the peculiar illumination which may be seen in studios and other well-lit rooms, and which itself affects the practice of painters and may be seen in their works, notably Rembrandt's, disguising or feebly showing the outlines or distinctions of things, but fetching out white surfaces and colored stuffs with a rich inward and seemingly self-luminous glow.

In this letter the palette of colors, the analysis of light in Rembrandt, and the precision of natural description generally indicate a sensibility attempting to render the phenomena of the world without forcing a distinction between science and art.

Throughout 1884 the controversy, in a continuous stream of letters, went on. Two moments, however, are sufficiently dramatic to merit special comment. G.J. Symons, a meteorologist, as the Chairman of the Krakatoa Committee, called for the collection of "the various accounts of the volcanic eruption at Krakatoa, and atttendant phenomena, " such as the fall of dust, pumice, and sulphurous vapors, unusual barometric readings, the "distances at which explosions were heard, and exceptional effects of light and color in the atmosphere" (February 14, 1884). Although this call for data would be routine in scientific circles in any modern era, here it surely gave the appearance of favoring the volcanic explanation for the unusual phenomena. Some negative reaction was therefore inevitable: *Knowledge* ran a column (June 6, 1884) stating that any data suggesting that Krakatoa was *not* the explanation for the spectacular sunsets should be sent to *Knowledge* instead. We should note at this point, however, that modern science has confirmed Symons' Committee's findings: according to Fred M. Bullard, a volcanologist, the 1883 eruptions—perhaps the "greatest" series of eruptions of "historic time"—provided for scientists then and now an excellent "opportunity to study the distribution of volcanic ash in an eruption."[11]

The second moment of note involved the entrance of Robert Leslie into the letters column of *Nature* and a review of the "storm-cloud" lectures that Ruskin delivered in London in February 1884. These two events were actually closely related, but only those who consistently read both Ruskin and *Nature* would have understood the relationship. We should note the critique of Ruskin first, since it appeared in the same issue as Symons' call for data and reports.

Under the heading of "Mr. Ruskin's Bogies," Ruskin's lectures were severely criticized and satirized by W. Clement Ley, a meteorologist who had already contributed a letter on *rayons du crépuscule* to the same issue of *Nature* in which Hopkins' first letter appeared. Ley recommended that "those who sympathize" with Ruskin's theory of the "storm-cloud" should "study some of those little books which are beginning to be the delight of our children"; Ley added that Ruskin was a representative of a Philistinism "which shows itself in oppostion to scientific culture" (February 14, 1884). Ley's critique was certainly mean-spirited, but it simply imitated Ruskin's own tone when baiting the scientists. Ruskin was vulnerable, as Ley perceived, because of the subjectivity of his scientific descriptions and his lack of records, since barometric gradients, wind force statistics, and other data favored by Ley were not cited. Ley's report was unquestionably accurate, for Ruskin had attacked the essence of scientific methodology in those lectures by arguing that "observations by instruments or machines were useless." Ruskin had gone to consult the anemometer at Radcliffe Observatory in Oxford, but he rejected its data: "What is the use of scientific apparatus," he asked, "when it can't tell you whether [the wind] is a strong medicine, or a strong poison."

Ruskin's approach to science at this stage of his career could scarcely have been welcomed in the pages of *Nature*, and for this reason, perhaps, Robert Leslie, his friend, would have to act as his stand-in. Leslie was no doubt present at the London lectures, where he would have seen Ruskin, in typically dramatic fashion, display one of his own drawings, done at Brantwood, "of a sunset in entirely pure weather, above London smoke"; it was a drawing, Ruskin asserted, of "one of the last pure sunsets" he had seen in 1876. Ruskin then remarked that, for contrast, he *could have* "blotted down...a bit of plague-cloud to put beside" this "last pure sunset," but all his listeners could see an ugly sunset every day. He added, somewhat illogically, that even his garden was not growing properly anymore. (In fact, in his despondency he had neglected it.) The rhetorical finish of these lectures was the phrase, "blanched sun, blighted grass, blinded man," a phrase which metaphorically linked the radical change in the sun's appearance, the decay in his garden, and the moral blindness in society. Citing biblical passages, Ruskin could only reiterate what "every seer of old predicted": that "physical gloom" mirrors the "moral gloom" of society.

When Ruskin delivered his lectures, he thought that the plague-cloud with its effects and the Krakatoa eruptions with their solar phenomena were no doubt related; a short time later, when he published the lectures, he added a note indicating that he realized that a body of opinion held that the Krakatoa eruptions had *not* caused the unusual solar effects. But for him, in any case, the matter of causation was not acute: in his lectures he had said that the plague-cloud had blotted out traditional English sunsets at least seven years before Krakatoa. In fact the two sets of phenomena *were* in one specific way related, for some of the Krakatoa sunsets did fulfill his prophecy of a plague-cloud choking off the sun's rays. Ruskin, it will be recalled, had told his audience that the plague-cloud "blanched" the sun "instead of reddening it"; he added that if they were "in a hurry to see what the sun looks like" behind such a cloud, they had "only to throw a bad half-crown [coin] into a basin of soap and water.

This description of a "blanched sun" corresponds to one of the special atmospheric phenomena associated with Krakatoa, called "Bishop's Ring," named after its first published observer, the Reverend Sereno Bishop of Hawaii. Another of Ascroft's "Sky Sketches" (FIGURE 3) from September 2, 1884 captures this "blanching" effect beautifully, and reveals, to use Bishop's own descriptions, the sun's "whitish corona" or "haze canopy" (*Nature*, April 10, 1884, and "Prize Essay," *The History and Work of the Warner Observatory*, Rochester, New York, 1887). Bishop described the "Ring" in this way: "A conspicuous object when the sun is high has been from the first the opalescent silvery glow around the sun. This occupies a circle of 25° radius or more. The outer part develops a pinkish hue, which against the blue sky shows lilac or chocolate tints" ("Prize Essay").

Bishop's description was matched by others in the pages of *Nature*, but Leslie's contribution concerning the "Ring" raised the issue of its definite connection with Krakatoa. Leslie wrote:

> I have watched the sky as an artist (out of London) for quite forty years, and feel sure that this corona, or blanching of the sun, has been a more presistent feature of late years than formerly....The last very mild winter and the preceding one could have had no connection with the Krakatoa eruption, and I think we must now seek for an explanation of the present and past atmospheric phenomena in some increase of solar energy, and consequent lifting of vapor higher than usual. (September 11, 1884)

Leslie's other letters to *Nature* (September 25, October 16, and December 4, all in 1884), like the one quoted above, differ markedly from those of other correspondents, because Leslie, like Ruskin, believed that the unusual solar phenomena had occurred for a number of years previous to the Krakatoa eruptions. Furthermore, he insisted on the use of "blanched," Ruskin's term, for the description of the sun; and finally, although Ruskin himself insisted that "the sun's going out,"[12] Leslie believed that there was "some increase of solar energy."

But other scientists, and Hopkins, disagreed. E. Douglas Archibald, a "halo observer for twenty-five years," believed that the "new" solar halo, that is, Bishop's Ring, had "never been seen here before, at any rate within the last twenty years." He went on to describe it briefly: "I remember noting the halo in November, and calling the attention of my assistant to the beautiful salmon color it showed in the interstices of a mackerel sky, which shut off the direct glare of the sun" (October 9, 1884). Hopkins' criticism of Leslie came in the poet's final letter to *Nature* on October 30, 1884, in which Hopkins directly addressed the question of Bishop's Ring. Hopkins wrote that the phenomenon should not be called a halo (which is a clearly defined circle of color, rainbow-fashion, around the sun) nor a corona (which implies that the phenomenon is part of the sun).[13] He noted that the Ring's color was "sometimes rose, sometimes amber or buff" and that "towards sunset it becomes glaring, and white and sallow in hue."

Hopkins also argued in this letter for the objective approach of science over the subjective judgments of Leslie (and, by extension, Ruskin). Even though Ruskin and Leslie disagreed about the changes in solar energy, both men believed that the sun was in fact changing in some dramatic way; Hopkins, like most of the astronomers and other letter-writers to *Nature*, believed that the issues at hand were meteorological or atmospheric, not cosmic:

> If there is going on, as Mr. Leslie thinks, an "increase of sun power," this ought to be both felt and measured by exact instruments, not by the untrustworthy impressions of the eye. Now Prof. Piazzi Smyth says that sunlight, as tested by the spectroscope, is weaker, not stronger, since the phenomena of last winter began. To set down variations in light and heat to changes in the sun, when they may be explained by changes in our atmosphere, is like preferring the Ptolemaic to the Copernican system. (October 30, 1884)

So much for Ruskin's trust in the eye over the scientific instrument! Removed from the context of Victorian science, Hopkins' remark about the "untrustworthy impressions of the eye" would seem very uncharacteristic of the precise and careful nature poet we know him to be.

Hopkins concluded his letter with an impressive summation of the principal issues that had been debated in the journal for two years:

> It is...right and important to distinguish between phenomena really new from old ones first observed under new circumstances which make people unusually observant. A sun seen as green or blue for hours together is a phenomenon only witnessed after the late Krakatoa eruptions (barring some rare reports of like appearances after like outbreaks, and under exceptional conditions); but a sun which turns green or blue just at setting is, I believe, an old and, we may say, ordinary one, little remarked till lately.

In addition to Hopkins' role as a disinterested evaluator, evident in the passage just quoted, Hopkins the experimentalist was also behind this letter, as he attempted to explain how sunlight can affect the eye: "it may be noticed," he wrote, "that when a candle-flame is looked at through colored glass, though everything behind the glass is strongly stained with the color, the flame is often nearly white." But Hopkins the poet (the glow was "bronzy near the earth; above like peach, or of the bluish color on ripe hazels") and Hopkins the speculative scientist ("It would seem as if the volcanic 'wrack' had become a satellite to the earth, like Saturn's rings, and was subject to phases, of which we are now witnessing a vivid one") were present as well.[14]

On scientific grounds, there is little doubt that Hopkins' criticism of Leslie was accurate and telling and that in general Hopkins' contributions to *Nature* were of a very high standard. With the exception of Ruskin, Leslie found himself without allies. The Royal Society excluded Leslie's role and in effect Ruskin's role in the debate by not using any of Leslie's letters in the Krakatoa report, even though *Nature* was a major source of data and even though it had allowed some space for alternative views of the origins of the unusual sunsets. The subsequent careers of Ruskin and Leslie were not often remarked in the public press, except in the case of Ruskin, with some regret: he soon passed into a terrible senility (and probably insanity), never reappearing in public after the "storm-cloud" lectures. Leslie himself continued painting and writing, but almost exclusively on marine subjects.

Our other figures passed, perhaps unfortunately, into temporary oblivion. Ascroft's and Dyason's works were rarely exhibited again, even in their lifetimes. Only Hopkins, after thirty years of silence after his death in 1889, gained fame from the posthumous publication of his poetry in 1918 (edited by the then poet laureate and Hopkins' old friend, Robert Bridges) and of his prose in the 1930s.

In the 1880s, however, all of these figures came into critical contact, and Victorian culture was the richer, as science and art joined—and sometimes contended—for the vision of the spectacular sunsets.

ACKNOWLEDGMENTS

For permission to reproduce FIGURES 1 and 2, I would like to thank the Royal Society of London; for FIGURE 3, I would like to thank the Science Museum, South Kensington. For invaluable aid in carrying out this research, I am grateful to Ms. Mary Ellen Ryan and the Reference Department of the Library of Northern Kentucky University; to Ms. Judy Turner of the Meteorology Department of the Science Museum, South Kensington; and to the librarians of the Royal Society.

NOTES AND REFERENCES

1. See David A. Roos, "Matthew Arnold and Thomas Henry Huxley: Two Speeches at the Royal Academy, 1881 and 1883," *Modern Philology*, vol. 74 (February 1977), pp. 316–324.
2. Jack Lindsay, *Turner: His Life and Work* (1966; St. Albans: Granada, 1973), pp. 246–249.
3. The letter by Ascroft is dated July 12, 1886, and is located in the files of the Krakatoa Committee, Royal Society of London.
4. The first quotation is from Dyason's letter, dated February 5, 1884, to the Secretary of the Royal Society (located in the Miscellaneous Correspondence file); the second quotation is from a letter to the Krakatoa Committee, undated but from early 1884 (located in the files of the Krakatoa Committee).

5. All four letters are, of course, available in *Nature*; the first three are reprinted in Appendix II, pp. 161–166, of *The Correspondence of Gerard Manley Hopkins and Richard Watson Dixon*, ed. Claude C. Abbott (London: Oxford University Press, 1935; 2nd ed., 1955). The fourth letter is reprinted in Patricia M. Ball, *The Science of Aspects: The Changing Role of Fact in the Work of Coleridge, Ruskin, and Hopkins* (London: The Athlone Press, 1971), pp. 148–150; Ball's book analyses the Krakatoa letters as part of the Romantic tradition of nature description, but not as part of the contemporary scientific scene. The third letter was reprinted, in part, in G.J. Symons, ed. *The Eruption of Krakatoa and Subsequent Phenomena: Report of the Krakatoa Committee of the Royal Society* (London: Trübner and Co., 1888), p. 172

6. For the subjective and objective factors involved in Ruskin's attitudes during this period, see John D. Rosenberg, *The Darkening Glass: A Portrait of Ruskin's Genius* (New York: Columbia University Press,1961), especially pp. 214–215. Ruskin's lectures, "The Storm-Cloud of the Nineteenth Century," are reprinted in *The Works of John Ruskin*, ed. E.T. Cook and A. Wedderburn (London: George Allen, 1903–1912), vol. 35.

7. The quotation is from Philip Henry Gosse, *Sea–Side Pleasures* (London: Society for the Propagation of Christian Knowledge, 1853), p. 8; see also David Elliston Allen, *The Naturalist in Britain: A Social History* (Harmondsworth: Penguin Books, 1978) for locating the role of naturalists and amateur scientists in the context of Victorian science.

8. See George P. Landow, "The Rainbow: A Problematical Image," in *Nature and the Victorian Imagination*, ed. U.C. Knoepflmacher and G.P. Tennyson (Berkeley: University of California Press, 1977), pp. 341–369.

9. Ruskin's letter on reflected rainbows is available in *The London Review* (May 16, 1861); it was reprinted once during his lifetime in *The Arrows of the Chace* (1880). Symons' "Extraordinary or Water-Reflected Bow," a reply to a letter and sketch by Charles Clouston, was published in *Symon's Monthly Meteorological Magazine*, vol. 10 (December, 1875), pp. 165–170.

10. For a discussion of the poetic efforts of Tennyson, Bridges, and Swinburne, and the prose of Hopkins concerning Krakatoa, see Richard D. Altick, "Four Victorian Poets and an Exploding Island," *Victorian Studies*, vol. 3 (March, 1960), pp. 249–260. Altick includes a brief discussion of the eruptions and reprints a diagram of the volcano from *Nature*.

11. Fred M. Bullard, *Volcanoes of the Earth*, 2nd ed. (Austin: University of Texas Press, 1976), p. 48.

12. John Ruskin, *Dilecta*, Chap. I (1886), in *Works*, vol. 35, pp. 579–580. (*Dilecta* was published concurrently with the autobiography *Praeterita* as a kind of ongoing appendix of miscellaneous letters and commentary.)

13. In the only reply (as far as I know) to Hopkins' letter, T.W. Backhouse on November 13, 1884 wrote that there is an acceptable use of "corona" to designate a meteorological phenomenon "classified by some observers with halos"; he did not answer Hopkins' objection in terms of the *color*, since Bishop's Ring was a distinctly colored "haze" and not rainbow-colored (which halos have as their distinguishing feature).

14. Bishop in his "Prize Essay" raised a similar speculation a few years after Hopkins' letter: Bishop wrote that future study of the Krakatoa "smoke-belt" may "furnish material aid in elucidating the still mysterious problem of the Belts of the planet Jupiter."

The Language of Science and Psychology in George Eliot's *Daniel Deronda*

SALLY SHUTTLEWORTH
Department of English
Princeton University
Princeton, New Jersey 08544

Daniel Deronda (1876) has frequently been regarded by critics as an anomaly in George Eliot's fiction. Although the portrait of Gwendolen Harleth is seen as one of the highest achievements of psychological realism, it is argued that in the Jewish sections of the book George Eliot relinquished firm control of her art and wandered off into the vagaries of mysticism.[1] I intend to challenge this critical assessment and to demonstrate that contemporary scientific ideas and theories of method provided a basis not only for the psychological theory, but also for the social and moral vision, and narrative methodology of *Daniel Deronda*.

With respect to narrative form and social theory, *Daniel Deronda* certainly differs from George Eliot's earlier fiction. Yet it is cast in terms of the moral framework common to all of George Eliot's previous works. The protagonists within each novel are shown struggling against social constrictions and the limitations of egoism, each striving to reconcile desire for individual fulfillment with the demands of social duty. Both the definition of the problem and its ideal resolution in the organic union of individual and society can be traced, I will argue, to the social and scientific theories of organicism that arose at the end of the eighteenth century.

It is impossible to assign definite temporal priority to either the social or scientific theories of organicism. What can be stated, however, is that the scientific organic theories developed at the end of the eighteenth century offered an organizational model to which social theorists turned in order to make sense of society following the upheaval of the French Revolution. The social theory espoused by the Revolutionary leaders was atomistic, based on the physical principle of association, which also dominated the natural sciences of the eighteenth century. Thus, Buffon viewed the organism as an association of parts whose movements could be interpreted according to the Newtonian mechanical laws of attraction, while social theorists similarly viewed society as an association of independently formed parts.[2] Like the earlier Hobbesian theory of social order, this model incorporated into its essential argument neither ideas of a reciprocal interaction between whole and part nor those of historical growth and change. Society, the revolutionary theorists believed, was simply a collection of individuals, all endowed with equal rights, and thus an artificial structure that could be transformed by the rational action of men.

In the wave of reaction that followed the Revolution, theorists throughout Europe turned to the newly developed principles of organic life to express an alternative social vision. Social theories of organicism arose at the same time that biology emerged as an autonomous science.[3] Cuvier and Goethe, who were among the founders of biology, were in fact directly influenced by contemporary philosophy. They translated into specific biological theory Kant's classic definition of the organism in The Critique of Judgement (1790): a whole in which each part is reciprocally means and end. Hitherto there had only existed the science of natural history, which included organic and inorganic phenomena alike and was primarily concerned with the recording and classification of details of external form. With the work of scientists such as Cuvier and Goethe, however, attention was turned to the formative processes of life itself. Interest was now focused on the internal principles of organization and the relationship between part and whole.

For social philosophy this organic conception offered a model that appeared to reconcile the conflicting ideas of individualism and social integration. On the historical plane, the idea of organic growth — development linked to stability of form — appeared to fulfill the demands of both historical change and continuity. In place of the atomism of eighteenth-century philosophy, organic theory stressed the interdependence of the whole, rather than the freedom of the parts; the necessity for gradual, cumulative growth, rather than the infinite potentiality for change. In the light of the organic analogy, the Revolutionary ideas of individualism were revealed to be dangerously related to a morally reprehensible, and socially disruptive, egoism, and the doctrine of rights associated with the upheaval of the Revolution was replaced by that of social duty. Thus, the Nature philosopher Johann Fichte argued in the

Principles of Natural Right (1791) that the distinction between an isolated man and a citizen was like that between the parts of an inorganic and organic body: "In the organic body each part constantly maintains the whole, and is in maintaining the whole thereby itself maintained, just so stands the citizen in relation to the state."[4] The doctrine of rights was replaced by that of duties, of individual interests by that of social functions. Despite the gulf that separated German Romanticism from French Positivism, Comte offered similar arguments. He observed that in the true organic society "the vague and stormy discussions of rights would be replaced by the calm and precise discussion of duties."[5] His proclaimed "science of society," being founded on the theory of an organic interdependence between part and whole, appeared to offer a scientific foundation for the moral conception of duty.

The ideal of organic social union defined by Comte underlies all George Eliot's novels and can clearly be traced in *Daniel Deronda*. Daniel's goal is to become "an organic part of social life, instead of roaming about like a yearning, disembodied spirit, stirred with a vague social passion, but without fixed local habitation to render fellowship real."[6] Furthermore, Gwendolen attempts to learn, through Daniel's influence, to transcend her egoism. Dislodged from "her supremacy in her own world," she develops an awareness of her membership in the social organism, acquiring "a sense that her horizon was but a dipping onward of an existence with which her own was revolving" (chap. 69, vol. III, p. 399). George Eliot's relation to the social and scientific thought of her day is not restricted, however, to the vague area of a general moral framework. Although all her works display the same preoccupation with questions of organic unity, there are very marked differences in the social theory and narrative form of the novels. George Eliot's first novel, *Adam Bede* (1858), conforms to what may be termed the traditional ideal of pastoral organicism: it portrays a stable hierarchical society. The cyclical structure of the narrative and the empiricist methodology upon which it is based sustain this static social vision. In *Daniel Deronda* (1876), however, empiricism is supplanted by scientific idealism and the organic ideal is focused not on a concrete society but on membership in the symbolic language community of Hebrew. The static, harmonious model of character and society of the earlier novel is replaced by a more fluid model which accentuates the possibility of conflict; and the unified, cyclical narrative structure is supplanted by a more open and fragmented form. These changes, I will argue, can all be correlated with the development of a specific form of organic theory in the science of the period.

Although George Eliot's thought was certainly affected by German Romanticism, she was primarily influenced by developments in organic theory stemming from the work of Comte. French and German theories of organicism converged only in their general moral orientation: the different biological principles upon which they were based ensured divergence in their social theory and scientific methodology. Comte rejected the vitalist ideas of German Romanticism. Avoiding the extremes of both vitalism and materialism, he defined life neither as a principle in itself, nor simply as the sum of chemical reactions, but as the continuous process of decomposition and recomposition that occurs in the interaction of organism and medium.[7] This biological theory laid the foundations for the later physiology of Claude Bernard and the psychological theory of G.H. Lewes, Eliot's close companion. Drawing on the work of these figures, I will show the significant consequences of this principle for the scientific methodology and the social and psychological theory of *Daniel Deronda*.

If *Daniel Deronda* is viewed purely in thematic terms, it appears to display the rather simplistic division attributed to it by critics. There is a polarity of value in the novel. On the one hand, there is a series of artistic figures, usually alien to English culture, who form the novel's center of positive value: the visionary Mordecai, his sister Mirah, herself a singer, the musician Klesmer, and the variously talented Meyricks. On the other hand, there are the figures concerned with the history of Gwendolen and her disastrous marriage to Grandcourt, who reveal, through their actions and speech, the decline of values, and the corruption prevalent in English high society. Between these two groups stands Deronda in, but not of, English social life; able, therefore, to mediate between the two cultures, to find, for himself, organic social union through the Jewish cause expressed by Mordecai and to act for Gwendolen as guide and mentor. Yet this simple thematic summary fails to capture the complexity of *Daniel Deronda*: the correlation between the value concerns, social theory, and methodology of the novel. *Daniel Deronda* is based on a bitter critique of English society: of its economic and social practices, and the restricted range of social and psychological understanding it permits. The idealism of Mordecai, the psychological analysis of Gwendolen's internal conflict, and on a methodological plane, the significant breaks with the accepted conventions of realism that occur in the narrative structure of the novel, can all be related to George Eliot's search for an adequate form to express this radical social vision.

Adam Bede conforms to George Eliot's programmatic statement of art contained in her essay "The Natural History of German Life," which was published in 1856, just fifteen months before she commenced the novel. The role she adopts as novelist is that of the traditional natural historian: the careful observer, devoted to classification. Her goal is to achieve a direct reflection of reality: "to give a faithful account of men and things as they have mirrored themselves in my mind."[8] *Daniel Deronda*, however, represents a radical break with this naïve realism, the social vision it sustains, and the narrative methodology upon which it is based. The opening words of the novel are a statement of the novelist's constructive role: "Men can do nothing without the make-believe of a beginning." The declaration draws attention to the status of the novel as written text, and thus effectively undermines the dominant assumptions of the traditional realist novel. The surety of the narrator of *Adam Bede* has disappeared; no longer is it assumed that the novelist's role is merely to outline a story that possesses a predefined beginning and end. In place of the solidity of objects portrayed in the opening pages of *Adam Bede*, *Daniel Deronda* starts with a series of questions: "Was she beautiful or not beautiful?" The role of the novelist or scientist in *Adam Bede* had been to classify, to arrest and fix; it was a methodological conception that could neither capture complexity nor contain contradiction. It had thus sustained a static vision of a unified society. In *Daniel Deronda*, however, George Eliot's goal as novelist is no longer to represent a fixed "reality," but to challenge the social assumptions, values, and practices that she saw operating in society. To express this social vision she turned, not to the empiricism of natural history, but to a more radical theory of scientific method.

The admission of the novelist's constructive role resembles Claude Bernard's conception of the experimental scientist as a "real foreman of creation."[9] The vitalist biologists had proclaimed that life was indivisible and experimental biology thus impossible. Comte, however, was able to overcome this objection by situating life in the process of interaction between organism and medium, thus preparing the way for Bernard's development of the science of experimental biology. Bernard contrasts the passive science of observation to active experimentation:

> In sciences of experimentation, man observes, but in addition he acts on matter, analyzes its properties and to his own advantage brings about the appearance of phenomena which doubtless always occur according to natural laws, but in conditions which nature has not yet achieved. With the help of these active experimental sciences man becomes an inventor of

phenomena, a real foreman of creation; and under this head we cannot set limits to the power that he may gain over nature through future progress in the experimental sciences.[10]

Bernard's theory of science casts light on George Eliot's conception of her authorial role in *Daniel Deronda*. In a letter of this period she observes:

But my writing is simply a set of experiments in life — an endeavour to see what our thought and emotion may be capable of — what stores of motive, actual or hinted as possible, give promise of a better after which we must strive.[11]

Like Bernard, George Eliot possesses a vision of future changes made possible by the experimental creation of conditions not yet achieved in nature. Within the novel the idealist Mordecai fulfills this creative role. His visions of an alternative form of social life are based on a firm belief in human potential and his thought processes are directly compared by George Eliot to those of an experimental scientist. On seeing Deronda row towards Blackfriars Bridge, in accordance with the mental image he had conceived of a disciple coming to him, Mordecai felt an exultation "not widely different from that of the experimenter, bending over the first stirrings of change that correspond to what in the fervour of concentrated prevision his thought had foreshadowed" (chap. 40, vol. II, p. 328).

George Eliot cannot, within the scope of the novel, demonstrate the validity of Mordecai's social visions. However, she deliberately breaks the conventions of realism to include within the novel both plot coincidences that strain credibility and individual visions that narrative events fulfill. Examples of the former include the relationship of Mordecai and Mirah, Daniel's Jewish ancestry, and his presence in Genoa when Grandcourt is drowned; and of the latter, Gwendolen's fear of the painting of a death figure at her mother's home which prefigures the scene of Grandcourt's death, and Mordecai's mental picture of Deronda as his savior. The narrative itself thus seems to suggest that there exist forms of mental and social association, and patterns of events, that normal social judgment excludes or denies. George Eliot uses the analogy with science to validate this judgment. Although she does not, as narrator, directly endorse Mordecai's visions, she allows Daniel to reflect that:

Even strictly-measuring science could hardly have got on without that forecasting ardour which feels the agitations of discovery beforehand, and has a faith in its preconception that surmounts many failures of experiment. (chap. 41, vol. II, p. 358)

This theory of scientific method, in which faith in preconception is not necessarily to be overruled by apparent scientific results, also coincides with that of Bernard and his predecessor Comte. Comte warns against the dangers of an unguarded empiricism for "no real observation of any kind of phenomenon is possible, except in so far as it is first directed, and finally interpreted by some theory."[12] In place of the quantitative description of isolated phenomena of natural history, Comte proposes a relational theory founded on the biological concepts of organic process and interdependence upon which he based his social philosophy. Thus, he argues that observation must be subordinated to what he terms the "statical" and "dynamical" laws of phenomena, for "no social fact can have any scientific meaning till it is connected with some other social fact."[13] Bernard similarly argues that experimentation must be guided by theory: "A fact is nothing in itself, it has value only through the idea connected with it or through the proof it supplies."[14] There are occasions, he concludes, when the scientist is justified in ignoring certain empirical "facts" that appear in contradiction to reason.[15] He observes that "blind belief in fact, which dares to silence reason, is as dangerous to the experimental sciences as the beliefs of feeling or of faith which also force silence on reason."[16] George Eliot is not arguing in *Daniel Deronda* that reason should be abandoned for a vague, implausible, mystical faith, but that man should not be content to accept appearances unchallenged. He should remain open to hitherto unexplored alternatives or different forms of thought.

In the novel's opening epigraph George Eliot formulates a comparison that points to the underlying methodological premises of the novel, and relates the practice of science to that of poetry:

> Even Science, the strict measurer, is obliged to start with a make-believe unit, and must fix on a point in the stars' unceasing journey when his sidereal clock shall pretend that time is at Nought. His less accurate grandmother Poetry has always been understood to start in the middle; but on reflection it appears that her proceeding is not very different from his; since Science, too, reckons backwards as well as forwards, divides his unit into billions, and with his clock-finger at Nought really sets off *in medias res*.

Science, like poetry, the epigraph states, cannot provide absolute surety; all findings are dependent on the constructive role of the scientist. Bernard, when defining experimental science, contrasted it with the science of "scholastics or systematizers [who] never question their starting point, to which they seek to refer everything; they have a proud and intolerant mind and do not accept contradiction since they do not admit that their starting point may change."[17] Experimenters, on the contrary, never accept an immutable starting point; their work can thus encompass contradiction. "Possessing absolute truth," Bernard concludes, "matters little to the man of science, so long as he is certain about the relations of phenomena to one another. Indeed, our mind is so limited that we can know neither the beginning nor the end of things; but we can grasp the middle, i.e., what surrounds us closely."[18]

George Eliot's opening epigraph contains a similar statement of relativity. For men, as for planets, neither origins nor history can be fully known: the comfort of certainty must be exchanged for an openness to the unknown. Taken as a methodological premise, the epigraph defines the underlying assumptions that govern the representation of Mordecai's idealism and the psychological analysis of Gwendolen's experience of self-conflict. Both are founded on an epistemology that can be related to Bernard's understanding of organic life, for his theory of the relativity of experimental science stems directly from his belief that the complex relations of determinacy in organic life do not conform to a linear sequence of cause and effect: there is no single point of origin or end. Organic interdependence, he argues, should rather be understood in terms of the old cyclical emblem of a serpent with a tail in his mouth.[19] George Eliot similarly rejects linear causality. Full authoritative knowledge, she asserts in the epigraph, cannot be obtained by tracing through a linear sequence of cause and effect. Through the figure of Mordecai she criticizes those forms of social thought in which relativity or contradiction cannot be accommodated, but which are predicated on an immutable starting point and refer everything to these value-laden premises. Mordecai's idealistic social vision and his apparently nonrational faith challenge accepted social definitions of rationality, revealing forms of connection and association excluded by the limited sequence of thought that passes for rational social judgment. Similarly, in portraying Gwendolen and the conflict and contradiction that characterize her psyche, George Eliot challenges the dominant social conception of the rational actor and the theory of causality upon which it is based. The narrative of *Daniel Deronda* does not follow the linear development of *Adam Bede*, but starts in *medias res* and jumps back and forth in time. Gwendolen, with all her conflicting impulses, is not a unified character, the sum of her previous experiences. Her history cannot therefore be represented through a simple temporal sequence of cause and effect.

It is impossible to divorce George Eliot's methodological reflections in *Daniel Deronda* from the narrative form of the novel or its social themes.

By drawing attention in the opening epigraph to the relationship between the continuous process of the "stars' unceasing journey" and the arbitrary division introduced by scientific classification or narrative intervention, George Eliot introduces as a methodological question the preoccupation with the relationship between part and whole that also underlies the social and moral vision of the novel. The comparison, in addition, establishes a chain of astronomical imagery, which is employed throughout the novel to define both the "science" of Mordecai's idealism and the fundamental values of the novel. Thus, when it next occurs, the function of astronomical imagery is to criticize the restricted nature of Gwendolen's social vision. In the epigraph to the chapter in which Klesmer delivers his devastating verdict on Gwendolen's singing, and the culture that produced it, imagery drawn from astronomy is employed to define Gwendolen's wounded egoism: "Croyez-vous m'avoir humiliée pour m'avoir appris que la terre tourne autour du soleil? Je vous jure que je ne m'en estime pas moins" (chap. 6, vol. I, p. 72).

The epigraph captures both Gwendolen's bravado and the relationship between self-assessment and cosmology. It is taken from Bernard Fontenelle's *Entretiens sur la Pluralité des Mondes*, which, published in 1686, explored the social and personal implications of the new Copernican astronomy. The contrast it draws between the closed self-assurance of the stable Ptolemaic universe and the admission of infinite space, of personal insignificance and of change in the Copernican universe is central to George Eliot's examination of nineteenth-century social experience. On the one hand, astronomy is employed to illustrate the traditional organic ideal of a stable, harmonious order. For the Meyricks the objects in their home seemed "as necessary and uncriticised a part of their world as the stars of the Great Bear seen from the back windows" (chap. 18, vol. I, p. 294). On the other hand, however, astronomy illustrates the relativity of Gwendolen's universe, its lack of absolute order or stability. The narrator intervenes to lament Gwendolen's lack of roots and a fixed home, for she cannot therefore perceive the stars as an unchanging order, since "the best introduction to astronomy is to think of the nightly heavens as a little lot of stars belonging to one's own homestead" (chap. 3, vol. I, pp. 26–27). Without this stable personal foundation astronomy inspires Gwendolen only with terror: "The little astronomy taught her at school used sometimes to set her imagination at work in a way that made her tremble: but always when some one joined her she recovered her indifference to the vastness in which she seemed an exile" (chap. 6, vol. I, p. 90). The sense of panic, of displaced centrality, is equivalent to that created by the discovery that the sun did not move around the earth.

Fontenelle's Marchioness was horrified by the picture of an infinite and changing universe. Her philosopher companion, by contrast, found inspiration in the new astronomy, experiencing in its vision of infinite space and constant movement a sense of liberty and release.[20] The two perspectives underlie *Daniel Deronda*, encapsulating the closed, arrogant social vision of the English ruling class, and the expansive possibility envisaged by Mordecai. Gwendolen moves between these two poles. First encountered by the reader in a foreign gaming saloon, where the only law is that of chance, she dwells in a world without apparent fixity, whether of place, social class, fortune, or religion. Her "Ptolemaic response" to life has been to see herself as the center around which all else revolves. The novel traces the progress of her education, through which she finally learns to accept the vastness of the universe, the earth's insignificance, and thus a horizon wider than the self. It is a process that is accomplished through her relationship with Deronda, and contact with the values represented by the "Jewish side" of the novel.

The Jewish mystic Mordecai, who offers through his social vision an alternative to the stultifying, enclosed life of English society, is explicitly compared to Copernicus and Galileo,[21] scientists who broke through the geocentric perspective of Ptolemaic astronomy and changed our conceptions of the universe. The scientific comparison is not restricted, however, to the revolutionary nature of Mordecai's visions; it has extended social and methodological implications. With reference to Mordecai's possible social impact, George Eliot observes that the appearance of fanaticism should not discredit a theory: responsibility for judgments rests with each individual:

> Shall we say, 'Let the ages try the spirits, and see what they are worth?' Why, we are the beginning of the ages, which can only be just by virtue of just judgment in separate human breasts — separate yet combined. Even steam-engines could not have got made without that condition, but must have stayed in the mind of James Watt. (chap. 41, vol. II, pp. 354–355)

The discussion associates Mordecai's visions, not with the vagaries of mysticism, but with the concrete realm of scientific practice. Furthermore, it reintroduces, as both a social and methodological question, the novel's dominant concern with the relationship between part and whole. The process according to which Mordecai's visions are to be assessed anticipates Deronda's conception of organic social union. In embracing Judaism, Daniel accepts his grandfather's belief that "the strength and

wealth of mankind depended on the balance of separateness and communication" (chap. 60, vol. III, p. 273). The organic ideal that underlies
the novel — the balance of individuality and social integration — also
determines the representation of Mordecai's idealism.

Imagery of astronomy is also extended to include that other element of
the Jewish story: Daniel's love for Mirah. George Eliot observes that even
in Romeo's discourse with Juliet his objections to Ptolemy would have
had an effect: "this passion hath as large a scope as any for allying itself
with every operation of the soul: so that it should acknowledge an effect
from the imagined light of unproven firmaments, and have its scale set to
the grander orbits of what hath been and shall be" (chap. 32, vol. II, p.
125). Even love is affected by breadth of vision. This perspective raises
questions concerning scientific method: the mind must accept not simply
a wider horizon but "unproven firmaments," that which cannot be
known, scientific hypotheses that cannot be proved. The observations
belong, not to the world of Paley and natural history or theology, but to
that of Darwin or Bernard; to a world in which the certainty of fixity was
renounced for the potentiality of constant change. They can be related
not to the traditional Baconian model of science rejected by Comte and
Bernard, but to a theory of scientific method that stresses deduction and
the role of the imagination. Thus G.H. Lewes, drawing his conclusions
from Comte and Bernard, argued in *The Foundations of a Creed* that
"fictions are potent; and all are welcome if they can justify themselves by
bringing speculative insight within the range of positive vision."[22] As a
methodological statement it defines George Eliot's practice in *Daniel
Deronda*.

Lewes insists in *The Foundations of a Creed* that ideal construction is
essential to the scientific method:

> It is through the manifold ideal constructions of the Possible that we learn
> to appreciate the Actual. Facts are mere letters which have their meaning
> only in the words they form; and the words again have their meaning not
> in themselves alone but in their positions in the sentence.[23]

Both the linguistic image and theory of scientific method are based on an
analogy originally drawn by Bernard to define organic interdependence.
Bernard argues that the function of an organ can only be comprehended
by relating it to the operation of the whole. To try to discover
physiological properties by isolating organs is like trying to determine
the difference between comedy and tragedy by seeing which has more a's
and which more b's:

En effet, les lettres ne sont rien par elles-mêmes, elles ne signifient quelque chose que par leur groupement sous telle ou telle forme qui donne un mot de telle ou telle signification. Le mot lui-même est un élément composé qui prend une signification spéciale par son mode de groupement dans la phrase, et la phrase, à son tour, doit concourir avec d'autres à l'expression complète de l'idée totale du sujet. Dans les matières organiques il y a des éléments simples, communs, qui ne prennent une signification spéciale que par leur mode de groupement.[24]

Bernard's theory of organic interdependence gives rise not only to a theory of scientific method, but also to an epistemology that challenges empiricism in science and naïve realism in art. There is no one-to-one correspondence between sign and signified, for meaning, like organic life, is a product of a total system.

The consequences of this theory can be traced in *Daniel Deronda*, not only in George Eliot's use of ideal construction, but in her approach to language. Her conception of science is no longer that of the natural historian, who limits his activities to the simple labeling and classifying of physical reality. In place of this confident authority in the power of definition, Geroge Eliot questions in *Daniel Deronda* the functions of language, and thus her own role as author. Of Gwendolen's eyes being "mysteriously" arrested by Grandcourt she comments:

Mysteriously; for the subtly-varied drama between man and woman is often such as can hardly be rendered in words put together like dominoes, according to obvious fixed marks. The word of all work Love will no more express the myriad modes of mutual attraction, than the word Thought can inform you what is passing through your neighbour's mind. (chap. 27, vol. II, pp. 38–39).

The observation expresses a rejection of modes of thought that presume that all life can be known and strictly defined or quantified. It arises directly from George Eliot's search for a language and narrative methodology that could express her radical social vision. Thus, her distrust of "fixed marks" finds a social and thematic reflection in the novel's social critique of those worldly figures who, following the rules of liberal, discursive reason, judge all issues according to neat, predefined categories. Deronda easily envisages Sir Hugo's traditional-minded assessment of Mordecai: "In such cases a man of the world knows what to think beforehand" (chap. 41, vol. II, p. 353). The phrase "man of the world," expressing the class assumptions of those who equate social position with authoritative knowledge, reveals the relationship between linguistic usage and social power.

The ruling classes' confidence in their power of definition is shown by their political callousness toward the Jamaican Negro. In supporting Governor Eyre's brutal massacre of the Jamaicans, Grandcourt dismisses the Jamaican Negro as "a beastly sort of baptist Caliban" (chap. 29, vol. II, pp. 80–81). The smug assumption of authority in definition is but the linguistic correlative of the expression of assertive power in colonial rule. The species identification, which absolves the speaker from confronting the horror of the murder, preserves ruling-class confidence in its own rule by blocking perception of alternative definitions.

Violence is institutionalized by language — whether that of imperialist murder or its more subtle forms of expression, which George Eliot traces in the life of English high society. Her critique of the social practices that led to Gwendolen's subjection to Grandcourt's tyranny is based on an analysis of the implications of the language employed by Gwendolen's relatives. The Reverend Mr. Gascoigne, Gwendolen's uncle, uses a monetary metaphor to quiet doubts about Grandcourt's past, thus illustrating the contemporary basis of the religion he represents: "All accounts can be suitably wound up when a man has not ruined himself, and the expense may be taken as an insurance against future error. This was the view of practical wisdom; with reference to higher views, repentance had a supreme moral and religious value" (chap. 13, vol. I, p. 207). All religious considerations or values are subordinated to the language of the cash-nexus. Rebellion against the dominant social values thus takes the form of a challenge to its language. The heiress Catherine, wishing to marry the musician Klesmer, struggles with her parents for the power of definition. She refuses her mother's conception of duty for "People can easily take the sacred word duty as a name for what they desire anyone else to do" (chap. 22, vol. I, p. 370). Her father's appeals to the "nation" and "public good" she exposes as the linguistic mask for his belief that an heiress should "carry the property gained in trade into the hands of a certain class" (chap. 22, vol. I, p. 371). Words, as Lewes and Bernard demonstrated, do not hold meaning in themselves; their meaning is dependent on the system of assumptions within which they are employed.

In place of the language and social vision of the English ruling class George Eliot proposes two alternative languages, that of the artist Klesmer and that of the visionary Mordecai. Klesmer, in his conversation with that representative of the English political system, the "aimably confident" Mr Bult, challenges the lack of idealism in English politics, its dependence "on the need of a market" (chap. 22, vol. I, p. 362). He asserts instead the artist's right to rule: "a man who speaks effectively through music is compelled to something more difficult than parliamentary eloquence" (chap. 22, vol. I, p. 363). But, within the novel, it is Mordecai who offers a truly alternative language: Deronda's movement away from the values of English ruling-class society is symbolized by his study of Hebrew.

George Eliot found within Judaism an autonomous yet historically based language and culture, open to imagination and not restricted to the rules of the barren, culture-bound reasoning of nineteenth-century English thought. Jewish culture represented for her the virtues of organic historical growth without the attendant disadvantages of the corruption of the English social organism. She took great care, therefore, to base Mordecai's speeches on those of the medieval scholar, Halevi, thus validating his visionary ideals through an authentic language. Mordecai asserts: "if I chose I could answer a summons before their tribunals. I could silence the beliefs which are the mother-tongue of my soul and speak with the rote-learned language of a system, that gives you the spelling of all things, sure of its alphabet covering them all" (chap. 40, vol. II, p. 343). His challenge is to the ruling class alphabet — the belief that all things may be defined by merely rearranging the sequence of predefined letters. It is a challenge that has its methodological issues in the narrative pattern of *Daniel Deronda.* The abrupt opening and fracturing of temporal and spatial continuity and the pattern of coincidences and fulfilled visions within the novel mark George Eliot's departure from the smooth sequence of cause and effect and the manipulation of predefined themes or figures that are traditionally associated with the realist novel. "Man finds his pathways," Mordecai declares, but "has he found all the pathways yet?" (chap. 40, vol. II, p. 343).

Mordecai questions whether the language of social rationality offers an adequate description of reality. The same underlying question governs George Eliot's subtle, psychological portrait of Gwendolen, who, with her conflicting impulses and self-division, does not conform to the dominant nineteenth-century model of the rational actor. The psychological consequences of the social practices criticized in the novel are shown in the history of Gwendolen. Like her uncle, Gwendolen has no form of spiritual idealism; she associates religion with economics: she had "always disliked whatever was presented to her under the name of religion, in the same way that some people dislike arithmetic and accounts." In a comparison that associates the political consequences of this attitude with imperialism, George Eliot observes that Gwendolen had no more enquired into religion than into the "conditions of colonial property and banking" (chap. 6, vol. I, pp. 89–90). The reference to imperialism is highly significant, for it relates not only to social practices, but also to Gwendolen's understanding of her own psyche. Gwendolen's self-image reflects the social imperialism of her class; she concentrates her ambitions on "the possibility of winning empire" (chap. 6, vol. I, p. 90). Social imperialism, however, cannot brook ambiguity, as Grandcourt's drawling dismissal of the Jamaican Negro illustrates. It effectively blocks perception of contradiction. In similar fashion, Gwendolen's imperialist goal rests on the assumption that she possesses a unified and noncontradictory ruling self. The inadequacy of her conception is revealed, however, by her "fits of spiritual dread," experiences that cannot be accommodated within her attempt to win empire, and for which the language of society could not account.

Gwendolen's image of herself as being in full control of her wishes and desires and in a commanding social role stems from a view of society and the self which ignores all divisive possibilities. She experiences, however, "subjection to a possible self, a self not to be absolutely predicted, . . . [which] caused her some astonishment and terror: her favourite key of life — doing as she liked — seemed to fail her" (chap. 13, vol. I, p. 201). George Eliot exposes the falsity of Gwendolen's self-image: in doing so she draws on contemporary psychological theory. In clinging to the idea of free will, Gwendolen has fallen victim to the theory of causality implicit in the social conventions of language, which, as the psychologist, James Sully, George Eliot's friend and contemporary, remarked, lead us to attribute actions to a dominating ego through the assumption "that the agent expressed by the subject of the verb is the adequate cause" — thus identifying the process of causation with the linear temporal ordering of language.[25] In his discussion of free will Sully argues that forms of speech that ascribe to a person the act of choosing between contending motives imply "not only that there exists quite apart from the processes of volitional stimulation some substantial *ego*, but that this *ego* has a perfect controlling power over these processes."[26]

Sully's challenge to theories of a directing controlling ego is founded on premises drawn from contemporary developments in physiology and psychology that were to undermine the models of man that had dominated early nineteenth-century thought — the Cartesian *cogito* or Bentham's rational actor — and the theories of social order they had sustained. His observations offer a linguistic basis for G.H. Lewes's argument, which Lewes draws from his own work in physiological psychology, that "Consciousness is not an agent but a symptom."[27] Lewes argues that the unity and simplicity assigned to a "Thinking Principle" cannot be upheld: "In any positive meaning of the term, that Principle is not an antecedent but a resultant, not an entity but a convergence of manifold activities."[28] Lewes's theories, which define the principles that lie at the heart of the social and psychological vision of *Daniel Deronda*, are based on a radical new conception of the organism drawn from the theories of Claude Bernard. Bernard developed Comte's earlier theories to produce a definition of life as a regulative process of interaction between an internal and an external milieu. He thus proposed a theory of causality that transcended the old division of materialism and vitalism and, in its vision of the complex causal chains in the interactive process, laid the foundations for the principles of homeostasis or feedback control and modern communications theory. His biological theories suggested a mode in which the split could be closed not only between materialism and vitalism, but also between mind and body and the organism and environment of earlier physiology. Lewes applied the same principles to psychology and challenged conceptions of individual autonomy, the dualism of subject and object, self and other, and the Cartesian division of mind and matter, which had sustained the identification of the self with conscious thought. Viewing the "Thinking Principle" solely as "the convergence of manifold processes," he replaced the conception of a unified, rational actor with a model of mind that stressed contradiction and the operation of the unconscious, and the conflict of different streams within the mind. As a psychological conception it had significant social and political implications.

Descartes' psychological theory had supported a conception of society as a mechanical association of autonomous, rational actors; a view whose economic foundations have been aptly summarized recently in Anthony Wilden's observation that "Every cogito was free to sell his disposable energy at the best price."[29] It was on this model of social order that Herbert Spencer's social organicism had been based. Gwendolen's "key of life — doing as she liked" conformed to Spencer's "first principle" of organic theory that "Every man has freedom to do all that he wills, provided he infringes not the equal freedom of any other man."[30] However, subjection to a self "not to be absolutely predicted about" exposes the falsity of such assumptions of individual autonomy and the economic model of free exchange to which they can be related.

With its emphasis on the principles of energy exchange, Spencer's organicism had remained within the epistemological atomism of social physics, whose premises Bernard had overturned. As Buckley, writing to define the basis of modern communication theory, observes: "after

Spencer it became clearer that whereas the relations of parts of an organism are physiological, involving complex physico-chemical *energy* interchanges, the relations of parts of society are primarily psychic, involving complex communicative processes of *information* exchange."[31] It is within this latter framework that *Daniel Deronda* primarily belongs: Mordecai's idealism, Gwendolen's experience of unconscious conflict and contradiction, Daniel's final incorporation not within the economic social organism of England, but a symbolic language community, all signal George Eliot's departure from the linear causality of Spencer's mechanistic interpretation of the social organism. Arguing that the government could not alter the sum total of injustice, Spencer had concluded:

> It is impossible for man to create force. He can only alter the mode of its manifestation, its direction, its distribution. . . . This is as true in ethics as in physics. Moral feeling is a force—a force by which men's actions are restrained within certain prescribed bounds; and no legislative mechanism can increase its results one iota."[32]

It is precisely against the crass use of such conceptions of energy transfer and their underlying economic and political assumptions that the moral force and epistemology of *Daniel Deronda* are directed.

Daniel dreaded "that dead anatomy of culture which turns the universe into a mere ceaseless answer to queries" (chap. 32, vol. II, p. 132), the belief that the universe could be parceled out into segments of certainty, the question "why" presuming only one chain of causation. Reflecting on Mordecai's faith in their relationship, Daniel cannot dismiss it as an illusion. For Daniel "the way seems made up of discernible links" (chap. 41, vol. II, p. 360); but he refuses to accept the principles of associationism that had underlain earlier mechanistic interpretations of the social organism, to assume that because one causal chain has been traced there cannot be another. Mordecai's conception of society cannot be expressed in terms of quantitative analysis; it is rather, as in Bernard's conception of the organism, a nonlinear homeostatic system: the same starting point may lead to many different ends, and different initial states may produce the same end. In terms of the symbolic system of society this may be expressed through the concept of overdetermination developed by Freud to explain the function of language in the Unconscious: because of the freedom of association that language permits, many causal chains may have led to the same symptom.[33] This principle of causation underlies both the representation of Mordecai's idealism and the psychological analysis of Gwendolen.

Gwendolen's social ambitions and her conception of a unified, directing self are synthesized in her Phaeton-like desire to "mount the chariot
and drive the plunging horses herself" (chap. 13, vol. I, p. 202). The
horses Grandcourt had brought to support his marriage proposal to her
had represented for Gwendolen the social control she would assume:
they were "the symbols of command and luxury, in delightful contrast
with the ugliness of poverty and humiliation at which she had lately been
looking" (chap. 27, vol. II, p. 43). Yet once she has entered her economic
bargain with Grandcourt, her engagement, the horses come instead to
represent her failure to command either her self or her destiny: "it was as
if she had consented to mount a chariot where another held the reins"
(chap. 29, vol. II, p. 76). George Eliot's analysis of Gwendolen's self-
division and of her loss of directing social power is founded on an implicit critique of theories of the self that stress rational, unified control,
and the concomitant theories of society that are based on a conception of
free interaction between autonomous entities.

The imagery George Eliot has chosen is drawn directly from contemporary psychological debate in which Lewes was involved. The analogy
between the soul and the charioteer and his horses dates back to the
Phaedrus of Plato, but in the nineteenth century the imagery of the horse
and its commander was recast to take account of a newly perceived problem in psychology. The discovery of the physiological basis of reflex action had threatened to disturb the model of a unified, controlling psyche.
William Carpenter, therefore, in *Principles of Mental Physiology* (1874),
which Lewes records reading in March of that year, used the analogy of
horse and rider to explain the relationship between the directing Will and
the body's automatic activity. Although the muscles furnish the power,
Carpenter argues, they are controlled and directed by a skillful rider: "the
role and direction of the movement are determined by the Will of the
rider, who impresses his mandates on the well-trained steed with as much
readiness and certainty as if he were acting on his own limbs."[34] In its
reliance on the concept of a dominating Will, Carpenter's argument is
based on a vitalist biology that Bernard in his biological theory, and
Lewes in his psychology, had transcended. Although Carpenter's argument is introduced purely in a physiological context, its political consequences can be traced in the work of Edward von Hartmann. Hartmann
argues in *Philosophy of the Unconscious* that the organization of the
human organism, a hierarchy dominated by a guiding intelligence, supplies a perfect model for political government. His political stance is
directly reflected in his theories of physiological psychology: reflex ac-

tion occurs in response to "commands coming from above," thus proving,

> the artistic and purposive organization of the nervous system, in which the lower energies are kept, it is true, prepared and always ready for action, but at the same time are held in check by the superior authorities as a squadron of skilful riders and snorting steeds by the will of the leader until the moment seems to have arrived for unchaining these energies by a nod.[35]

In rejecting the conception of a dominating Will, the self as a rider in firm control of his steed, George Eliot is also challenging the theory of social order it sustains: the conception of society as a harmonious system, wonderfully coordinated by the guiding intelligence of its rulers.

It should not be assumed, however, that George Eliot employed physiological psychology in *Daniel Deronda* solely in support of radical argument. Like her previous novels, *Daniel Deronda* was written within the moral framework of organicism; George Eliot employs the resources of physiological psychology to give an apparent scientific force or validity to her moral propositions. Thus, Mordecai's plans for the Jewish race are couched in language that combines the sentiments and expressions of the medieval visionary Halevi with the precision of contemporary physiological theory. He proposes that the Jews draw on "the heritage of Israel" beating in their veins, the "inborn half of memory" (chap. 42, vol II, p. 393). Physiologically the Jews represent a cultural and religious unity, a historical continuity that is lacking in fragmented European society: European culture was "an inheritance dug from the tomb. [The Hebrew culture] is an inheritance that has never ceased to quiver in millions of human frames" (chap. 42, vol. II, p. 394). "Where else," Mordecai exclaims, "is there a nation of whom it may truly be said that their religion and law and moral life mingled as the stream of blood in the heart and made one growth" (chap. 42, vol. II, p. 385). The theory of psychic and cultural evolution proposed is that of Spencer's Lamarckian vision: "Every one of the countless connections among the fibres of the cerebral masses answers to some permanent connection of phenomena in the experiences of the race."[36] It is a theory that underlies the opposition dramatized in the text between Mirah, whose "religion was of one fibre with her affections and had never presented itself to her as a set of propositions" (chap. 32, vol. II, p. 128), and Gwendolen, whose lack of hereditary roots is associated with her psychic disunity, her lack of an unquestioned center of value. Yet it is this reliance on purely biological factors that creates the division experienced by the reader in the novel's conclusion, the potentiality of Deronda's future contrasted with the barrenness of Gwendolen's.

George Eliot also employs physiological concepts to reinforce the moral distinction between Mordecai and Grandcourt. Deronda is made to reflect on the role of ardent men: "the men who had visions which, as Mordecai said, were the creators and feeders of the world—moulding and feeding the more passive life which without them would dwindle and shrivel into the narrow tenacity of insects, unshaken by thoughts beyond the reaches of their antennae" (chap. 55, vol. III, p. 213). The power of vision is related to the ardor of its possessor; an opposition in thus proposed between creative force and insect-like tenacity and this is enacted in the text by Mordecai and Grandcourt. Grandcourt is defined by his "small expense of vital energy," his lack of "regulated channels for the soul to move in" (chap. 15, vol. I, p. 232). On Lush, Grandcourt's servant, puzzling over Grandcourt's actions, George Eliot observes: "Of what use, however, is a general certainty that an insect will not walk with his head hindmost, when what you need to know is the play of inward stimulus that sends him hither and thither in a network of possible paths?" (chap. 25, vol. II, p. 9). Without strong directing motives, Grandcourt's energy, like that of an insect, is dissipated in response to immediate stimuli: "How trace the why and wherefore in the mind reduced to the barrenness of a fastidious egoism, in which all direct desires are dulled, and have dwindled from motives into a vacillating expectation of motives. . .?" It is a condition that belongs to a life "unmoulded by the pressure of obligation" (epigraph, chap. 25, vol. II, p. 3). Grandcourt refuses the responsibilities of social duty. His actions are therefore unpredictable, merely a listless response to the changing demands of egoism. Eliot correlates Grandcourt's egoism with vacillation, and Deronda's sense of obligation with directed energy. The contrast between these two sets of correlations reveals an underlying organic ideal that informs Eliot's use of physiological metaphors.

Early in the novel, Daniel, questing for organic union, had been paralyzed by "a too reflective and diffusive sympathy." He longed for "some external event, or some inward light, that would urge him into a definite line of action, and compress his wandering energy" (chap. 32, vol. II, p. 132). Unlike Grandcourt, his problem is not lack of energy but misuse of that precious resource, a consequence of his lack of a definitive role within the social organism:

> But how and whence was the needed event to come?—the influence that
> would justify partiality and make him what he longed to be yet was unable

to make himself — an organic part of social life, instead of roaming in it like a yearning disembodied spirit, stirred with a vague social passion, but without fixed local habitation to render fellowship real?" (chap. 32, vol. II, pp. 133).

The desired organic union is given to him unproblematically through biological filiation. Following the discovery of his Jewish birth "there was a release of all the energy which had long been spent in self-checking and suppression because of doubtful conditions" (chap. 63, vol. III, p. 307). George Eliot's desire to resolve the social questions raised by organicism, to fulfill the role normally associated with the realist novel, and to offer a concrete moral solution to the problems posed, results in an overreliance on ideas of energy transmission and biological inheritance. The divided structure of the text is thus a consequence both of conscious experimentation and the final breakdown of the organic ideal.

Physiological psychology is not only employed, however, to sustain moral argument; it is also used to undermine the conception of a unified, integrated character, which had underpinned the traditional linear development of the realist novel. "We mortals," George Eliot observes of Grandcourt, "have a strange spiritual chemistry going on within us." The comment introduces her reflections on the absence of any correlation between intelligence and rational action; the description of mind that follows conforms to the theory of Lewes: "Grandcourt's thoughts this evening were like the circlets one sees in a dark pool continually dying out and continually started again by some impulse from below the surface" (chap. 28, vol. II, p. 63). In Lewes's favourite image the mind is like a lake: consciousness is the interaction between the stationary waves below the surface and fresh incoming waves.[37] In both illustrations it is the ceaseless action of the waves below the surface of consciousness that determines behavior.

Jean Sudrann in her study of *Daniel Deronda* has argued that whereas twentieth-century psychologists have given contemporary novelists a vocabulary by which to describe the descent into the self, George Eliot was forced to use the idiom available to her: that of melodrama.[38] George Eliot did have available, however, the idiom of Lewes's psychology, which anticipated in many ways twentieth-century developments in psychological theory. An alternative image of the mind employed by Lewes is that of the palimpsest:

> The sensitive mechanism is not a simple mechanism, and as such constant, but a variable mechanism which has a history. What the senses inscribe in it are not merely the changes of the external world, but these characters are

> commingled with the characters of preceding inscriptions. The sensitive
> subject is no *tabula rasa:* it is not a blank sheet of paper, but a palimpsest.[39]

The metaphor of the palimpsest, indicating the coexistence of many psychic levels, was to be employed by Sully to describe the dream process, and later borrowed by Freud to define his distinction between manifest and latent content.[40] In all three writers the image signifies a conception of mind that is neither that of conscious control nor a hierarchy of levels dominated by a directing will. Gwendolen's history is not that which can be captured by the chronology of social time: "there is a great deal of unmapped country within us," the narrator observes, "which would have to be taken into account in an explanation of our gusts and storms" (chap. 24, vol. I, p. 416). The cartography of that region, however, cannot be that of visible surface, but must represent the simultaneity of conflicting desires. In the work of Lewes, George Eliot found a theory adequate to her purpose.

George Eliot, in her descriptions of Gwendolen's psyche, skillfully interweaves terminology from the Romantic-Gothic literary tradition with an interpretation of the mind's processes that accords with Lewes's psychological theory: "Fantasies moved within her like ghosts, making no break in her more acknowledged consciousness and finding no obstruction in it: dark rays doing their work invisibly in the broad daylight" (chap. 48, vol. III, p. 94). Lewes argued in *The Physical Basis of Mind* (which he was working on while George Eliot wrote *Daniel Deronda*) that the operations of the mind were composed equally of sensations within consciousness and the unconscious; there was no absolute distinction between the two levels, merely one of gradation, such as that between light and dark: "The nervous organism is affected as a whole by every affection of its constituent parts . . . . the thrill which any particular stimulus excites will be unconscious, sub-conscious in proportion to the extent of the *irradiated* disturbance."[41] In George Eliot's description, the "rays" of the unconscious operate concurrently with conscious processes, never rising to the surface of consciousness.

At issue is the question of polar opposition: whether of a mode of judgment that would divide all into black and white ("in such cases a man of the world knows what to think beforehand") or of a theory of the self that would institute a rigid division between rational thought and other psychic phenomena. Gwendolen is far from being Carpenter's rider in firm control of his steed. Conscious and unconscious thought patterns coexist without forming one unified stream, while the rigid division between the self and the external world predicated upon the notion of a rational actor or controlled rider is also dissolved. Gwendolen recalls that seeing Grandcourt drowning in the water, "I only know that I saw my wish outside me." In her jump into the water beside him, she was "leaping from my crime, and there it was—close to me as I fell—there was the dead face—dead, dead" (chap. 56, vol. III, pp. 231–32). The external action, the leap away from self, only becomes a new entry into the self, the external embodiment of her crime. It functions as a symbolic demonstration of the fact that the self cannot be identified with the narrow limits of rational consciousness.

George Eliot describes Gwendolen when on the yacht as being "at the very height of her entanglement in those fatal meshes which are woven within more closely than without, and often make the inward torture disproportionate to what is discernible as outward cause" (chap. 54, vol. III, p. 188). The image of the web is one employed by Lewes to define the psyche to demonstrate the impossibility of a division between organism and environment, self and other, consciousness and the unconscious.[42] In George Eliot's illustration the complex interlocking social and psychic meshes do not exhibit linear causality, one-to-one correspondence between external cause and inner effect, but rather overdetermination. The individual cannot be comprehended, in Lockean fashion, as the *sum* of his previous experiences—a quantitative mode of assessment that cannot account for complexity or contradiction. George Eliot contrasts the linear, social time of language with the actual simultaneity of conflicting experience. The contradictory feelings Gwendolen inspires Eliot attributes to:

> the iridescence of her character—the play of various, nay, contrary tendencies. For Macbeth's rhetoric about the impossibility of being many opposite things in the same moment, referred to the clumsy necessities of action and not to the subtler possibilities of feeling. We cannot speak a loyal word and be meanly silent, we cannot kill and not kill in the same moment; but a moment is room wide enough for the loyal and mean desire, for the outlash of a murderous thought and the sharp backward stroke of repentance. (chap. 4, vol. I, p. 57).

Rational, syntactic progression cannot readily capture both conscious and unconscious complexity.

George Eliot describes Gwendolen's psychic conflicts using a variety of techniques. Gwendolen's speech after Grandcourt's death is punctuated by gaps which, like the lapses in speech investigated by Freud, signify an underlying unarticulated disturbance: "for I was very precious to my mother—and he took me from her—and he meant—and if she had known—" (chap. 65, vol. III, p. 341). The literal transcription is also accompanied by explicative commentary as George Eliot seeks not merely to present, but to offer as well theoretical explanations of Gwendolen's behavior: "the question had carried with it thoughts and reasons which it was impossible for her to utter, and these perilous remembrances swarmed between her words, making her speech more and more agitated and tremulous" (chap. 65, vol. III, p. 341).

George Eliot, as author, confronts the same problem as Daniel when faced with the inadequacy of language to aid Gwendolen:

> Words seemed to have no more rescue in them than if he had been beholding a vessel in peril of wreck—the poor ship with its many-lived anguish beaten by the inescapable storm. How could he grasp the long-growing process of this young creature's wretchedness?—how arrest and change it with a sentence? (chap. 48, vol. III, p. 100).

Language appears mere surface gloss when called upon to transform the overdetermined effects of a lifetime. Deronda's problem is that of explication or articulation, and as such is identical with the project of the novel: that of rendering in language the complex network of determinate processes that constitute the conflicts and contradictions of the psyche. Like Mordecai, who challenged the language of "a system that gives you the spelling of all things, sure of its alphabet covering them all," George Eliot is seeking in *Daniel Deronda* a narrative structure and language that could encompass complexity and contradiction: a form that would disrupt the association of the temporal sequence of language with a theory of causality. Linguistic convention, as Sully demonstrated, had led men to attribute actions to a dominating ego through the assumption "that the agent expressed by the subject of the verb is the adequate cause." George Eliot attempts in *Daniel Deronda* to break this convention, to undermine, through manipulation of the language and structure of the novel, conceptions of the self as a unified directing force, and of society as the free interaction of autonomous entities.

Lewes, applying Bernard's theories to psychology, broke down the Cartesian dualism of mind and matter, consciousness and the unconscious, and self and other. Furthermore, in questioning ideas of individual autonomy he also explicitly challenged the correlated social theories of free exchange. The story of Gwendolen is a critique of both dominant conceptions of the self and of correlated social practices. The marriage market into which Gwendolen enters is, George Eliot stresses, one of commodity exchange:

1ST GENT: What woman should be? Sir, consult the taste
of marriageable men. This planet's store
In iron, cotton, wool, or chemicals —
All matter rendered to our shaping skill,
Is wrought in shapes responsive to demand:
The market's pulse makes index high or low,
By rule sublime. Our daughters must be wives,
And to be wives must be what men will choose:
Men's taste is woman's test.

(epigraph, chap. 10, vol. I, p. 144).

Women, like raw materials, are shaped to the needs of the market; while social relations are reduced to relations between isolated objects or commodities. This process receives symbolic expression in the recurrent mirror imagery of the novel.

In chapter 2 we see Gwendolen posing before the mirror, reducing the fullness of her being to the dimensions of her bodily image. Moved by her own reflections she "leaned forward and kissed the cold glass which had looked so warm" (chap. 2, vol. I, p. 21). The opposition between cold and warm suggests the loss entailed for Gwendolen in her identification of herself purely in terms of her isolated physical image.

Yet the mirror in *Daniel Deronda* is but a secondary form of reflection, a physical representation of social practices. Gwendolen was "A girl who had every day seen a pleasant reflection of that self in her friends' flattery as well as in the looking glass" (chap. 2, vol. I, p. 20). Friends and family confirm her identification of self with bodily image. Gwendolen aspires to self-determination, to be unlike others who "allowed themselves to be made slaves of, and to have their lives blown hither and thither like empty ships in which no will was present" (chap. 4, vol. I, p. 53). She desires to "achieve substantiality for herself and know gratified ambition without bondage" (chap. 23, vol. I, p. 378), but she is doomed to failure. Enslaved by the idea of self as reflected image, the only career she can conceive, acting, is based on a marketing of her beauty. Far from achieving freedom, she is merely following the market's pulse, treating herself as a commodity to be shaped according to social demand.

The world Gwendolen enters on her marriage to Grandcourt is one of mirrors. At Ryelands, in their London home, or on the yacht, the walls are hung with mirrors, their endless repetition of images reflecting the enclosed system in which Gwendolen is entrapped. Mirrors offer only an illusion of depth and truthful representation; restricted to the field of surface appearance they cannot capture contradiction. The self-referentiality of their endless repetition is designed to deny the possibility of conflict, of a world outside that cannot be captured with such symmetry. Grandcourt's judgment is of a piece with this world of mirrors: "And, in dog fashion, Grandcourt discerned the signs of Gwendolen's expectation, interpreting them with the narrow correctness which leaves a world of unknown feeling behind" (chap. 54, vol. III, p. 202). Grandcourt exists only within the world of the sign, incapable of comprehending the complexity of symbolic language. In a naïve theory of signification, which George Eliot was attempting to transcend in *Daniel Deronda*, he believes that, mirror-like, there is a one-to-one correspondence between sign and signified. Like Spencer in his mechanistic social theory, Grandcourt does not allow for the introduction of a higher level of complexity than simple energy exchange in social communication. His mode of judgment thus reinforces the psychological atomism of a mirror world.

The figure of Grandcourt is counterbalanced by Gwendolen's "terrible-browed angel," Daniel. Like Klesmer, Daniel refuses to enter the circle of exchange, to treat Gwendolen as a commodity, and to accept the false values and distortions employed as currency there. He offers "not one word of flattery, of indulgence, of dependence on her favour" (chap. 54, vol. III, pp. 195–196) and, stimulating Gwendolen's impulse to complete honesty and confession, he holds up a mirror capable of capturing not just surface illusion, but the multileveled layers of psychic confusion as well. This trait enables him to break through the barriers of self and other. Thus Gwendolen "had learned to see all her acts through the impression they would make on Deronda" (chap. 54, vol. III, p. 195), who becomes for her "an outer conscience" and, in an image that captures the organic process, the "breathing medium" of her joy (chap. 64, vol. III, p. 334). "In this way," George Eliot observes, "our brother may be in the stead of God to us" (chap. 64, vol. III, p. 335). In the transfer and interdependence that characterize their association, Gwendolen and Deronda epitomize the "I and Thou" relationship that Feuerbach saw as the foundation of true religion — a religion that would not constitute a self-alienation by situating itself within God, but one firmly founded in humanity. In *The Essence of Christianity* (a work translated by George Eliot), Feuerbach had criticized the "exaggerated subjectivity of Christianity," which, knowing only individuals and not species, is forced to resort to God as supernatural mediator. The more natural reconciliation, he suggested, is that in which "my fellow-man is per se the mediator between me and the sacred idea of the species."[43] Deronda, breaking down

Gwendolen's sense of autonomous individuality, fulfills this function. Through his influence Gwendolen "was for the first time feeling the pressure of a vast mysterious movement, for the first time being dislodged from her supremacy in her own world, and getting a sense that her horizon was but a dipping onward of an existence with which her own was revolving" (chap. 69, vol. III, p. 399). Like Mordecai, Deronda has played the role of Copernicus, displacing Gwendolen's Ptolemaic sense of cosmology. Feuerbach was to be criticized by Marx for abstracting the human essence from the historical process, situating it in contemplation rather than sensuous activity, but George Eliot's study corrects this deficiency. The relationship between Gwendolen and Daniel is placed firmly within the context of social and economic relations of power in nineteenth-century England.

In a letter to Frederick Harrison, who was to read a paper on Lewes's psychology, George Eliot observed: "It is melancholy enough that to most of our polite readers the Social Factor in Psychology would be a dull subject. For it is certainly no conceit of ours which pronounces it to be the supremely interesting element in the thinking of our time."[44] More clearly than in any of her previous novels George Eliot demonstrates in *Daniel Deronda* the social determination of psychic currents, and reveals the correlation that exists between models of the self and society. Both the idealism of Mordecai and the psychological intricacy of Gwendolen stem, as I have argued, from the same source. Through Mordecai, George Eliot questions atomistic and inductive science, and theories of linear causality; through Gwendolen, she questions social and psychological atomism, the opposition between self and other, consciousness and the unconscious. In both cases Eliot challenges dominant modes of social categorization—the identification of thought with a restricted form of rationality. Her social analysis and theory of psychology and scientific practice are no longer those of *Adam Bede;* they derive, not from natural history, but from the biological and psychological theories of Bernard and Lewes. The ideals of organicism are those that have determined George Eliot's earlier work—the protagonists still strive for the intergration of individual fulfillment and social duty, historical change and continuity. But the underlying model is now firmly that of Bernard: the organism conceived not as fixed structure, but as interactive process. It is a conception whose radical social and psychological dimensions George Eliot explores in *Daniel Deronda*.

NOTES AND REFERENCES

1. The most famous example of this form of assessment is to be found in F.R. Leavis's, *The Great Tradition* (London: Chatto, 1948).
2. See Georges Canguilhem, *La Connaissance de la Vie* (Paris, 1969), p. 57 and Walter Buckley, *Sociology and Modern Systems Theory* (Englewood Cliffs: Prentice Hall, 1967), p. 8.
3. For these arguments concerning the origins of biology, I am indebted primarily to T.H. Huxley, "On the Study of Biology" in *The Scientific Memoirs of T.H. Huxley*, ed. M. Foster and E.R. Lankester, 4 vols. (London: Macmillan, 1898); Ernst Cassirer, *The Problem of Knowledge*, trans. W.H. Woglom and C.W. Hendel (New Haven: Yale University Press, 1950); François Jacob, *The Logic of Life: A History of Heredity*, trans. Betty E. Spillman (New York: Pantheon, 1973); Michel Foucault, *The Order of Things: An Archaeology of the Human Sciences*, trans. (London: Tavistock Publications, 1970); and Oswei Temkin, "Basic Science Medicine and the Romantic Era," in *The Double Face of Janus and Other Essays in the History of Medicine* (Baltimore: Johns Hopkins University Press, 1977).
4. Quoted in F.W. Coker, *Organismic Theories of the State*, eds. Faculty of Political Science of Columbia University (New York: Columbia University, 1910, [Longmans, Green agents]), p. 22.
5. Auguste Comte, *The Positive Philosophy of Auguste Comte*, freely translated and condensed by H. Martineau, 2 vols. (London, 1855), vol. II, p. 472.
6. George Eliot, *Daniel Deronda*, Cabinet Edition, 3 vols. (Edinburgh: W. Blackwood, 1878–80), chap. 32, vol. II, p. 133. References to this edition will be cited hereafter in the text.
7. Comte,[5] vol. I, pp. 361–362.
8. George Eliot, *Adam Bede*, Cabinet Edition, 2 vols. (Edinburgh: W. Blackwood, 1878–80), chap. 17, vol. II, p. 265.
9. Claude Bernard, *An Introduction to the Study of Experimental Medicine*, trans. Henry Copley Green (U.S.A., 1949), p. 18. Although there is no direct record of George Eliot's reading Bernard, he was a major influence on the work of George Eliot's companion, G.H. Lewes. Lewes's diaries (in the Beinecke Rare Book and Manuscript Library, Yale University) reveal his extensive reading of Bernard's works throughout his career. He records reading *An Introduction to The Study of Experimental Medicine* in March 1871, and his copy (in the original French edition), which is kept in Dr. Williams' Library, Gordon Square, London is extensively marked.
10. Bernard,[9] p. 18.
11. *The Letters of George Eliot*, ed. Gordon Haight, 7 vols. (New Haven: Yale University Press, 1950–54), vol. VI, pp. 216–217. To Dr. Joseph Frank Payne, 25 January 1876.
12. Comte,[5] vol. II, p. 97.
13. Comte,[5] vol. II, p. 97.
14. Bernard,[9] p. 53.
15. Bernard,[9] p. 54: "It follows from the above that if a phenomenon in an experiment had such a contradictory appearance that it did not necessarily connect itself with determinate causes, then reason should reject the fact as non-scientific."
16. Bernard,[9] p. 53.
17. Bernard,[9] p. 50.
18. Bernard,[9] p. 50.
19. Bernard,[9] p. 88.

20. Bernard Fontenelle, *Entretiens sur la Pluralité des Mondes*, 4th ed. (Paris, 1866), This edition is in the George Eliot and George Henry Lewes Collection in Dr. Williams' Library, London, and is the one used by George Eliot, containing her annotations.

21. See chap. 41, vol. II, p. 354. of *Daniel Deronda*.

22. G.H. Lewes, *The Problems of Life and Mind, First Series; The Foundations of a Creed*, 2 vols. (London: Trübner, 1874), vol. I, p. 47.

23. Lewes,[22] vol. I, p. 296.

24. Claude Bernard, *Leçons de Physiologie Expérimentale Appliquée à la Médicine*, 2 vols. (Paris: Bailièrre, 1855–56), vol. II, pp. 12. This passage was marked by Lewes in his copy (now in the Dr. Williams Library). In *Seaside Studies* (London: W. Blackwood, 1858) he refers the reader to it, observing that Bernard "warns us against attempting to deduce a function from mere inspection of the organ, without seeing that organ in operation and applying to it the test of experiment" (p. 153).

25. The quotation is from a passage underlined in George Eliot's and Lewes's copy of James Sully's *Sensation and Intuition: Studies in Psychology and Aesthetics*, (London: H.S. King, 1874), p. 138, in Dr. Williams' Library, London. It is taken from the essay entitled "The Genesis of the Free-Will Doctrine," which has been heavily annotated. In his record of their reading, Lewes lists Sully's *Sensation and Intuition* on July 12, 1874, and then on subsequent dates (manuscript diary of 1874, Beinecke Rare Book and Manuscript Library).

26. Sully,[25] p. 138.

27. G.H. Lewes, *Problems of Life and Mind*, 3rd series, 2 vols. (London: Trübner, 1879), vol. II, p. 365.

28. Lewes,[22] vol. I. p. 144. A strong bond existed between George Eliot, Lewes, and Sully, giving rise to close interconnections in their work. In *Sensation and Intuition* Sully evinces admiration for both George Eliot and Lewes. In the essay "On Some Elements of Moral Self-Culture," he observes: "However systematic his ethical studies may have been, one will pretty certainly discover a new grandeur in virtue and a new foulness in base living after reading a play by Shakespeare or a story by Eliot" (p. 152). In "The Relation of the Evolution Hypothesis to Human Psychology," he states that he hopes to solve "a question recently raised by Mr. Lewes in his *Problems of Life and Mind*." The question he addresses is one crucial to the present argument: the relationship between the individual and the social medium. Sully was to aid George Eliot in preparing the third series of *Problems of Life and Mind* for publication after Lewes's death.

29. Anthony Wilden, *System and Structure* (London: Tavistock Publications, 1977), p. 214.

30. Herbert Spencer, *Social Statics* (London: Chapman, 1851), p. 103.

31. Buckley,[2] p. 42.

32. Spencer,[30] pp. 267–268.

33. For an analysis of the theory of causality underlying Freud's concept of overdetermination, see Wilden,[29] p. 35.

34. William B. Carpenter, *Principles of Mental Physiology* (London, 1874), p. 24. Lewes's diary for 1874 is in the Beinecke Library, Yale University.

35. Edward von Hartmann, *Philosophy of the Unconscious: Speculative Results According to the Inductive-Method of Physical Science*, trans. W.C. Coupland, 2nd ed., 3 vols. (London: K. Paul, Trench Trübner, 1893), vol. III, p. 284. Lewes records reading Hartmann's *Philosophy of the Unconscious* (in the original German edition) on December 15, 1869 and again in April 1872 (manuscript diary, Beinecke Library).

36. Herbert Spencer, *The Principles of Psychology* (London: Longmans, 1855), p. 581.

37. Lewes,[22], vol. I, p. 150n.

38. Jean Sudrann, "Daniel Deronda and the Landscape of Exile," *English Literary History*, vol. 39 (1970), p. 446.

39. Lewes,[22] vol. I, p. 162.

40. Sigmund Freud, *The Interpretation of Dreams, The Standard Edition of the Complete Psychological Works of Sigmund Freud*, trans. J. Strachey, vol. IV (London: Hogarth Press, 1953–66). Freud quotes the following passage from Sully to illustrate this distinction: "The chaotic aggregations of our night-fancy have a significance and communicate new knowledge. Like some letter in cypher, the dream-inscription when scrutinized closely loses its first look of balderdash and takes on the aspect of a serious intelligible message. Or to vary the figure slightly we may say that like some palimpsest, the dream discloses beneath its worthless surface character traces of an old and precious communication" (p. 135n).

41. G.H. Lewes, *The Physical Basis of Mind; The Problems of Life and Mind*, 2nd series (London: Trübner, 1877), p. 359.

42. Lewes,[22] vol. I, p. 189. Lewes employs the theory of organism and medium interaction to challenge the traditional division drawn between subject and object. He concludes: "Out of the general web of Existence certain threads may be detached and rewoven into a special group — the Subject — and this sentient group will in so far be different from the larger group — the Object; but whatever different arrangement the threads may take on, they are not different threads."

In the final volume of *Problems of Life and Mind*[27] Lewes, using the image of a web, expresses his theory that the processes of the mind do not obey the seriation of conscious thought: "We have seen that the discriminated experiences comprised under Attention must be regarded as a *series*; but the true comparison for sensorial reaction is that of a *web*. The attitude of the Sensorium is a fluctuating attitude which successively traverses and retraverses all the portions of the sensorial field, and which thus successively brings now one and the other point into daylight; leaving the others momentarily obscured though still impressing the sentient organism" (p. 217).

43. Ludwig Feuerbach, *The Essence of Christianity*, trans. Marian Evans (London: Chapman, 1854), p. 158.

44. *The Letters of George Eliot*,[11] vol. VII, p. 161.

Facts and Constructs: Victorian Humanists and Scientific Theorists on Scientific Knowledge

DONALD R. BENSON
Department of English
Iowa State University
Ames, Iowa 50011

As a society dedicated to the stimulation of popular interest in science, the British Association for the Advancement of Science could hardly have made a more appropriate choice for its president in 1859 than the Prince Consort. This was the year that brought forth *The Origin of Species*, the single most powerful stimulus to popular interest in science of the nineteenth century, and in retrospect Prince Albert's presidential address to the Association might almost be taken as an official inauguration of science into a central role in British life. The address itself reinforces a sense of such an occasion in its account of the matter and method of science, an account that epitomizes the popular conception of those subjects in Victorian England. The domain of the inductive sciences, the Prince told the Association, is the domain of facts. Unlike feelings, which are "subjective. . . facts are 'objective' and belong to everybody — they remain the same facts at all times and under all circumstances. . . . It is with facts only that the Association deals." Inductive method he describes as a "reasoning upwards from the meanest fact established, and [a] making every step sure before going one beyond it, like the engineer in his approaches to a fortress. We thus gain ultimately a roadway, a ladder by which even a child may, almost without knowing it, ascend to the summit of truth. . . ."[1] Modern cultural historians have tended to characterize Victorian science in much the same way. One historian says that in the era of the Great Exhibition science conceived "a world of external nature, sharply distinguished from subjective illusions, whose course, observable by the senses, is the result of the operation on inert matter of

299

universal unalterable laws deducible by reason."[2] Another writes that the Victorian defender of " 'the scientific method' . . . had behind him the known and the knowable, the tangible evidence of the laboratory. . . . Confident that he had found the key to all things relevant to human life, he could enter controversy with a buoyant assurance denied to those who asked some sanction beyond sense experience."[3]

The historian who had not encountered this conception of science in the Prince Consort's address would certainly have found something like it in the works of the major Victorian humanists, in the assumptions they make about scientific fact. John Henry Newman, for example, in rejecting Sir Robert Peel's natural theological argument for the religious value of scientific study, held that science simply "brings before us phenomena, and it leaves us, if we will, to call them works of design, wisdom, or benevolence. . . . We have to take its facts, and to give them a meaning . . . ." Thus, although Newman rejected the loose claims made for science as an instrument of transcendent knowledge, he did not doubt its capacity to provide a factual description of nature: "The whole framework of nature is confessedly a tissue of antecedents and consequents . . . . What the physical creation presents to us in itself is a piece of machinery . . . ." These are the givens, and we must decide whether to "refer all things forwards to design, or backwards on a physical cause."[4]

Matthew Arnold grounded his critique of science on the same assumptions. In "Literature and Science" Arnold endorsed the discipline that the practice of natural science enforces through "the habit of dealing with facts" and admitted the appeal of its authority: "Not only is it said that the thing is so, but we can be made to see that it is so." This kind of visual confirmation assures the "reality of natural knowledge," the "knowledge of things." Science satisfies "our instinct for intellect and knowledge" by providing "pieces of knowledge," which, according to Arnold's notion of scientific method, are added one to another until we come to propositions as "interesting" as Darwin's about our ancestors or Huxley's about the inexorable order of nature. Much as he was disturbed by it, Arnold no more suspected the certainty of science in the realms of factual knowledge than did Newman, sharing with him what has been called the "touching faith in 'scientific method' " current among Victorian humanists.[5] Scientists, Arnold admitted, "will give us other pieces of knowledge, other facts, . . . and they may finally bring us to those great 'general conceptions of the universe, which are forced upon us all,' says Professor Huxley, 'by the progress of physical science.' " These were for Arnold, as they were for Newman, the givens. The humanist's task was

to put what is "only" knowledge "into relation with our sense for conduct, our sense for beauty, . . . touched with emotion by being so put."[6]

Walter Pater's commitment to an objective scientific method was quite
explicit. Pater, expressing an ontological skepticism not evident in
Newman and Arnold, maintained that a growing awareness of solipsistic
isolation was the dominant experience of the nineteenth century, evident
in its art, its science, and its life. The science of such an age, he said, must
be "a science of crudest fact," and the development of such a
science—along with corresponding modes of art and life—constitutes
"the mental story of the nineteenth century." Lamenting, as Arnold did,
the age's loss of all sense of wholeness in human experience, Pater went
beyond Arnold's acceptance of scientific fact as a foundation from which
a morally and emotionally tenable life might be structured; he suggested
the pursuit of fact itself—scientific as well as aesthetic fact—as a way of
life: "One resource of the disabused soul of our century . . . would be the
empirical study of facts, the empirical science of nature and man, surviving all dead metaphysical philosophies. [Prosper] Mérimée, perhaps,
may have had in him the making of a master of such science, disinterested, patient, exact. . . ."[7] Disinterested observation of fact, even emotional and aesthetic fact, on the scientific model and communication of
that observation had become for Pater the most meaningful of activities.
In his Preface to *The Renaissance* he defined the task of the aesthetic
critic as knowing his own impression of a given work or experience "as it
really is." To this end the critic regards his objects, "like the products of
nature," as "receptacles of so many powers or forces" that produce "plea-
surable sensations" whose "influence he feels, and wishe[s] to explain, by
analysing and reducing it to its elements." The critic has finished his work
when he has "disengaged that virtue" or property by which the work
makes its unique impression "and [has] noted it, as a chemist notes some
natural element, for himself and others. . . ."[8]

The popular conception of Victorian science, shared at least implicitly
and disseminated by Victorian humanists, and still credited by some
historians of the period, appears to be of a discipline built upon passive
and objective observation of physical phenomena, precise recording of
these observations as facts, and manipulation of the facts by induction
(or possibly addition) to yield general truths about nature. The exact
meaning of *fact*, a term so frequently but uncritically used in expressions
of this view, is problematic; the evidence seems to point to the assumption of an exact correlation—if not an identification—between fact and
phenomenon. In sum, as one historian puts it, "the situation [for Vic-

torian science] was beautifully clear-cut and precise. On one side lay the field of inquiry, the material world, or nature. . . preserving an essential and unalterable 'reality' independent of the thoughts and wishes. . . of the minds that contemplated it. On the other side were those minds, whose scientific function was solely to discover, first by observation and experiment and then by rational deduction, what the material world contained and how its course was ordered."[9]

It is not my purpose to assess the general validity of this conception of Victorian science. Relevant studies by historians of science would not, I think, support a claim that it accurately reflects what scientists themselves believed,[10] although there is a line of theorists stretching from John Herschel through T. H. Huxley and Herbert Spencer whose conceptions of science share significant features with it. Herschel's influential *Preliminary Discourse on the Study of Natural Philosophy* (1830) grounds scientific knowledge on absolutely objective, if not wholly passive observation and experiment, on a commitment by scientists "to stand and fall by the result of a direct appeal to facts in the first instance, and of strict logical deduction from them afterwards."[11] Herschel believed that impressions are trustworthy as "the sensible results of processes and operations carried on among external objects" (pp. 84, 85). While knowledge of the ultimate causes of these observed phenomena is beyond our ken, science can, by inductive procedures, determine their laws and predict their future occurrence. This discussion of the "principles" of natural science is completed by extended treatments of the main Baconian categories — controlled observation and collection of facts, classification, induction, and formation of theories. Herschel does not fully resolve the ontological issues implied in this account of scientific method: The business of the scientist, he says, is to discover "what *are* [the] primary qualities originally and unalterably impressed on matter, and . . . the *spirit* of the laws of nature " — or if these qualities are "really *occult*" to comprehend them as far as possible and "devise such forms of words as shall include and *represent* the greatest possible multitude and variety of phenomena" (p. 39).

Huxley and Spencer were members of a younger generation who espoused a scientific naturalism in some ways akin to the popular conception of science. The naturalists were epistemological positivists, content simply to describe external and mental phenomena and the laws of their succession and coexistence, while remaining uncommitted as to the underlying reality of these phenomena. Their objective method required verification "by observable empirical facts and correspondence with the known laws of nature." This prevented the confusion of "ideas with objects or . . . hypothetical concepts with reality" and assured correspondence between "the materials of experience and the ideas of the mind." Their positivism obviated any formal ontology, and in practice they ranged from "near Berkeleian idealism" to "naive realism."[12] It has been suggested, for example, that for Huxley scientific naturalism was simply "a critical principle," the assertion of reason against supernaturalism, whereas for Spencer it was "a theory of reality, an ontology."[13]

Even such limited agreement with the popular conception of science was not shared by all Victorian theorists. I propose to consider here a set of theoretical statements that, among them, call into question the passivity and objectivity of scientific observation, the certainty of scientific facts and their basis in physical reality, and the adequacy of the inductive method itself. These theorists recognize the active participation of mind in the construction — as opposed to the simple discovery — of scientific knowledge and the vital role of imagination and its instruments, specifically analogies and models, and of language and other symbols in this process. Certain of them find in this participation a basis for the integration of scientific with other modes of knowledge. The authors of these statements were not detractors of science, but its advocates; all were more closely conversant with its specific problems and directions than were the humanists who embraced the popular conception. They were in no sense members of a school or of a tradition, disagreeing among themselves in fundamental ways. Indeed their diversity itself seems significant.

William Stanley Jevons and Karl Pearson, both with roots in positivist thought, were rigorous students of scientific method.[14] Jevons treated science primarily as mental activity, Pearson exclusively so, and both were skeptical about the possibility of knowing any reality outside the mind. William B. Carpenter and John Tyndall were experimentalists, but deeply committed to popular scientific education.[15] In their accounts of scientific method both severely modified the inductive element, Carpenter by insistence on the holistic character of all thought and Tyndall by emphasis on the essential role of imagination in the creation of knowledge. Arthur James Balfour and John Theodore Merz were even more deeply interested in the relation of science to the broader spectrum of Victorian thought, Balfour as a theologian, and Merz as an intellectual historian.[16] Both outlined paradigms of scientific knowledge that emphasized dynamic interrelations between mind and observed evidence — paradigms that they believed science shared with other modes of knowledge.   ⋅§ ξ⋅

Jevons in his *Principles of Science* (1874) and Pearson in his *Grammar of Science* (1892) undertook definitions of the method of science and of the limits of its validity. Both placed active minds, constructing scientific law if not the phenomena themselves, at the center of the scientific enterprise. Both limited the validity of scientific knowledge to these constructs — that is, in keeping with positivist principles, they denied any ontological authority to scientific law, and in Pearson's case at least to scientific description. Jevons argued further that all scientific knowledge is approximate.

Jevons announces in his preface to *The Principles of Science* an emphasis on the limits of scientific knowledge, promising "that before a rigorous logical scrutiny the Reign of Law will prove to be an unverified hypothesis, the Uniformity of Nature an ambiguous expression, the certainty of our scientific inferences to a great extent a delusion."[17] He begins his scrutiny with a denial of the adequacy of strict Baconian induction, offering in its place an ampler logical process (which accounts for hypothesis) and probability theory. Although his account of scientific observation and inference is grounded in a Lockean-associationist psychology, Jevons does not construe the "nature" from which basic sense impressions derive as a machine impressing its order on the mind, but as "a spectacle . . . of endless variety and novelty." The mind responds to this spectacle with a "wonder" that fixes perception and draws its inferences by an active process, which Jevons terms "substitution" (I, pp. 1–3). The actual relation between impressions and external phenomena cannot be determined by logic or science, but a significant relation must be assumed. Logic, and by extension scientific inference, treat "ultimately of thoughts and things, and immediately of the signs [including language] which stand for them. Signs, thoughts and exterior objects may be regarded as parallel and analogous series of phenomena, and to treat one series is equivalent to treating either of the other series" (I, p. 10).

The validity of scientific conclusions deriving from these series is further limited in several respects by the laws of probability. Induction itself by definition yields no more than probability (the evidence never covers all cases), the assumption of the constancy of nature is only an assertion of probability, and the notion of cause is an expression of probable relationship. So complicated is nature that "the simplest bit of matter, or the most trivial incident . . . offers infinitely more to learn than the human intellect can fathom. The word *cause* covers just as much untold meaning as any of the words *substance, matter, thought, existence"* (I, p. 255). If physical events can only be taken as moments in a continuous state-change — as the theories of the conservation of matter and energy suggest — then *cause* can only mean "the group of positive or negative conditions which, with more or less probability, precede an event" (I, p. 260). Even physical measurements are based on probability, as described by the Law of Error. Thus Jevons' understanding of scientific "fact" is far from the simple and unitary one of the popular conception. There are empirical facts, observed but unrelated to any hypothesis; "explained facts," which are so related and brought into harmony with other facts; facts predicted by theory and confirmed by observation; and finally those "facts" known — and in the case of microphysical phenomena knowable — *only* to theory (for example, Joule's calculation of the velocity of gas molecules or William Thomson's of the limits of the size of basic particles).

Jevons' notions of the fundamental analogy among thoughts, things, and signs indicate the central place analogy holds in his epistemology of science. "The whole structure of language" itself, he maintains, "and the whole utility of signs, marks, symbols, pictures, and representations of various kinds, rest upon analogy." The significant analogies, however, are not between individual entities but between *relationships*. Of language, for instance, Jevons explains that "there is no identity of nature between a word and the thing it signifies . . . but there is analogy between words and their significates" — between word-series and thing-series. The reductive tendency in this view is evident in his illustration, the analogy between the physical substances iron/iron-carbonate and the corresponding words *iron/iron-carbonate*. His illustrations of the function of metaphor, however, suggest a more organic conception of language: "There would be no expression for the sweetness of a melody, or the brillance of an harangue, unless it were furnished by the taste of honey and the brightness of a torch" (II, p. 284). The claim here, at any rate, is that the sweetness and the brilliance can only be expressed metaphorically.

Most fruitful for scientific discovery is the exploration of analogies among different classes of phenomena. Jevons deals here with the process Tyndall had identified as imaginative construction, a process that was in fact enormously productive in Victorian science,[18] and to illustrate it he chooses the same set of analogies that Tyndall had used, the phenomena that can be construed under the pattern of wave motion. Jevons traces the successive elaborations of these phenomena, from water to sound to heat and light waves, from the palpable to the obscure, with different media all exhibiting the same essential characteristics: reflection, refraction, interference. This set of analogies, Jevons says, has been one of the most important in science, providing a basis for large mathematical generalizations (II, pp. 293–97).

Jevons sums up what he calls "the results and limits of scientific method" in three general points. The "laws of nature" are propositions, based on probability, about correlations among observed phenomena; they do not determine the location of matter in space (II, pp. 431, 433–36). The constancy of nature is not only an assumption, but a questionable one at that, as witness thermodynamic theory, which requires "a discontinuity of law" in the remote past, and thus opens the possibility of others in the future (II, pp. 438–39). The universe is "infinite in extent and complexity," a belief that leads Jevons to propose a "divergent" theory of knowledge (the more we learn, the more problems we raise) as against Tyndall's "convergent" theory (that science is reducing the amount of ignorance in the universe) (II, pp. 433, 499–51).

Karl Pearson's *Grammar of Science* represents for Victorian theory the extreme extension of positivism and the ultimate idealization of scientific fact. Science is rigorously elaborated by Pearson as essentially a process by and within the mind. He nonetheless claims that scientific method is applicable to all fields of knowledge, which by his definition excludes metaphysics, theology, and poetry. Pearson claims Baconian induction as the basis of scientific method (specifically rejecting Jevons' criticism of Bacon here), but his account of the method goes beyond induction: it is the classification of facts which are perceived by imagination as flowing from law. "The discovery of some single statement, some brief *formula*" from which the facts can be seen to flow "is the work not of a mere cataloguer, but of the man endowed with creative imagination. . . . The discovery of law is therefore the peculiar function of the creative imagination."[19] Significant as Pearson may have thought this role to be, it is clear that the original accurate classification of facts is the crucial phase in scientific investigation and that the important contribution of imagination is economy rather than discovery. The facts themselves, the "external" objects and phenomena, are constructs of the mind, built out of immediate and stored sense impressions. Pearson insists on a solipsism of the imprisoned "ego," conditioned by a nervous system. "We *know* ourselves, and we *know* around us an impenetrable wall of sense-impressions." We cannot say that these are produced by external objects, only that behind them lies *"sensation"* — thus expressing our agnosticism about the stuff of an external world (p. 82). Logically "we ought to use the word *know* only for conceptions, and reserve the word *believe* for perceptions" (p. 180). The validity of both facts and formulas for Pearson is grounded in their "universal" acceptance by "all normally constituted and duly instructed minds" (p. 30).

Pearson's definition of "the Scientific Law" is framed within a brief history of scientific ontology. As Pearson reads this history, it began in a materialism — that of the Greeks, most notably — which construes the phenomenal world as dead matter subject to nonrational laws that determine the order of our perceptions. The Stöics projected a rational character akin to human reason into this matter, the natural theologians removed the source of this rational character to an external lawgiver who orders material events by it, and most recently the metaphysicians have hypothesized a will and consciousness in matter (p. 131). The scientist, on the other hand, recognizing that our perception is limited to sense impression, rejects an external world of dead matter altogether and locates the rational character of our sensory routine and its reducibility to law in the human mind itself. "Natural law" appears to the scientist as "an intellectual product of man, and not a routine inherent in 'dead matter.' The progress of science is thus reduced to a more and more complete analysis of the perceptive faculty — an analysis which unconsciously and not unnaturally we project into an analysis of something beyond sense-impression. Thus both the material and the laws of science are inherent in ourselves rather than in an outside world" (p. 132). The scientist makes no affirmation about whatever may lie beyond sense impressions. In attributing the order of our perceptions to our own minds, he has "withdrawn from the beyond the last anthropomorphical element, and left it that chaos behind sense-impression, whereof to use the word knowledge would be the height of absurdity" (p. 130).

For Pearson in a more radical way than for Jevons, not only the method and the laws of science but its materials as well are constructions of the mind: matter, as we have seen earlier, and also space, time, and motion. Interestingly, it is ether, that insubstantial substance so alluring to nineteenth-century science as a means for resolving the anomalies of radiant energy transfer, that most excites Pearson's own imaginative powers and nearly tempts him into the forbidden region of metaphysics. In his discussion of matter he speculates that within a quarter century there could develop an adequate concept of ethereal motion as the substratum of all sense perception, under which matter would become "non-matter [ether] in motion" (p. 312). The character of ether as perfect fluid or perfect jelly would resolve the anomaly of atomism, that conceptual discontinuity yields perceptual solidity. Pearson outlines his own concept of ether as perfect fluid in motion, the "ether-squirt" theory, which involves an inflow of ether from something like a fourth dimension and the repulsion of "negative matter" beyond our portion of the universe (pp. 318–19). This emboldens him to speculation, under the heading "A Material Loophole into the Supersensuous," on the implications of such a theory for an ultimately verifiable *meta*physics (pp. 319–23) — speculation from which, however, he quickly draws back (pp. 323–25).

Carpenter and Tyndall dealt with scientific method, somewhat less rigorously than Jevons and Pearson, in public lectures in the 1860s and 1870s. Both denied the adequacy of Baconian induction (Carpenter came close to rejecting induction altogether) and insisted on the importance of intuition and imagination to the individual scientist, who after all carries on the work of science. Carpenter further emphasized the provisional character of even the most firmly established scientific conceptions, for example, Newtonian cosmography.

Carpenter's presidential address to the British Association in 1872 was a wide-ranging challenge to what he perceived as science's one-sided development of the Baconian program — a preoccupation with "Nature in her Relation to Man" at the expense of what is more basic, "Man as the 'Interpreter of Nature.' "[20] In a critique that foreshadows a standard twentieth-century interpretation of the unrecognized metaphysical implications of Newtonian method,[21] Carpenter chides "those who set up *their own conceptions* of the Orderly Sequence which they discern in the Phenomena of Nature, as fixed and determinate *Laws*, by which those phenomena not only *are* within all Human experience, but always *have been*, and always *must be*, invariably governed." They are guilty, he charges, of the very "Intellectual arrogance they condemn in the Systems of the Ancients" (p. 416). But in fact, even such a cumbersome system as Ptolemaic astronomy "did intellectually represent" everything the astronomer of its time could actually observe, and it supported accurate predictions. One can ask no more of the science of any age, since even in the most exact sciences "we cannot proceed a step, without translating the actual Phenomena of Nature into Intellectual Representations of those phenomena." We now accept the Newtonian "Scheme of the Universe," Carpenter argues, only because as the simplest and most universal summation of the observations made so far it "satisfies our Intellectual requirements" (pp. 420–21).

Although Carpenter shares with the positivists he is attacking in this address the assumption of mind's separation from the phenomena of nature, he treats this less as a limitation than as simply the context for man's interpretive role. As the scientist undertakes "the Intellectual Representation of Nature" so the poet, "rendering into appropriate forms those deeper impressions made by the Nature around him on the Moral and Emotional part of his own Nature," represents "what he *feels* in Nature; and to each true Poet, *Nature is what he individually finds in her*" (pp. 417–18). Because it is based on verifiable facts elaborated by a commonly accepted reasoning process and because its results are widely agreed upon, there is a "general belief . . . that the Scientific interpretation of Nature represents her not merely as she *seems,* but as she *really is,*" a belief that the scientist's interpretation is "less individual" than the poet's. But closer consideration shows that the scientific interpretation no less than the poetic "*is a representation framed by the Mind itself* out of the materials supplied by the impressions which external objects make upon the Senses; so that to each Man of Science, *Nature is what he individually believes her to be*" (pp. 418–419). Carpenter even questions the exclusive claims of inductive method itself, insisting that scientific interpretations are often "clearly matters of *judgment . . . a personal act.*" There is validity in such interpretations, but it is not dependent on the reductive certainty of method, rather on "Common Sense," consensus, whose trustworthiness in turn "arises from its dependence, not on any one set of Experiences, but upon *our unconscious co-ordination of the whole aggregate of our Experiences,* —not on the conclusiveness of any one train of Reasoning, but on *the convergence of all our lines of thought towards this one centre*" (pp. 422–423). Pushed far enough, this insistence on individual experience leads to recognition of what A.N. Whitehead was to call the fallacy of misplaced concreteness, that is, the attribution of greater reality to the objects of sense experience than to sense experience itself: Carpenter challenges the positivist dogma that "we know nothing but Matter and the Laws of Matter, and that Force is a mere fiction of the Imagination." Is it not actually the other way around, he asks, "that while our notion of *Matter* is a Conception of the Intellect, *Force* is that of which we have the *most* direct —perhaps even the *only* direct —cognizance . . . from our own perception of *exertion*?" (p. 432)

John Tyndall spoke to the British Association in 1870 on "Scientific Uses of the Imagination," making strong claims for "the prepared imagination" of Newton, "the constructive imagination" of Dalton, the rich imaginative faculty of Davy, and for the imaginative bases of such powerful scientific concepts as force and causal relationship.[22] Tyndall exemplifies the operation of imagination by describing how ether was imaginatively constructed on the model of tangible wave phenomena — water and sound waves — and he chides chemists who deny the reality of the elementary sources of these phenomena, atoms and molecules: "Ask your imagination if it will accept a vibrating multiple proportion — a numerical ratio in a state of oscillation," that is, a mere abstraction, as the source of ethereal vibrations. "The scientific imagination, which is here authoritative, demands . . . a particle of vibrating matter quite as definite . . . as that which gives origin to a musical sound" (II, p. 109). In an earlier lecture (1865) Tyndall had described the scientist's "picturing atoms, and molecules, and vibrations, and waves, which eye has never seen nor ear heard, and which can only be discerned by the exercise of the imagination. This, in fact, is the faculty which enables us to transcend the boundaries of sense, and connect the phenomena of our visible world with those of an invisible one" (I, p. 71).

In his address to the British Association of 1868 Tyndall envisions the generation of scientific knowledge as an individual, highly creative, even mysterious activity. He describes the individual scientist carrying "the light of his private intelligence a little way into the darkness by which all knowledge is surrounded," maintaining that "the force of intellectual penetration into this penumbral region which surrounds actual knowledge is not . . . dependent upon method, but upon the genius of the investigator." Even "the brightest flashes" of insight, however, must be "proved to have their counterparts in the world of fact. Thus the vocation of the true experimentalist may be defined as the continued exercise of spiritual insight, and its incessant correction and realisation. His experiments constitute a body, of which his purified intuitions are, as it were, the soul" (pp. 76–77). In an 1877 address Tyndall suggests that not only the universe itself, but also the science that interprets it may be organic: "Great discoveries grow. . . . Theories sometimes float like rumours in the air before they receive complete expression" (p. 340). Progress in science is, as Emerson characterized all intellectual progress, "rhythmic" (p. 342).

Balfour in his *Foundations of Belief* (1895) and Merz in his *History of European Thought in the Nineteenth Century* (1896–1914) agreed that mind participates actively and imaginatively in basic scientific observation, but avoiding positivist solipsism, they emphasized the interplay between observation and its changing theoretical context. On such lines they developed comprehensive paradigms for the construction of knowledge which provided a basis for the integration of science into the larger

realm of Victorian thought. Merz further suggested a means for resolving
the separation of mind and physical nature.

Balfour's *Foundation of Belief* is first of all a critique of the prevailing
scientific naturalism, in which he asserts a realism like that assumed in
the popular conception. His strategy is to turn the solipsistic implications
of positivism — essentially those spelled out by Pearson — back on the traditional and still widely held realist assumptions of scientific practice to
show how far abstruse theory has outrun experience. Having cleared the
ground, he might then offer his own account of the reasoning process
and of the development of knowledge, both general and scientific.

The point of departure for Balfour's critique is the deep commitment
of the naturalists to the distinction between real and permanent primary
qualities and merely contingent secondary qualities. To apply this
distinction to aesthetic experience, for instance, as naturalist thinkers
have done, is to reduce it absurdly, for our experience of the beautiful
only assures us that it "is not the object as [from science] we know it to
be — the vibrating molecule and the undulating ether — but the object as
we know it not to be — glorious with qualities of colour or of sound."[23]
To apply the primary-secondary distinction, critically, to science itself is
to reveal the paradox of a discipline dedicated to experimentation in a
solid physical world that it cannot affirm to exist. The attempt to build
solid physical inferences on immediate mental facts, Balfour charges,
contradicts science's own premises, since historically its propositions are
not "about states of mind, but about material things," about a nature independent of the observer (pp. 115–16). Had scientists known from the
beginning that they were observing their own sensations, they would
scarcely have bothered to "invent" an independent nature (p. 117). Just as
Pearson had denied any metaphysical extension of scientific constructs,
Balfour makes clear that it is not the empirical practice of science he objects to, but its *negative* metaphysical extension by the naturalists. The
empirical philosophers — Hume, Mill, Huxley, Spencer — have played
"unconscious havoc" with the sound results of empirical practice by
replacing "the atoms and motions and forces" of physics with a "private
reality" of their own, for example, Mills' "permanent possibilities of sensation" or Spencer's "unknowable" (pp. 123–24).

In Balfour's scheme the grounds of knowledge and of belief are not
rigidly fenced off as they are for the naturalists, and to a large extent for
humanists like Arnold. Knowledge, including that cultivated by science,
and belief both develop out of a complex interplay of active reason and
intellectual and social authority. Sense experience itself is conditioned by

environmental factors beyond the mere equipment and occasion and even disposition for perception. Authority of various sources and a tendency to rationalize preconceptions are intimate and inescapable factors in reasoning, as is personal volition. Balfour elaborates these basic premises into a paradigm for the development of knowledge, again including scientific knowledge, which bears little resemblance to the naturalist and popular paradigms and which anticipates certain modern views, such as that of Thomas Kuhn, in a remarkable way.[24] It is not — as Arnold was willing to concede to Huxley — "by the steady addition of tier to tier . . . not by mere accumulation of material, nor even by a plant-like development, that our beliefs grow less inadequate to the truths which they strive to represent," Balfour argues. Rather the structure of knowledge is like some "ancient dwelling" under continuous alteration, one part decaying while another is rebuilt. Or perhaps more accurately, it is like "a plastic body" constantly altered by internal (rational) and external (authoritative) forces, both adding and destroying. The "whole mass" changes configuration and balance, "settling towards a new position of equilibrium, which it may approach, but can never quite attain." The main framework of the structure or body is provided by hypotheses, theories, generalizations, explanatory formulas. The interaction between these and their specific contents is "the most salient, and in some respects the most interesting, fact in the history of thought. Called into being, for the most part, to justify, or at least to organise, pre-existing beliefs, they can seldom perform their office without modifying part, at least, of their material" (pp. 252–53). Eventually this equilibrium is destroyed by materials that can no longer be "held in check," and "a new theory has to be formed, a new arrangement of knowledge has to be accepted, and under changed conditions the same cycle of not unfruitful changes begins again" (p. 254).

Merz's impressive integration of the main elements of nineteenth-century thought in *A History of European Thought in the Nineteenth Century* assumes a similar paradigm. By "thought" Merz intends something more fundamental and comprehensive than either philosophy or science, although both are organic to it. Thought reaches from "Reason, i.e., defined Thought, to the obscurer regions of Feeling and Imagination, to the unconscious world of Impulse," and it is the structure of thought that gives facts, external events all their significance.[25] Knowledge develops from man's invention of methods and theories and his guessing at results — "in fact, erecting scaffoldings with the help of which he raises the structures of Society, Art, and Science," scaffoldings that must continually be "remodelled on new principles" and for new demands (I, pp. 55–56). What Merz calls the "unmethodical" aspect of thought, intuition broadly construed, "is a reflection of the knowledge of science or the light of philosophy, but, like all reflected light, it not only follows, it also precedes the real and full light. . . . In it lie hidden the germs of future thought, the undeveloped beginnings of art, philosophy, and science yet unknown and undreamt of . . ." (I, p. 66). It is less the assertion of these broad claims for intuition — they had been heard frequently in the nineteenth century — than the close application of them to the history of nineteenth-century thought, science in particular, that gives Merz's work its significance.

If he does not manage to bring all, or even most, of nineteenth-century scientific practice and theory under his paradigm, Merz does nevertheless identify significant instances of the constructive view of scientific description and explanation in both. The atomic conception, for example, one of the strong candidates during the century for acceptance as the fundamental description of nature, was widely treated by chemists until the 1860s as no more than a convenient symbolic means of dealing with observed phenomena. Another such candidate, the kinetic or mechanical conception, had its most comprehensive exposition in the work of Maxwell, who took its central physical assumption, ether, as a theoretical tool only. Advocates of the physical or energy-transfer conception also increasingly questioned the substantial reality of ether, and certain of them the reality of matter itself. The statistical conception of nature, which was making an increasingly credible claim to be fundamental, assumed that our basic knowledge of physical phenomena was not particular, but rather summary and approximate.

Merz identifies several more general trends in the science of the latter half of the nineteenth century that reflect the constructive view of scientific knowledge. The first of these is the shift from the explanatory conceptions of force and cause to the positivist descriptive conceptions of energy and concurrence, which undercut the century's typical dualistic systems that assume a rigidly causal material realm separate from any metaphysical realm. Second is the decline of confidence not so much in the results as in the "processes and contrivances of mathematical and mechanical reasoning," a growing realization of the hypothetical character of "mechanical models of . . . elementary motions and mechanisms," which can "no longer be considered as describing the real processes of nature," but only as "convenient and helpful means by which to start a train of reasoning . . ." (III, pp. 401–402). Finally, against the basic assumption of continuity in nature, there has been the growing conviction that, as one of Merz's contemporaries puts it, "we are everywhere confronted with discontinuities, with new beginnings, with breaks in what we would fain consider the orderly development of things," a conviction that promises to lead thought from its traditional emphasis on uniformity and continuity in nature to a new emphasis on the continuity and coherence imposed by the observer of nature (III, p. 412).

Two contrasting attitudes characterize the century's ontological thought for Merz. He takes Rudolph Hermann Lotze as a point of departure for the development of one of these, phenomenalism. Lotze's scheme required "some central idea in the light of which the phenomenal world can not only be described and analysed, but also interpreted and understood. . . ." But with a phenomenalist such as Spencer, Lotze's "underlying conception is reduced to the empty form of a mere affirmation" and "with Wundt no outlying or underlying conception exists at all," only a highest abstraction from phenomenal experience (III, pp. 516–517). However, Merz finds Lotze and, among later thinkers, F.H. Bradley to be ultimately more representative of nineteenth-century thought. Lotze's underlying "idea," which unifies phenomena, and Bradley's repudiation of "the separation of feeling from the felt, or of the desired from desire, or of what is thought from thinking, or of the division . . . of anything from anything else" represent for Merz an attitude fundamental to later nineteenth-century scientific and philosophical thought, the refusal to consider facts in isolation. Indeed Merz maintains that a "synoptic" (holistic) attitude has been evident in much nineteenth-century thought about nature, beginning with Goethe.

Merz does admit that a powerful tension persists between, on the one hand, the demand of science for external verification and its tendency toward a psychophysical reduction of mind and, on the other, the insistence of philosophy on inner verification and its idealization of mind. He proposes to resolve this conflict by means of critical introspection and a recognition of the constructive function of mind and of the role of language in the development of knowledge. Merz observes that attempts by nineteenth-century scientists to erect universal systems all rest on assumptions of real space and geometric order. Mind is an inescapable embarrassment to these physical systems, being at once nowhere and everywhere in space. If, however, as much nineteenth-century thought suggests, space is a construct of mind not vice versa, then not only "that cluster of sensations" that we recognize as the external world, but also the whole of the mind demand our critical attention. "The explorer of the

firmament of the soul will have to recognize as equally real those regions
in the field of consciousness which are less fixed," irregular, not externalized, not exactly defined by us (IV, pp. 780–781). Merz admits that this
construction of space may be essential to exact knowledge (science and
philosophy), but insists that it be recognized for what it is—an "edifice"
that has been built up over the course of human history and transmitted
person-to-person by language, not a simple and original perception of
primitive consciousness. Further, he warns, "the construction of the
geometric world of physical reality, together with logical thought, forms
the pattern upon which we are always tempted to model an explanation
of the larger field of consciousness . . ." (IV, p. 782). To accede to this
temptation is simply to shrink reality. Merz, then, affirms the validity of
scientific knowledge, but as a human construct, subject to constant
reshaping, even in its most fundamental aspects, within the larger construction of human consciousness itself.

It is more than a passing irony that these scientific theorists should
have treated science as an essentially human and even humane activity
while the humanists I discussed earlier resignedly accepted the reductive
popular conception of it. The foundations for a truly critical interpretation of scientific method and scientific knowledge had been laid by
earlier nineteenth-century humanists, Goethe, for example, as Merz
observed, and Coleridge.[26] It was Arnold who defined the humanist's responsibility as the criticism of life, yet who was unable to extend that
criticism to science in a serious way.[27] As we have seen, he regarded
"facts" as firm, objective, and ordered. The world presents us "the spectacle of a vast multitude of facts awaiting and inviting [our] comprehension," and our part is to "observe [these] facts with a critical spirit; to
search for their law. . . ."[28] The determination to see these facts as they
are, so fundamental to Arnold's notion of culture and of the criticism of
life and art, is precisely "the scientific passion, the sheer desire to see
things as they are. . . ."[29] However, Arnold may have found this commitment to scientific ontology and scientific method, as he understood
them, irreversible, for he was later to warn that "our religion has
materialised itself in the fact, in the supposed fact; it has attached its
emotion to the fact, and now the fact is failing it," whereas poetry has attached "its emotion to the idea; the idea *is* the fact." But Arnold seems to
have concluded that the all-important realm of *real* fact is the domain of
science. What then is left for poetry? "Without poetry, our science will

316 ANNALS NEW YORK ACADEMY OF SCIENCES

appear incomplete. . . . For finely and truly does Wordsworth call poetry 'the impassioned expression which is in the countenance of all science'; and what is a countenance without its expression?"[30] The rhetoric is confident as Arnold predicts that poetry will replace a "hollow" religion and philosophy, but the admission is clear: Poetry has become an appendage to science, the knowledge of fact.

Surely such a conception of science, overstating the solidity, the certainty, and the conclusive power of facts and understating the creative activity of mind and the provisional character of its constructions, transmitted from Victorian humanists to their twentieth-century inheritors has been an ideal matrix for the myth of the Two Cultures. The six theorists discussed here, taken together at least, were perhaps working toward a conception that might accommodate everyone in the same culture.

NOTES AND REFERENCES

1. George Basalla, William Coleman and Robert H. Kargon, eds., *Victorian Science: A Self-Portrait from the Presidential Addresses of the British Association for the Advancement of Science* (Garden City, N.Y.: Doubleday, 1970), pp. 52–53.

    Havelock Ellis, looking back from 1900, would likewise perceive Victorian science as a grand set of engineering problems when he called it "merely mechanical aptitude, the aptitude to make and to measure," important as that was. (Quoted by Frank Miller Turner, *Between Science and Religion: The Reactions to Scientific Naturalism in Late Victorian England* [New Haven: Yale University Press, 1974], p. 6.)

2. Herbert Dingle, "The Significance of Science," in Herbert Dingle, ed., *A Century of Science* (London: Hutchinson's, 1951), p. 309.

3. Jerome Hamilton Buckley, *The Victorian Temper: A Study in Literary Cultures* (Cambridge: Harvard University Press, 1951), pp. 186–187.

4. Charles F. Harrold and William D. Templeman, eds., *English Prose of the Victorian Era* (New York: Oxford University Press, 1938), pp. 541, 544–545. Newman's comments here were made in a series of letters to *The Times* in 1841 on Peel's Tamworth reading-room proposal.

5. R.H. Super, "The Humanist at Bay: The Arnold-Huxley Debate," in U.C. Knoepflmacher and G.B. Tennyson, eds., *Nature and the Victorian Imagination* (Berkeley and Los Angeles: University of California Press, 1977), p. 239.

6. The quotations from "Literature and Science" (1885) are from R.H. Super's edition of *The Complete Prose Works of Matthew Arnold* (Ann Arbor: University of Michigan Press, 1960–77), vol. X, pp. 60, 62, 65.

7. Walter Pater, *Miscellaneous Studies* (London: Macmillan, 1924), pp. 13, 16–17. This lecture on Mérimée was first published in 1890.

8. Walter Pater, *The Renaissance: Studies in Art and Poetry* (London: Macmillan, 1919), pp. viii, ix–x.

9. Dingle,[2] p. 305. Dingle is perhaps less disturbed than he should be about a basic difficulty with this interpretation: "This view of things appeared so evident that it is difficult to find a passage in the writings of the time that even openly implies it; the implication lay so

deep that detailed analysis is necessary to show how inevitably the whole system of thought rested on it." Even "up to the close of the century . . . the incompatibility [between scientific practice and such an ontology] lay too far beneath the surface to be observed . . ." (p. 309).
10. See, for example, David B. Wilson, "Concepts of Physical Nature: John Herschel to Karl Pearson," in Knoepflmacher and Tennyson,[5] Nature and the Victorian Imagination, pp. 210–15; David M. Knight, Atoms and Elements: A Study of Theories of Matter in England in The Nineteenth Century (London: Hutchinson, 1967); and George M. Fleck, "Atomism in late Nineteenth-Century Physical Chemistry," Journal of the History of Ideas, vol. 24(1963), pp. 106–14.
11. John Herschel, A Preliminary Discourse on the Study of Natural Philosophy (London, 1830; rpt. New York: Johnson Reprint Corp., 1966), p. 80. Citations of Herschel will be made parenthetically by page number to this edition.
12. Turner,[1] Between Science and Religion, pp. 18–19. Other scientists Turner identifies as naturalists are Charles Darwin, W.K. Clifford, Francis Galton, and John Tyndall. Turner's main interest here is in six writers who challenged naturalism on different scientific and intellectual grounds—Henry Sidgwick, A.R. Wallace, Frederic W.H. Myers, George John Romanes, Samuel Butler, and James Ward.
13. James G. Paradis, T.H. Huxley: Man's Place in Nature (Lincoln: University of Nebraska Press, 1978), p. 180.
14. Jevons (1835–1882), who spent most of his rather short professional life in Manchester, made substantial contributions to economic thought and logic as well as to scientific theory. Pearson (1857–1936) served as professor of mathematics, mechanics and eugenics at University College, London, while maintaining wide cultural interests in socialism, Darwinism, German culture, Christian history, and the status of women. He pursued in effect two careers, first as a theorist of scientific method and later as the virtual founder of modern statistics.
15. Carpenter (1813–1885) took medical training at Edinburgh and held teaching and research posts at the Royal Institute and University College, London. His major interest was in the physiology of the nervous system and in what he believed to be the mind's active contribution to sense perception. His work in this area, in microscopy, in marine zoology, and in science education has led one biographer to nominate him the last "complete naturalist" (K. Bryn Thomas in The Dictionary of Scientific Biography, ed. C.C. Gillispie [New York: Scribner's, 1973]). Tyndall (1820–1893), after several years' practical engineering experience, studied mathematics and physics at Marburg and was appointed professor of natural philosophy and later superintendent at the Royal Institute. His significant experimental work was in electromagnetism, thermodynamics, and bacteriology, and he was widely known as a popularizer of science, especially after his Belfast lecture of 1874, which antagonized religious traditionalists.
16. Balfour (1848–1930), too, pursued a double career: as a speculative thinker concerned with the relations among physics, metaphysics, and religious faith, and as a politician and statesman (Member of Parliament, party leader and Leader of Commons, cabinet member, and Prime Minister). Merz (1840–1922), educated in Germany, was a chemist and successful industrialist, with a wide circle of intellectual acquaintants. Beginning with his study of Leibniz in 1884, Merz published in the areas of philosophy, psychology, religion, and the history of science throughout his lifetime.
17. The Principles of Science: A Treatise on Logic and Scientific Method (London, 1874), vol. I, p. ix. Citations of Jevons will be made parenthetically by volume and page number to this edition.

18. Maxwell was certainly the most successful (and articulate) practitioner of this method. Robert Kargon, "Model and Analogy in Victorian Science: Maxwell's Critique of the French Physicists," *Journal of the History of Ideas*, vol. 30 (1969), pp. 423–436, cites Maxwell's description of this heuristic method as one in which "a partial similarity between the laws of one science and those of another . . . makes each of them illustrate the other," whereas a merely mathematical reduction of the phenomena in question, "though true, would be deficient in 'vividness' and 'fertility of method' " (pp. 432–33). Harold I. Sharlin, *The Convergent Century: The Unification of Science in the Nineteenth Century* (New York: Abelard-Schuman, 1966), pp. 80–98, analyzes in detail Maxwell's development of electromagnetic theory by means of analogical modelling.

19. Karl Pearson, *The Grammar of Science* (London: W. Scott, 1892), p. 37. Citations of Pearson's work will be made parenthetically by page number to this edition.

20. Basalla et al.,[1] *Victorian Science*, pp. 415–416. Subsequent citations of Carpenter will be made parenthetically by page number to this edition.

21. E.A. Burtt, *The Metaphysical Foundations of Modern Science* (Garden City, N.Y.: Doubleday, 1954), summarizes this central theme on page 10.

22. John Tyndall, *Fragments of Science* (London: Longmans and Co., 1879), vol. II, p. 104. Citations of Tyndall will be made parenthetically by volume and page number to this collection.

23. A. J. Balfour, *The Foundations of Belief, Being Notes Introductory to the Study of Theology* (London: Longmans & Co., 1895), pp. 61–62. Citations of Balfour will be made parenthetically by page number to this edition.

24. Thomas Kuhn, *The Structure of Scientific Revolutions* (Chicago: University of Chicago Press, 1970).

25. J. T. Merz, *A History of European Thought in the Nineteenth Century*, 3rd ed. unaltered (London, 1907), vol. I, p. 5. Citations of Merz will be made parenthetically by volume and page number to this edition (vol. III, 1912, vol. IV, 1914).

26. Owen Barfield outlines the basic critique of scientific ontology implicit in Coleridge's writings in chap xi of *What Coleridge Thought* (Middletown, Conn.: Wesleyan University Press, 1971).

27. Arnold was not alone in this, as I have indicated. I single him out because of his influence and his specific concern with science.

     Hayden White, "The Fictions of Factual Representation," in Angus Fletcher, ed., *The Literature of Fact* (New York: Columbia University Press, 1976), pp. 21–44, identifies some of the symptoms I have ascribed to Victorian literary humanism in Victorian historiography. Truth is equated with fact only at the beginning of the nineteenth century, partly in reaction against the dangers of myth-based history, dramatized by the French Revolution, and partly in pursuit of scientific certainty. White excepts from this pattern the age's great philosophers of history — Hegel, Marx and Nietzsche — who recognized "that all original descriptions of any field of phenomena are *already* interpretations of its structure" (p. 32). White holds that some sort of "poeticizing" is "the immediate base" of all cultural activity, science included (p. 29). He exemplifies this with an analysis of *The Origin of Species*, "that *summa* of the 'literature of fact,' " as "a history of nature meant to be understood literally but appealing ultimately to an image of coherence and orderliness which it constructs by linguistic 'turns' alone" (p. 43).

28. Arnold,[6] "On the Modern Element in Literature" (1869), *Complete Prose Works*, vol. I, pp. 20, 24.

29. Arnold,[6] "Culture and Anarchy" (1869), *Complete Prose Works*, vol. V, p. 91.

30. Arnold,[6] "The Study of Poetry" (1880), *Complete Prose Works*, vol. IX, pp. 161–162.

Bernard Shaw and Science: The Aesthetics of Causality

THOMAS POSTLEWAIT

Department of Humanities
Massachusetts Institute of Technology
Cambridge, Massachusetts 02139

> If the fool would persist in his folly he would become wise.
> —WILLIAM BLAKE,
> *The Marriage of Heaven and Hell*

SCIENCE is knowledge by means of causes (*Scientia est cognitio per causas*): this essential proposition, which supports both the order and the purpose of basic science, is the foundation on which Bernard Shaw mounts his challenge to modern science. While his intention was to question the presuppositions of scientific theory and methodology, his reputation during his lifetime (1856–1950) was that of a clever but often misinformed critic. His attacks on materialism, Darwinism, doctors, vivisection, and immunization earned him popular eminence as a delightful caricaturist, but not a sympathetic and understanding observer of scientists and scientific ideas. Anyone who claims that the moon and the sun are just a few miles away from the earth must be a clown. Thus, his three-time biographer Archibald Henderson wrote: "Shaw's writings exhibit . . . little preoccupation with science except as a subject for satire and ridicule."[1] And Prince Kropotkin, one of Shaw's acquaintances in socialist circles, accused Shaw of arguing against science in the manner of the old Catholic priests. There is, however, something so absurd in this statement that it sounds appropriately paradoxical when applied to Shaw, but it yokes him with the wrong group. On the surface it is wrong because Shaw's historical sympathy was with Descartes and Galileo rather than with the Inquisitors. Neither censorship nor dogma had his support. In a deeper sense it is misleading because Shaw put the old Catholic priests and the defenders of "materialist science" in the same camp, thereby casting himself in the heroic role of rebel-prophet. "The disciples of Pavlov would burn me if they had Torquemada's power."[2] Kropotkin accused Shaw of being an opponent of truth, but Shaw accused him and the many "enlightened" critics of religion of being "the dupes of science,"[3] making of it another inadequate faith.

Illusions and self-deceptions rule the world, not logic, not reason, not even natural law. As "a prophet new inspired" Shaw challenges scientists to recognize that they are prone to illusions, as are all other human beings. He accuses them, often with devilish hyperbole, of being irrational true believers or professional impostors. And he counters them, quite seriously for all his irony, with his own "science of metabiology" (V, p. 337), which serves as a theology of Creative Evolution.[4] "The Agnostics and Atheists and Determinist Diehards who have thrown over the creeds whipped into them in their childhood are surprised and alarmed to see Creative Evolution reviving much that they have discarded as superstitious" (V, p. 698). Shaw goes them one better by making his own "natural" religion,[5] thus apparently abandoning science at just the time that its triumph seemed securely warranted to all progressive thinkers.

Why does Shaw, one of the leading spokesmen in England of the progressive causes, attack science so roundly? Was it simply because his waspish humor, his puritan conscience, made him distrust all forms of authority? In a drama review of 18 January 1896 he wrote: "It is instinct with me personally to attack every idea which has been full grown for ten years, especially if it claims to be the foundation of all human society."[6] Obviously, Shaw had a healthy disregard for idols and cant, no matter what shape they took. But this "instinct," however well developed, was not a blind reflex triggered only by social conditioning or moral outrage. (Shaw, in fact, would call this instinct a Lamarckian habit acquired by will power.) "You must not think," Shaw wrote to H. M. Hyndman, the leader of the Social Democratic Federation, "that I disagree with you merely because disagreement is my pose. I am not a mere intellectual anti-gravitation man. . . . I am a moral revolutionary, interested, not in the class war, but in the struggle between human vitality and the artificial system of morality, and distinguishing, not between capitalist & proletarian, but between moralist and natural historian."[7] By "moralist" Shaw means here someone who forces reality into ready-made moral categories and "flattering illusions." By "natural historian," which he applies to himself, Shaw means someone who looks at nature and human nature with the precision of a thinker who can see through illusions, hypocrisy, and false sentiment. "I assure you I am as sceptical and scientific and modern a thinker as you will find anywhere" (IV, p. 459). He claims to be a pragmatic naturalist,[8] not confined within the prison of systematic thought.

Now it is well to acknowledge this combative instinct of Shaw's, for obviously his nay-saying is a genuine part of his character. But it is not strictly just to describe, in G. K. Chesterton's words, Shaw's "great game of catching revolutionists napping, of catching the unconventional people in conventional poses," as the motivations of ". . . a merely destructive person."[9] Nor is it finally to the point to call Shaw "antiscientific,"[10] as did Bertrand Russell in 1951. Nor, I would argue, can we fully explain Shaw's argument with science as a disapproval of scientific rationalism and "amorality," although this is clearly an aspect of the issue, as Eric Bentley has well argued.[11] Ironically, we have to come to fair terms with Shaw's repeated attacks on science—all those repudiations of scientific method, logic, and theory—because such statements mask a rather complex and often positive (but not positivistic) idea of science in Shaw's writing. Because he seeks nothing less than ". . . the revival of religion on a scientific basis" (V, p. 332), we need to see how he attempts to get beyond both "pseudo-Christianity" and "scepticism." He establishes a critique of both science and religion that denies "either/or" dichotomies because such "choosing up sides" derives from an invalid division of life and mind into such opposed pairs as matter versus spirit, mechanism versus vitalism, natural versus supernatural, reason versus imagination.

> Both our science and our religion are gravely wrong; but they are not all wrong; and it is our urgent business to purge them of their errors and get them both as right as possible. If we could get them entirely right the contradictions between them would disappear: we should have a religious science and a scientific religion in a single synthesis. Meanwhile we must do the best we can instead of running away from the conflict as we are cowardly enough to do at present.[12]

One of the greatest challenges for Shaw's readers and audiences is to keep pace with his ironic turns, his deeply paradoxical mind. He is a moralist who attacks moralists, a socialist who attacks socialists, an idealist who attacks idealists. His irony is not merely a comic technique but an intellectual way of perceiving life's apparent contradictions. He is the quintessential Victorian giant—novelist, essayist, music critic, drama critic, social economist, politican, lay philosopher, metabiologist, playwright—and the greatest debunker of Victorian culture. Influenced by Thomas Carlyle, Arthur Schopenhauer, and John Ruskin (among various nineteenth-century writers who criticized their culture), but not sharing their anguish and pessimism over social divisions and philosophical dichotomies, Shaw uses his profound genius for paradoxical understanding to turn antithesis into synthesis. Yet he rejects, with a clear-sighted irony, the impulse toward system building, the great Victorian vice. To understand him is to see how his ironic temper is his primary mode of apprehension, a mode that allows him to challenge the divisive forces in his culture, to reveal their contradictions and half-truths, and to show, befitting a dramatist's vision, the harmony in conflict (and the division within order). In Shaw's perspective all genuinely intellectual work is ironic and all genuine irony is intellectual work. With this in mind, let us see how he incorporates a criticism of science into his "scientific criticism."

[I]

The common image we have of Shaw as a critic of science is that he was both "a violent and reckless supporter of scientific lost causes, like Lamarckian Evolution,"[13] and a persistent opponent of materialistic science because of its reductive causality and its denial of spiritual values or qualities. In both cases Shaw's argument is against any theory that, in the words of Samuel Butler, "banishes Mind from the universe" (V, p. 696).[14] Following Butler, Shaw argues that science is incapable of explaining the immaterial nature of thought: "The physiology of thought and action, so far, gives no explanation of consciousness."[15] The Shavian idea of consciousness includes such powers of will as self-control and self-judgment — that is, self-consciousness (which implies also the potential for self-irony, a quality of mind that plays havoc with mechanistic and materialistic theories of mind).

Shaw rejects Darwinism, or what he calls "Neo-Darwinism," not because he disbelieves in evolutionary theory, for he readily accepts its concept of historical change and adaptive modification, but because its causal principle of natural selection, as he understands it, is too dependent upon purposeless behavior or random circumstances. Shaw sides with Lamarckian theory because he wanted — or rather needed — a theory of evolution that placed emphasis upon the mind's control over nature. How else can we account for moral judgment? When Lamarck argues in the *Zoological Philosophy* (1809) that man's adaptive abilities derive from intelligent choice and feeling, he establishes a line of argument that Shaw adopts and blends with the Schopenhauerian thesis about the will to live. In fact, Shaw praises Schopenhauer's *The World as Will and Idea* (1818) because it is the "metaphysical complement to Lamarck's natural history" (V, p. 282).

The weakness in Darwin's theory of natural selection for Shaw is that it locates change outside of mind, in a deterministic environment that shapes our instinctual drive to survive. While Shaw was willing to admit that outside forces play a significant part in evolutionary adaption, he refused to accept the "hideous fatalism" (V, p. 294) of life reduced to blind forces. To do so is "a ghastly and damnable reduction of beauty and intelligence, of strength and purpose, of honor and aspiration, to such casually picturesque changes as an avalanche may make in a mountain landscape, or a railway accident in a human figure" (V, p. 294). Arguing that natural selection lacks the status of a law, Shaw turned instead to moral and metaphysical ideas as the basis for both his critique and his alternative to Darwinism.[16] His idea of a Life Force requires that evolutionary change, as it progresses in human beings, be an agency of consciousness that is in turn its own agent. The Life Force is not only the efforts of self-preservation, self-development, and self-awareness — all of which are important aspects of life — but also self-knowledge. As Don Juan states in *Man and Superman*: "Life was driving at brains — at its darling object: an organ by which it can attain not only self-consciousness but self-understanding" (II, pp. 662–63). Shaw's conception of a Life Force operating in nature is, in the final analysis, more metaphysical than physical; but for this very reason such an idea, with its emphasis on will power and purpose, offers a moral alternative for Shaw to the theory of the survival of the fittest, even when Shaw's rather messy but dramatically important idea of the Superman gets thrown into the argument.

Quite simply, Shaw's celebration of the Superman — be he Caesar, Napoleon, Don Juan, Wagner, Ibsen, Shelley, or, of course, G.B.S. himself — is a highly dramatic, if often ironic, way of asserting the value of intelligence, purpose, self-control, and will power. In the historical order of things, mind matters. Change occurs by means of the Life Force, which is Shaw's metaphor for a universal will that individuals discover within themselves and follow or direct towards a desired goal. Causality is thus teleological, at least in the sense that human will is capable of striving progressively "upward," like Don Juan leaving hell and moving toward "something better than [him]self" (II, P. 679). Working within us is "Life's incessant aspiration to higher organization, wider, deeper, intenser self-consciousness, and clearer self-understanding" (II, p. 680).[17]

The hero manifests the power and purpose of the Life Force, yet he may often misuse his talent. So Shaw remains suspicious of the Superman while celebrating him. Because Shaw was attracted to the idea of the hero, he was acutely aware of the myths and dangers of heroes.[18] Not surprisingly, he looked with reservation on one of the most successful myths of the nineteenth century: the scientist as hero. Shaw thus rejects a popular sentiment expressed by T. H. Huxley: "Men of science will always act up to their standard of veracity, when mankind in general leave off sinning; but that standard appears to me to be higher among them than in any other class of the community."[19] This is sentimentalism for Shaw. He is quite prepared to praise great scientists and he does acknowledge his "great Makers of Universes": Pythagoras, Aristotle, Ptolemy, Kepler, Copernicus, Galileo, Newton, and Einstein. In a speech honoring Einstein in 1930 Shaw called him the greatest of his contemporaries. But scientists in general receive no special praise from Shaw.

And doctors, who serve as Shaw's convenient symbol of the scientist in his social role, are a favorite target of Shaw's. In attacking them he can use understatement: "I have not much faith in doctors in the present condition of their science."[20] Or overstatement: "The physician is still the credulous impostor and petulant scientific coxcomb whom Molière ridiculed" (II, p. 768). Doctors, like the rest of us, serve their own interests, yet often pretend to us and to themselves that they are serving noble aims: "Doctors are no more proof against such illusions than other men" (III, p. 240). "As to the honor and conscience of doctors, they have as much as any other class of men, no more and no less" (III, p. 228). So much for Huxley's proud claim. To believe in a special providence for truth among scientists is an illusion, a common one, but then "Science is the mother of credulity" (VI, p. 747).

Science's claim to being a norm of Truth failed to impress Shaw. He also refused to bow down in homage to the specialized expert — the professional — as this exchange between two doctors in *The Doctor's Dilemma* (1906) illustrates:

> RIDGEON: We're not a profession: we're a conspiracy.
> SIR PATRICK: All professions are conspiracies against the laity. (III, p. 351)

As a member of that laity, Shaw refused to believe that knowledge belongs only to the expert. Even when wrongheaded, as in his criticism of immunization, he did not argue out of ignorance. Our present knowledge of viruses reveals Shaw's mistakes, yet the eminent crystallographer J. D. Bernal notes that "a great deal of Shaw's objections were in his days reasonable, valid ones."[21] Often Shaw is not so much entering a scientific debate on particular facts as attacking the common faith in panaceas and the scientific belief in a reductive causality. In this sense his criticisms of bacteriology in the preface to *The Doctor's Dilemma* represent a cogent analysis of the tendency to confuse symptoms with causes. And in the play itself the doctors are satirized because they have become true believers in their pet theories, even in disregard of the immediate evidence that contradicts their hypotheses.

> B.B.: If you have been scientifically trained, Mr. Dubedat, you would know how very seldom an actual case bears out a principle. In medical practice a man may die when, scientifically speaking, he ought to have lived. I have actually known a man die of a disease from which he was, scientifically speaking, immune. But that does not affect the fundamental truth of science. (III, pp. 389–90)

Such fundamental truths both outrage Shaw and appeal to his sense of comic irony: "I presume nobody will question the existence of a widely spread popular delusion that every doctor is a man of science" (III, p. 247). Shaw's ironic criticism is directed against our own credulity, the doctors' self-delusion, and the heroic "man of science."

Shaw's outrage over vivisection results from his refusal to believe that scientific methods are beyond reproach. In the words of Roger Boxill, "Shaw's case against vivisection rests on the principle that science is not beyond good and evil. . . . Like all human endeavor, it is the product of will, and the will can be either cruel or humane. The pursuit of knowledge, like the pursuit of wealth, power, or happiness, is bound by moral law."[22] Shaw challenges scientists not to retreat into a brutalized disregard of the pain of animals in the name of research. "All we insist on," he wrote in 1900, "is that science must be pursued under the same moral and legal restraints as any other civilized activity. . . . There are hundreds of paths to scientific knowledge."[23] Of course, the search for knowledge must continue, but within the constraints of moral deliberation. Responsibility—a key word for Shaw—must be taken.

Shaw believed that human knowledge cannot properly be separated from the full realm of moral, aesthetic, and metaphysical issues that engage us in our response to existence and our search for meaning.

> A laboratory may be a fool's paradise or a pessimist's inferno: it is made to order either way. Its doors may be shut against metaphysics, including consciousness, purpose, mind, evolution, creation, choice (free will), and anything else that is staring us in the face all over the real world. It may assume that because there is no discovered chemical difference between a living body and a dead one, only a difference in behavior, there is no difference. It may rule out all the facts that are incompatible with physicist determinism as metaphysical delusions. In short, it may reduce itself to absurdity in the name of science with a large S. Fine art is allowed no such licence.[24]

Does this mean then that Shaw is giving the "artist-philosopher" an heroic role denied to the scientist? Not really, except that Shaw honors both artists and scientists who avoid, as well as they can, the various kinds of reductions that plague human thought. At the same time Shaw acknowledges, quite willingly, that the artist can be as dangerous a fool as a laboratory researcher. Both are equally subject to prejudices, ignorance, blunders, and corruption. Thus, idolatry of artists is as foolish as idolatry of scientists. Nevertheless, some artists and some scientists do begin to suggest provisionally the fullness of life, however contradictory it may be:

> Shakespear raised a biological problem when he set up 'a divinity that shapes our ends, rough-hew them how we will.' It was certainly not solved by Darwin; but then it was not solved by Shakespear. Goethe led the way to the evolutionary solution, but got no farther than a guess. Scientific advances mostly begin with guesses, jests, paradoxes, fictions, superstitions, quackeries, accidents, and apparent irrelevancies of all sorts.[25]

Shaw understands something important here, something paradoxical about the search for order, systems, and causality. It is not just that reason needs the imagination, although this is lost sight of too often when scientific method and practice assume an unreflective stance of dogmatic standardization. More than that, true understanding of the nature of so much of life may very well *require* fictions, irrelevancies, jests, and paradoxes. As William Blake wrote: "Without Contraries is no progression." Our earth is the marriage of heaven and hell, of necessary order and just as necessary disorder.

[II]

Contraries mutually exist. And Shaw does have his contraries. For example, he seems to use the words "science" and "scientific" in rather contradictory ways. The word "science" can mean "materialism," "mechanism," "positivism," "physicist determinism," "simplemindedness," "reduction," self-delusion," "witchcraft," and "credulity"—all basically terms of disparagement for Shaw. But just as likely the word "science" can mean "common sense," "reality," "realism," "truth," "natural," "clear-sightedness," "intellectual penetration," and "moral vision." In this positive sense, Shaw praises "scientific history," "economic science" (meaning, of course, socialism), and the "scientific criticism" of the arts (which he practices). He calls for a "science of appearances" in acting that will reform the stage, for a scientific education that teachers truth not lies, and for a scientific psychology that dispels illusions. By following the "lights of science" we are directed to a fuller understanding of existence. So two sciences, negative and positive, exist in Shaw's writings. He can castigate "rationalized physical science" as a *reductio ad absurdum* because of its life-denying "materialism."[26] But he can also embrace science as the way to truth. In fact, in his early life Shaw was swept up by the advance of science.

He tells us that he "had been caught up by the great wave of scientific enthusiasm which was then passing over Europe as a result of the discovery of Natural Selection by Darwin, and of the blow it dealt to the vulgar Bible worship and redemption mongering which had hitherto passed among us for religion. I wanted to get at the facts. I was prepared for the facts being unflattering: had I not already faced the fact that instead of being a fallen angel I was first cousin to a monkey."[27] In the 1880s he was comfortably anticlerical, whacking parsons with the stick called Natural Selection. And in his second novel, *The Irrational Knot* (written 1880; serialized 1885 to 1887; published 1905), Shaw makes his case for both Rationalism and Materialism by creating a hero who, as an engineer, approaches vocation, sex, and society with an analytic sensibility that is dispassionate and logical. He proves himself superior to almost everyone else who lives by illusions and false sentiment.

For young Shaw science was playing a crucial role in the overthrow of irrational beliefs. In 1896 he wrote: ". . . I know that the real religion of to-day was made possible only by the materialistic-physicists and atheists-critics who perform for us the indispensable preliminary operation of purging us thoroughly of the ignorant and vicious superstitions which were thrust down our throats as religion in our helpless childhood."[28] The key word here is *preliminary*, for like Auguste Comte and John Stuart Mill, Shaw saw human knowledge as moving through progressive stages of development. In Comte's theory these stages are historically the theological, followed by the metaphysical, and, triumphantly as mankind throws over superstition and idealism, they culminate in the positive or scientific era. Shaw takes over the general idea of three stages, but in *The Quintessence of Ibsenism* (1891) he changes the progressive order by placing the positive era before the still-to-come metaphysical era. By doing so he intends to challenge both materialism and positivism.

While Shaw sided with Mill and Huxley in their dismissal of traditional religious explanations of life and history, he refused to accept a new metaphysics based upon the denial of metaphysics.

Positive science has dazzled us for nearly a century with its analyses of the machinery of sensation. Its researches into the nature of sound and the construction of the ear, the nature of light and the construction of the eye, its measure of the speed of sensation, its localization of the function of the brain, and its hints as to the possibility of producing a homunculus presently as the fruit of its chemical investigations of protoplasm have satisfied the souls of our atheists as completely as belief in divine omniscience and scriptural revelation satisfied the souls of their pious fathers. The fact remains that when Young, Helmholtz, Darwin, Haeckel, and the rest, popularized here among the literate classes by Tyndall and Huxley, and among the proleteriat by the lectures of the National Secular Society, have taught you all they know, you are still as utterly at a loss to explain the fact of consciousness as you would have been in the days when you were instructed from The Child's Guide to Knowledge. Materialism, in short, only isolated the great mystery of consciousness by clearing away several petty mysteries with which we have confused it; just as Rationalism isolated the great mystery of the will to live.[29]

The will to live is a passion beyond reason, so that, for example, even in pain and suffering we hold to life. For Shaw, following Schopenhauer, will must be separated from reason in our understanding of ourselves.

This does not mean that Shaw rejects the uses of rational discourse. "I need hardly add, I hope, that though I am a mystic and not a rationalist, I am not an irrationalist either. I do not expect facts to be reasonable (they are almost all miraculous); but I do expect facts to be reasonably acted on, and arguments to be logical."[30] Shaw the mystic? That does not fit our image of him as the Fabian economist and lecture hall debater. The paradox here is that reason reveals the irrational for Shaw, but the irrational nevertheless confronts us with the need to be rational. In Shaw's plays this apparent contradiction is the motive power in the characterizations of John Tanner in Man and Superman, Andrew Undershaft in Major Barbara, and of course, St. Joan. At her trial, for example, St. Joan reasons with great skill about the irrational nature of her will power, her voices, and her beliefs. In Shaw's gallery of characters she consummates his life-long refusal to reduce understanding to an either/or choice between metaphysical idealism or scientific rationalism. These are not necessary opposites. And his added irony in the case of St. Joan is to choose a character who is supposedly the model of Idealism, but who proves to be a greater "Realist" in some ways than all her doubters.

This said, we still have a problem with Shaw here because it is both easy and difficult to understand what he means by "rationalism." Easy, in that the word for the Victorians is generally associated with someone like Voltaire or Mill, spokesmen for the Age of Reason, for Enlightenment, for natural law and the logic of scientific materialism. Rational thought mirrors, therefore, the order of nature. Difficult, however, because "rationalism" also means a theory of knowledge that locates truth beyond sense perceptions, beyond empirical method, in the significant domain of universal ideas. On the one hand, this latter meaning of rationalism can refer to the deductive reasoning of mathematics, while, on the other hand, it may refer to the metaphysical idealism or authoritarianism of scholastic philosophy with its deductive systems of *a priori* logic that assume a universal design and harmony that reveals God as First and Final Cause. This theoretical union of Aquinas and Aristotle dominated science until the seventeenth century. Thus, A. N. Whitehead in *Science and the Modern World* calls modern science since Galileo and Newton essentially "anti-rationalistic" because it is committed to stubborn facts. Of the scientific revolution, he writes: "It is a great mistake to conceive this historical revolt as an appeal to reason. On the contrary, it was through and through an anti-intellectualist movement. It was the return to the contemplation of brute fact; and it was based on a recoil from the inflexible rationality of medieval thought."[31] This is a simplification, of course, ignoring as it does the mathematical basis for this revolt and the rationalism of Descartes, Leibniz, and Kant as philosophers of modern science. The seventeenth and eighteenth centuries may have been a time of revolt against medieval rationalism, but they were also a time when Reason became a new faith. Our problem is that the word "rationalism" gets used to describe both of these periods and notions of Reason.

While Shaw does have an important metaphysical strain to his argument that can be called Idealistic (linked backward through Hegel, Carlyle, Shelley, and Blake to an older tradition), he normally uses the word "rationalism" in its contemporary meaning (and not usually in the Hegelian sense of Reason in history). "I had better warn students of philosophy that I am speaking of rationalism, not as classified in the books, but as apparent in men."[32] That is, the rationalism of someone like Mill. Shaw equates rationalism with the Age of Reason, the post-Newtonian era of mechanical interpretations of nature, of Deism, of "Natural Law" and the Science of Man. Shaw is arguing against this new rationalism, whether it be scientific materialism, mechanism, empiricism, or positivism. For the sake of convenience he tends to bunch them together as the new faith, no matter that they are not the same thing.

But Shaw's strategy serves more than convenience. He is denying that such rationalism, drawn as it is from certain ideas about causality in science, provides a sufficient model for understanding nature or human nature. In nineteenth-century science, causality had come to mean primarily certain immediate and forceful determinants of action that are logical, not ontological; phenomenal, not spiritual. In fact, following the leads of Bacon and Descartes, modern science reduced the Aristotelian concept of the four causes — material, efficient, formal, and final — to basically efficient and material causality because of the new emphasis upon the natural laws of force. In Bacon's words:

> I divided natural philosophy into the inquiry of causes and the production of effects. The inquiry of causes I referred to the theoretical part of philosophy. This I subdivide into physics and metaphysics. It follows that true difference between them must be drawn from the nature of the causes that they inquire into. And therefore to speak plain and go no further about, physics inquires and handles the material and effective causes, metaphysics the formal and final.[33]

This division had major benefits, which Shaw acknowledges, but it also cut science off from the important issues about life and meaning that are contained in all concerns with formal and final causality.

As William A. Wallace has demonstrated in his masterful history of scientific philosophy and methodology, *Causality and Scientific Explanation* (Vol. I: Medieval and Early Classical Science; Vol. II: Classical and Contemporary Science), the triumph of classical science since Newton, with his emphasis upon the *vera causa*, stems in part from a dissociation of science from the traditional search for causes. This tradition, which flourished during the medieval and early period of classical science (sixteenth and seventeenth centuries), was still concerned with the four causes. But under the influence of Descartes, science banished questions about "why" and settled more readily for questions about "how." Not only did Descartes banish forms, or formal causes (which may be defined variously as the *eidos* or structural idea or form, the *paradeigma* or pattern, the *logos* or reason or creative principle, the *schema* or shape), but he also "saw final causality as beyond human understanding; and though he endorsed matter, and in this sense subscribed to material causality, he severely restricted its scope over the interpretations of his predecessors. Thus, in effect, there is only one type of cause for Descartes, and this is the active or efficient cause, which henceforth would be at the base of all scientific explanation."[34]

Although Galileo and Newton continued to search for "true causes" and were still influenced by the Aristotelian idea of four causes, they helped lead science toward a methodology and philosophy that narrowed the meaning of causality to such determinants that could be described or demonstrated phenomenonally and quantitatively. Newton sometimes describes the *vera causa* as a type of material cause and at other times rather as an efficient cause. Whatever the case, it is to be understood as an explanation of natural appearances by means of uniform "laws" of behavior that are physical, not metaphysical (even though Newton himself did not deny the possibility of metaphysical causes). By the end of the eighteenth century, under the influence of David Hume's skepticism, the idea of causality was narrowed further to the idea of correspondence between things, not even necessary connection. Various attempts were made during the eighteenth and nineteenth centuries to provide science with an epistemology and an ontology that were not limited to mechanism and materialism, but for the most part what Shaw calls "rationalism," with a scientific face, dominates the age. This produces what Shaw terms "the anti-metaphysical temper of nineteenth-century civilization" (VI, p. 67). In essence, Shaw is arguing that Newtonian physics provides a false analogy for understanding life and action. For Shaw, this operational or mechanistic idea of causality ignores, as it did in a somewhat different sense for Bishop Berkeley in his criticism of the new rationalism, the mind and its acts.

Existence must be seen from within and not merely as an interplay of outside forces. The Life Force, which is Shaw's metaphor for the motive power that animates our existence, provides the drive for action and re-action, conflict and resolution. It is dramatic in this manner because it has an active, motivational character. At least with human beings we must consider intentions as a source of causal action, and these intentions or motivations, which express more than instinct, need, and conditioning, are to be valued on their own terms as self-willed acts because they express values. Thus, Shaw states that there is a "Kantian moral law within us" (IV, p. 468) — that is, both intelligence and freedom — by which we are able to judge and act. Will power, however irrational and passionate, has purposes (final causes) not dreamed of in mechanistic philosophy. And rationalism, however logical and dispassionate, has patterns and ideas (formal causes) shaping life and action that go beyond any theory of effectual determinism.

We should note, however, that in spite of Shaw's attack on rationalism, his ideas derive in part from the rationalist tradition. Unavoidably, his critique of reason is quite impure, partly because he is not a systematic thinker. His historical ideas, for example, partake of the rationalist emphasis on progress, even though he claims in "The Revolutionist's Handbook" of *Man and Superman* that "progress is an illusion." His belief in "Creative Evolution" is not a rejection but a modification of evolutionary theory, by way not only of Lamarck but also of the German system of natural philosophy put forward by such theorists as Lorenz Oken (1779–1851), whose idea of the law of causality as a law of polarity and opposition is strikingly applicable to Shaw. And Shaw's assumptions about reality and the natural world are shaped in large part by modern science, even though he accuses it of being too reductive. These contradictions do not nullify his criticisms, but they do show, as he knew, that the scientific revolution was shaping his understanding, his values, and his culture. Such contradictions are, of course, the raw material for Shaw's self-irony and dramatic paradoxes. They are also the basis for his argumentative method. He did not so much stand in full opposition to materialism and rationalism as work his way through them quite dialectically. In this important sense, to quote Blake again, "opposition is true friendship."[35] In the process of arguing against the age's rationalism Shaw attempted to move beyond the nineteenth-century dichotomy of materialism and idealism, science and religion.

[III]

Emile Zola, in the preface to his play *Thérèse Raquin* (1873), argues "that the experimental and scientific spirit of the century will enter the domain of the drama, and that in it lies its only possible salvation."[36] To become modern the drama must become realistic, revealing with scientific precision the true causes determining human action. The conclusion of the action must reject melodrama and the tricks of the well-made play as written by Eugène Scribe and his followers, and instead should follow the mathematical logic of the problem proposed. In this scientific method, based upon close observation and direct analysis, the power of reality will give new life to the drama.

Seven years later in "Naturalism in the Theatre," Zola presented his dramatic manifesto:

> Naturalism is the return to nature; it is that operation which the scientists made the day they decided to start with the study of bodies and phenomena, to base their work on experiment, and to proceed by means of analysis. Naturalism, in letters, is equally a return to nature and to man; it

is direct observation, exact anatomy, the acceptance and depiction of what is. The writer and the scientist have had the same task.[37]

As in his long essay, "The Experimental Novel," in which he makes extensive use of Claude Bernard's *Introduction to the Study of Experimental Medicine* (1865) in order to equate the methods of the novelist with those of the doctor, Zola argues here that science is the motive power behind social advancement and understanding. Its methodology is thus the guide for the writer who wishes to reveal the true causes of human motivations and actions, of social events and forces. Human behavior, like that of the rest of the universe, can be shown to follow laws of causality. Thus literature should be determined by science. The writer should avoid all romantic sentiment, all moral idealism, all personal comment. The naturalist writers "teach the bitter science of life; we give the lofty lesson of the real."[39]

How lofty or scientific the naturalist drama was can be easily debated. But nevertheless it did help bring about a renewal of drama in terms of "the real." For Shaw, first as a dramatic critic and then as a dramatist, this renewal was vitally needed and worth fighting for. Especially during the 1880s and 1890s he advocated a scientific realism in literature, arguing for example in *The Quintessence of Ibsenism* (1892) and "A Dramatic Realist to His Critics" (1894) that "positive science" is a key instrument in the overthrow of romantic idealism and melodrama. Shaw praises Ibsen, Eugene Brieux, and Zola regularly during this period for their realistic and scientific artistry. Zola, whom he commends for his "scientific spirit" and dedication to truth, "wanted to tell the world the scientific truth about itself."[39] It is this "scientific quality" that Shaw attempts to capture, he tells us, in *Widowers' Houses* (1892), his first play. He aims to ". . . be as realistic as Zola."[40] The artistic and scientific qualities of his art must go together because "no point in a drama can produce any effect at all unless the spectator perceives it and accepts it as a real point. . . ."[41] To deviate from the real world is "a damnable sin" Shaw declares in a letter to Henry James, criticizing the unrealistic characterization and plot of James' play *The Saloon*: "WHY have you done this? If it were true to nature — if it were scientific — if it were common sense, I should say let us face it, let us say Amen. But it isn't."[42]

Shaw repeatedly defends his early plays in terms of their realistic and scientific authenticity (just as he attacks, in his dramatic reviews, most contemporary plays for their lack thereof):

Scientific natural history is not compatible with taboo; and as everything connected with sex was tabooed, I felt the need for mentioning the forbidden subjects, not only because of their importance, but for the sake of destroying taboo by giving it the most violent possible shocks. The same impulse is unmistakably active in Zola and his contemporaries.[43]

Zola gains great notoriety and censure for writing about prostitution in *Nana* (for example, Henry James proclaims: "Never surely was any other artist so dirty as M. Zola!"). Not to be outclassed, then, Shaw will match him by writing *Mrs Warren's Profession*, blaming prostitution on capitalist economics and treating Mrs Warren as a successful and even charming businesswoman. Shocking? Hardly, but the play was denied public performance by the British censor for three decades and caused a police raid when first performed in New York City in 1905.

Of course *Mrs Warren's Profession* is not another *Nana*, providing the subterranean lesson of the real, and Shaw is not another Zola. Also, while neither writer avoids melodrama, Shaw depends on its conventions, at least ironically, far more extensively than Zola. Shaw's commitment to scientific realism never attains to the naturalism of Zola's works. One important reason for their differences is suggested by a comment on Zola by Henry James:

> What will strike the English reader of M. Zola at large . . . is the extraordi-
> nary absence of humor, the dryness, the solemnity, the air of tension and
> effort. M. Zola disapproves greatly of wit; he thinks it is an impertinence in
> a novel, and he would probably disapprove of humor if he *knew* what it is.
> There is no indication in all his works that he has a suspicion of this; and
> what tricks the absence of a sense of it plays him![44]

James is correct: Zola explicitly attacked the literary use of wit. In fact, Zola claimed that a man of genius, such as himself no doubt, is by definition not witty. Scientific realism rules out such frivolous attitudes and techniques. Scientific methodology has no place for humor. The mission of the naturalist artist requires that only an objective, impersonal voice be used. Who wants — or believes in — a witty scientist?

Shaw's true genius, however, derives from an almost intuitive understanding of the world in ironic terms. And this irony operates as both wit and philosophy in tandem, because the mind plays counterpart harmony and discord with what it perceives. Reality is not merely an object of perception; it is also a manner of perception. That is, Shaw argues that the real cannot be separated from our individual idea of the real. While being committed to a "scientific realism," he denies that empirical methods and behaviorist psychology (such as dominated English science and philosophy from Locke to Mill) reveal adequately the rather equivocal relationship between mind and world.

The naturalist principle is thus too naive in its assumptions, too unreflective about the mind's own anticipatory role in shaping the objects of perception into the content of understanding. Shaw's irony, which by its nature depends upon the projection of personal values and judgments onto what is observed, is a subjective process of thought, a phenomenology of perception that translates immediate observation into reflective expression. Irony is self-consciously opposed to pure empiricism, which presupposes an objective mind that receives impressions from the outside world and accurately records them. Now admittedly such an empirical ideal implies a rigorous theory that few scientists worked out with any consistency in the nineteenth century, but it is a model of thought that became a popular assumption, aim, and even canon for many thinkers, including many practicing scientists. Noting this, and inflating such a frame of mind into a type, Shaw uses irony not only to make fun of such thinking but also to reveal how an ironic perspective, as a way of seeing double, is a more realistic and fuller perception. The ironic temper records the ambiguous tension between subject and object, and thereby more likely avoids the reductive traps, as Shaw sees things, of objectivism.

For the ironic mind reality is an obscure object of desire, and this desire, or passion as Shaw calls it ("Thought is a passion"), engages the world dynamically by means of a reflective process that is not value-free. This does not mean for Shaw that the world is mere appearance or that the mind is solipsistic. Such relativism does not attract him. His irony needs both the world in all its substantiality and the mind with all its formality. The mind's shaping power, which includes both formal and final causes (that is, idea and purpose), interacts with the world's shaping power, which includes both material and efficient causes. Together they perform strophe and antistrophe, turn upon turn, in dramatic tension. The mind takes the objects of perception and reformulates them so that they express not just themselves but also the perceiver's attitude toward them. How this reformulation occurs is a difficult question of epistemology that Shaw does not bother to answer, not being a philosopher. In general terms, however, his argument is that "the human mind is like the human hand in being able to grasp things only when they are shaped in a certain way. . . . And a logical theory, with its assumptions of cause and effect, time and space, and so on, is just such a mental handle and nothing else. Without a theory, natural occurrences may be put to use; but they cannot be thought out."[45] In the ratios of the mind things get converted. And the scientist, whom Zola elevates, is for Shaw just "a

professional thought-carpenter" who "fits a theory" to natural occurrences.[46] This theory has no absolute validity, whatever its sufficiency.
In time it will be found wanting.

In other words, things are not necessarily what they seem, which is a
good working definition of irony. The scientist's theory may be what
Shaw calls a necessary illusion. "What, then, is a necessary illusion? It is
the guise in which reality must be presented before it can rouse man's interest, or hold his attention, or even be consciously apprehended by him
at all."[47] Shaw's ironic perspective is thus tied to his important idea about
illusions, which the mind seems capable of holding in great number and
with masterful dedication. Because of this talent we locate reality not in
objective facts, but in the mindful (or sometimes mindless) way we
observe those facts. For Shaw this means that reality and illusion are
necessarily combined, in different configurations of conscious and
unconscious duplicity. So his irony is the quintessential element in his
humor and satire; it is his methodology and his philosophy for representing reality, just as naturalism is Zola's.

In Shaw's world we do not merely record reality in our minds, we perform it. We act out a role in which we convert the things of this world to
our uses. We may be conscious of our roleplaying — as Shaw was in
creating his public persona as "G.B.S." — or we may be quite absorbed in
our illusions and self-deceptions. In either case we are possible subjects
for comic irony. In fact, such irony may be the fullest way of picturing
our nature and our place in nature. To this point we should keep in mind
that Shaw called *Don Quixote* "the greatest realistic novel in the world."[48]
Upon serious consideration, who is prepared to deny this fundamental
irony?

Shaw's realism, like Cervantes', is based in a sympathetic yet ironic
understanding of the role of illusions in human behavior. When Shaw
describes his writing as scientific or realistic, therefore, he means that it
clearly represents such illusion in action:

> To me the tragedy and comedy of life lie in the consequence, sometimes ter-
> rible, sometimes ludicrous, of our persistent attempts to found our institu-
> tions on the ideals suggested to our imaginations by our half-satisfied pas-
> sions, instead of on a genuinely scientific natural history. And with that
> hint as to what I am driving at, I withdraw and ring up the curtain. (I, p.
> 385).

In this case the curtain rises on *Arms and the Man*, a less than naturalistic
play, obviously, but still one that provides a realistic picture of the
romantic illusions people maintain about heroism and war. Shaw well

knew that a tragedy lurks beneath his comedy, that human slaughter and evil follow from such illusions. He does not need a Zolaesque naturalism to be "realistic," especially since he perceived reality far more ironically than Zola. Thus, although he seems to set up his concept of a "genuinely scientific natural history" as the opposite of the false illusions that we hold, he in fact presents such illusions in his drama as an unavoidable element in our "natural" life. Our history includes our illusions.

Shaw's dramatic wit, with its ironic uses of old melodramatic plots and comic characters, thus distances him from Zola's naturalism, but his themes often derive nevertheless from the social issues of his time. During the 1880s and 1890s, in fact, no one was more involved in speaking out against the social ills of industrial society. These concerns contributed to his conception of his drama as discussion or problem plays. As he states in 1901: "What is a modern problem play but a clinical lecture on society; and how can one lecture like a master unless one knows the economic anatomy of society?"[49] The medical metaphor is not as self-consciously "scientific" as Zola's justification of literature in terms of medicine, but no less clinical and realistic in its appeal to science is Shaw's use of economics as both the basis for analyzing the true nature of society and the means for overcoming social illusions. Shaw's common bond with naturalism is in the broad indictment of capitalism and social injustice. By means of a rigorous "economic science" Shaw delivers his own "lesson of the real."

Shaw's economic masters were Henry George, who wrote *Progress and Poverty* (1879), and Karl Marx, whose *Capital* Shaw read, in part, in its French translation in the 1880s (before it became available in English). Upon hearing George lecture in the early 1880s, Shaw realized how he could move beyond his early intellectual "belief in Rationalism and Materialism."[50]

> It flashed on me then for the first time that the 'conflict between Religion and Science'—you remember Draper's book?—the overthrow of the Bible, the higher education of women, Mill on liberty, and all the rest of the storm that raged round Darwin, Tyndall, Huxley, Spencer, and the rest, on which I had brought myself up intellectually, was a mere middle-class business. Suppose it could have produced a nation of Matthew Arnolds and George Eliots—you may well shudder. The importance of the economic basis dawned on me: I read Marx, and was exactly in the mood for his reduction of all the conflicts to the conflict of classes for economic mastery, of all social forms to the economic forms of production and exchange.[51]

Shaw took to the Marxian dialectic—it makes good drama of history at the very least — because it provided a moral alternative to capitalist and liberal sentiment. For the next two decades, until he developed his meta-physics of creative evolution, economics provided the base for his in-vestigation of the causes of social ills.

Shaw's use of Marx, however, was always selective. The Marx of "scientific materialism" appealed to him no doubt, especially when he was most active as a Fabian lecturer and writer. But while Shaw was deeply concerned with specific social issues — with property rights, public revenue, welfare, wages, working conditions, labor unions, sewers, edu-cation — he often takes an unMarxian view on these concerns. His more substantial bond with Marx exists in a certain shared attitude they have toward human behavior. Both of them see their "scientific" mission as revealing the illusions that order and disguise "reality." In Marx's case, as Robert L. Heilbroner has pointed out, this scientific dialectic between il-lusions and reality is central to his analysis:

> In Vol. III of *Capital* Marx says that if appearances were like essences — if surface manifestations were all there was to reality — we would not need science. Science is therefore for him a penetrative task. It is not, however, the penetration of the natural scientist who tries to peer through the essen-tially *random* disturbances of nature to discover the workings of eternal laws. It is rather a penetration through a *systematic* distortion introduced into the social universe by the prescriptions ground into our social spec-tacles. These are distortions of which we are normally quite unaware. The object of Marx's scientific task is thus not to discover eternal laws in history (he is generally very guarded in his larger historical pronouncements), but to correct our social vision by discovering the prescription through which we have been looking.[52]

While Shaw does not build a new system to correct the old systematic distortion, he does share Marx's basic idea that pervasive illusions rule the world. Social institutions, whether they be traditional religions, capitalist organizations, political parties, or the like, warp understand-ing. By their very nature such institutions guide thought into distorted channels. And Shaw, with the help of Marx, sees how these distortions shape all of society:

> Take from the activity of mankind that part of it which consists in the pur-suit of illusions, and you take out the world's mainspring. Do not suppose, either, that the pursuit of illusions is the vain pursuit of nothing: on the contrary, there can no more be an illusion without a reality than a shadow without an object. Only, men are for the most part so constituted that realities repel, and illusions attract them.[53]

To be "scientific" is to be able to see through these distortions, to see the essential truth beneath the surface manifestations. Then change can occur.

Although Shaw turns to biology by the beginning of the twentieth century for the source of his theory of how change will be achieved, he never abandons the economic base that he established in the 1880s as a social critic. His realism and historicism develop out of his commitment to a "science of sociology" that will help provide rational programs for irrational man. His first "unpleasant plays" were an illustration of society's direct responsibility, whether or not acknowledged, for such evils as slum housing and prostitution. "Nothing," he writes, "would please our sanctimonious British public more than to throw the whole guilt of Mrs Warren's profession on Mrs Warren herself. Now the whole aim of my play is to throw that guilt on the British public itself" (I, p. 254). As a Fabian playwright Shaw uses economic science and its causal principles for his criticism of liberal, capitalistic society. This adaptation of Marxian ideas created a major problem for Shaw, however, because it led him towards certain deterministic propositions about history, change, and social organization. When Shaw writes in the preface to *Mrs Warren's Profession* that ". . . society, and not any individual, is the villain of the piece" (I, 264), he is apparently adopting a theory of causal efficacy that would place him in the same basic camp with the social determinists of his age. At this point he seems at one with Zola.

If in fact this is the thrust of Shaw's drama and social criticism, then we must say that his ideas on the hero, will power, moral choice, individual responsibility, causality, and science are all at odds with this idea of social determinism. Even when we recognize that Shaw's paradoxical thinking yokes opposites in dynamic relationship, we cannot logically accept an ironic resolution of this basic contradiction between determinism and free will. The two concepts are mutually exclusive. Furthermore, Shaw's drama itself becomes hopelessly muddled intellectually in its design and argument if this conflict exists. So, in the following section, we must ask whether Shaw stumbles at this crucial point into a deep confusion. Much of his argument with science rides on this issue. Specifically, two moral concerns arise: Does Shaw believe that social forces control one's destiny? And is his idea of the Life Force essentially deterministic?

[IV]

In his entry on "Destiny" in his *Philosophical Dictionary* Voltaire writes:

> Everything is performed according to immutable laws, everything is or-
> dained, everything is, in fact, *necessary*. . . . There are, some people say,
> some events which are necessary, and others which are not so. It would be
> comic for one part of the world to be arranged and the other not; that one
> part of what happens should happen inevitably, and another fortuitously.
> When we examine the question closely, we see that the doctrine opposed to
> that of destiny is absurd; but many men are destined to be bad reasoners,
> others not to reason at all, and others to persecute those who reason well or
> ill.[54]

The idea of destiny working itself out in human affairs appears in various
forms in Shaw's writings, embodied in a man of destiny such as
Napoleon or Caesar, a superwoman such as St. Joan, or a universal
power such as the Life Force. At least upon first consideration this idea
seems consistent with a belief in biological determinism. In turn, one
might argue that scientific determinism, as expressed in theories of
physics and biology, could serve as the theoretical base for such an idea
of destiny. Or, if this seems inappropriate for Shaw, one might compare
Shaw's idea of destiny to the Greek idea of fate or *moira*, since both
Shaw and the Greek dramatists wish to place Idea at the center of
causality, in contrast to believers in materialism, who attempt to remove
idea from action. But this fateful family of deterministic ideas does not
provide a proper home for Shaw. His "Man of Destiny," as he titled a
short play on Napoleon, is neither a modern Oedipus, caught in a fate
that measures his moral character and intellect, nor a victim of environmental and hereditary forces, enmeshed within a Zolaesque state of affairs. Shaw reasons that it is unreasonable and unnecessary, maybe even
absurd, to expect everything in the world to be prearranged. It is indeed
comic that one part of the world is arranged and the other part is not.
And we reason badly about which is which.

Yet what are we to make of Shaw's apparent acceptance of social determinism? In comparing capitalism with socialism, for example, he
states that "socialism is equally secular, and more materialistic and
fatalistic, because it attributes more importance to circumstances as a
factor in personal character and to industrial organization as a factor in
society."[55] Thus, in these terms money and social attitudes seem to determine action and character in Shaw's *Widowers' Houses*. The characters
are manipulated apparently by social forces (and Shaw's thematic purposes), so that we get little sense of them as people with personal control.
Shaw's argument determines their actions, rather than their actions,
developing out of their own choices, determining events. In like manner,

Shaw claims, *Mrs Warren's Profession* was written to show that prostitution is caused, not by female depravity and male licentiousness (faults of moral character), but by a social order that forces some women to resort to prostitution in order to survive. In a 1912 preface to Dickens' *Hard Times* Shaw writes that "until Society is reformed, no man can reform himself except in the most insignificantly small ways."[56]

What these statements reveal is that when Shaw's topic is Victorian capitalism, he tends to attack it in terms of the convenient socialist categories of economic determinism. He is drawn to hyperbole, as he acknowledges, in order to get his readers' attention. But when our attention is focused too exclusively on this aspect of Shaw's attack, we are led, misleadingly I believe, to argue that events determine character in his drama, that the conflicts are between social forces not individuals, and that individuals lack self-control. This is the argument of Friedhelm Denninghaus in *The Shaw Review*. According to him Shaw's characters are "shaped by history" in a "social-deterministic" world, wherein *"every personal responsibility is taken from them"* [his italics].[57] All social institutions are thus a form of destiny, imprisoning one in a social role that is imposed, not chosen.

Can we fit Shaw on this Procrustean bed? I think not. In the first place, this division of dramatic conflict and human events into either personal or social categories is too reductive for even the most "social" of Shaw's plays. When, for example, in *Widowers' Houses* a character such as Cokane (whose name suggests both someone who drugs himself and the rich idleness of the land of Cockaigne) excuses the fact of slum housing by blaming the situation on the "increase of the population" (I, 95) or when Sartorius (the capitalist well-clothed in deceptive words and false appearances) tells Trench that they both are "powerless to alter the state of society" (I, 94), we must recognize that Shaw is attacking them, not agreeing with them. Their Malthusian explanations appeal to a deterministic concept of causality that locks individuals in their social situation without the possibility of change. But as Shaw wrote in a letter of 1890: "the world does not now want to be told that society is rotten; it wants to know how it came to be rotten and what it should do to get sound."[58] One source of the rottenness is the irresponsible and rationalizing logic of a Cokane or a Sartorius. Their self-serving realism about capitalist society is no better (nor worse) than Trench's self-esteeming idealism. Both responses are inadequate.

Even during the 1880s and 1890s, his period of deepest involvement in socialist programs and theory, Shaw remained committed to the idea of moral responsibility as both a personal and a social imperative. In the *Quintessence of Ibsenism* he argues that history is moving from the age of reason to the age of individual will, so individuals must take responsibility for their destiny and not rationalize it as being determined. In one of the Fabian essays of 1889 Shaw wrote: "It is to economic science — once the Dismal, now the Hopeful — that we are indebted for the discovery that though the evil is enormously worse than we knew, yet it is not eternal — not even very long lived, if we only bestir ourselves to make an end of it."[59] This hope is crucial; it is an expectation, a purpose, a final cause that allows human beings to command their own destiny. The past is not fate; the future is not prearranged. By making an "end" for our actions — a hope and cause — we make an "end" of social suffering. Or as The Statue in *Don Juan in Hell* says: "For what is hope? A form of moral responsibility" (II, p. 642).

Will power and social responsibility are Shaw's "scientific" problems of moral character and human destiny. The effects of circumstances are obviously real, but so too are the powers of character: "I, as a Socialist, have had to preach, as much as anyone, the enormous power of the environment. But I never idolized environment as a dead destiny. We can change it; we must change it."[60] Or as Vivie says in *Mrs Warren's Profession*: "Everybody has some choice, mother. The poorest girl alive may not be able to choose between being Queen of England or Principal of Newnham; but she can choose between ragpicking and flowerselling, according to her taste. People are always blaming their circumstances for what they are. I dont believe in circumstances. The people who get on in this world are the people who get up and look for the circumstances they want, and, if they cant find them, make them" (I, pp. 309–10). Shaw may be gently satirizing Vivie's self-sufficiency, her cold independence, but her quality of character also has his support.

While Shaw did not believe that we have absolute free will, he attacked determinism as "a soulless stupidity" (IV, p. 533). Implied in this criticism is an argument against materialism as the scientific base for determinism. Indeed, Shaw wants to reject both physical and psychosocial theories of determinism,[61] while holding on to a metaphysical theory of an indeterminate Life Force or Will of God. This argument was an aspect of his desire to reinterpret religion in the light of science, but without having to accept determinism.

> God . . . is will. But will is useless without hands and brains. . . . The evolutionary process to me is God: this wonderful will of the universe, struggling and struggling, and bit by bit making hands and brains for himself, feeling

that, having this will, he must also have material organs with which to grapple with material things. And that is the reason we have come into existence. If you don't do his work it wont be done.[62]

Shaw takes the will of God and places it in each of us, thus resolving the conflict between determinism and free will by dispersing, as it were, the omnipotence and omniscience of God into the long history of the evolutionary process: "Now to admit that God can err, or that He is powerless in any particular, is to deprive Him of the attributes that qualify Him as God; but it is a very healthy admission for the strongminded. It increases our sense of responsibility for social welfare, and is radiant with boundless hopes of human betterment" (V, pp. 695–96).

Shaw's essential optimism, his belief in change and improvement of the human condition, arises partly from his acceptance of the basic thrust of evolutionary theory. That is, "the great principle of evolution," as Darwin phrases it, provides a law of causality which Shaw accepts in principle, while denying the deterministic implications of Natural Selection. Where Darwinism fails for Shaw is in its inability to comprehend the place of self-control and moral choice in human consciousness. The Neo-Darwinists are blind:

What is self-control: It is nothing but a higher developed vital sense, dominating and regulating the mere appetites. To overlook the very existence of this supreme sense; to miss the obvious inference that it is the quality that distinguishes the fittest to survive; to omit, in short, the highest moral claim of Evolutionary Selection, shewed the most pitable want of mastery of their own subject, the dullest lack of observation of the forces upon which Natural Selection works. (V. p. 309)

Biology transcends its own theory here; instinct becomes idea. The mind frees us from biological determinism and confronts us with moral choice. Value is thus a measure of something beyond survival instinct, beyond just biological advantage, beyond even reason. Shaw's "supreme sense" of self-control is an *inner* force, capable of development in each individual as purposeful decision. It is the direction of energy toward a specific end that is an idea. Human beings can hope, not just desire. They can conceive of what is not (for example, the hope for a good, even utopian society) and work for that end (even for others, not just themselves).

The Life Force, then, is not fateful, nor deterministic, nor fatalistic. In spite of Shaw's tendency to capitalize the phrase, thus giving it the aura of a Law or Principle of existence, the Life Force is for him a vital energy of will that we tap, a reservoir of evolutionary purpose or intention that we draw on. But Shaw is not always consistent about this. Sometimes in rather shorthand form in his letters and essays he represents the Life Force as a sexual power that sweeps us up in its drive toward procreation. As Ana says in *Don Juan in Hell*: "I believe in the Life to Come. A father! a father for the Superman!" (II, p. 689). As the sex drive, especially in women, the Life Force for Shaw is biological passion, subverting reason and idealism. But Shaw usually enlarges his meaning of it by showing that this biological drive is in the service of our will to understanding. We steer it. Or as Don Juan says: "The philosopher is Nature's pilot" (II, p. 685).

The Life Force is a power *within* us, yet this Shavian idea is itself a possible subterfuge if it implies causal action simply in terms of an obscure vitalist principle. Internal power, as Kant argues, matters not at all in the argument for free will if we have simply shifted from mechanical to psychological determinism. An internal chain of ideas is no more self-controlled than an external chain of events. Shaw attempted to avoid such a trap by drawing on the "creative evolution" of Henri Bergson, who argued that our intelligence, as opposed to instinct, confronts us with choice. The mind represents to itself various possible actions, then acts out the postulated idea. The mind is thus a symbolic staging ground for selfhood and purpose. By means of reflection and expectation we shape the autonomy of the will. We act, or can act, on principle. Our freedom then is an article of intelligence that we value because it confronts us with choice. Our judgment becomes a measure of value, an aspect of moral being.

By separating ethics from biological determinism and thereby proclaiming the value of self-control as an attribute of human consciousness, Shaw is extending the argument that T. H. Huxley reluctantly made at the end of his life in his important lecture of 1893 on "Evolution and Ethics." Despite his earlier attempts to solve the dualism of mind and matter by embracing purely naturalistic theories of human existence, Huxley had to admit finally that ethics cannot be derived from Darwinian theory, that ethics in fact exists somehow outside of all naturalistic theory. Huxley's rationalism and naturalism were unequal to the philosophical task of deriving civilization and its complex social arrangements from biology. The realm of law and order that science was extending, and of which Huxley sung well the praise, could not explain adequately man's ethical place in nature. This Huxley acknowledged in the end.

In like manner, more recently, the geneticist Theodosius Dobzhansky has argued that the agents of human history are contained within that history itself. In the great historical process man steps out of the purely biological into the ethical:

> Attempts to discover a biological basis of ethics suffer from mechanistic oversimplification. Human acts and aspirations may be morally right or morally wrong, regardless of whether they assist the evolutionary process to proceed in the direction in which it has been going, or whether they assist it in any direction at all. . . . Moral rightness and wrongness have meaning only in connection with persons who are free agents, and who are consequently able to choose between different ideas and between possible courses of action. . . . This new evolution, which involves culture, occurs according to its own laws, which are not deducible from, although also not contrary to, biological laws. The ability of man to choose freely between ideas and acts is one of the fundamental characteristics of human evolution.[63]

Human action, like dramatic action, is moral, and cannot be reduced to social Darwinism or "sociobiology." The will is not without various biological, psychological, sociological, and economical forces shaping it, but in the final analysis it is an internal force that turns consciousness into conscience. To think otherwise is to let science con us. "Conscience," Shaw wrote, "is just as much a force in humans as electro-magnetism is in the physical universe" (VI, p. 29).

Both mechanism and materialism are thus false analogies for understanding human behavior. Neither efficient cause nor material cause provides sufficient explanation. Whereas the classical scientific idea of causality makes motive power an action determined by natural laws of relation, Shaw makes relations a dramatic operation of motive powers, following the dictates of individual will. Conscience provides the basis for action; action is the test of conscience. Thus, in Kenneth Burke's words, ". . . insofar as ethics is treated in its own terms, as a special context of inquiry, rather than being reduced to nonethical terms, one is pledged in advance to discourse on the subject of action and passion. For that is what the study of ethics is."[64] No deterministic idea of causality, however much it may satisfy an appeal to Occam's razor as a principle of analysis and action, is appropriate for explaining human behavior. Unfortunately, the authority of the simple too often has served as the absolutism of the simpleminded.

As Shaw understood, the powerful desire for a single law of causality is in part admirable, but it is also a seductive danger that has plagued human history. Such idealism, which seeks universal consistency and rationality, transfigures our view of the world into a theory that reaches

beyond necessity into a logical contraction of life. Categorical impera-
tives thus become not Occam's razor, but a *reductio ad absurdum*. Such
is the destiny of the idea of destiny. While Voltaire states that the doctrine opposed to that of destiny is absurd, Shaw says in reply that doc-
trinaire thinking is absurd. Or, from the comic writer's point of view—
which the writer of *Candide* well understood in his satire on Pangloss's
simpleminded philosophy (which served as a necessary illusion)—such
classified thinking with its departmental disposition, its predictable pattern, its quintessential temperament, is the prototype for comic characterization. It is a humor: a repetitive cause or principle of behavior that
becomes ludicrous. Function is reduced to repetition; repetition turns
into habitual conduct or action. Thought becomes mechanical. And as
Henri Bergson noted: "The attitudes, gestures, and movements of the
human body are laughable in exact proportion as that body reminds us
of a mere machine."[65] As with the body, so with the thought. Physical
and mental action become comic reductions. For Shaw one such reduction to the mechanical was the theory of mechanism when applied to
humans. A "reality" in one field has become an illusion in another. So,
once again we are in the realm of irony, wherein things are not what they
seem (especially when what we see becomes a mechanical thing). Such
simpleminded causality is an illusion, and illusions, as we have seen, are
the shape of the world's body for Shaw. As he wrote in a letter of 1891:
"Every man sees what he looks for, and hears what he listens for, and
nothing else."[66] In the sense that such is our destiny, a fate of illusion and
self-deception, this central idea is Shaw's own singleminded law of
causality, his own *reductio ad absurdum* as a comic moralist. Another
paradox? Another necessary illusion?

[V]

Scientia est cognitio per causas. So too is drama. In Shaw's words: "the
dramatist's . . . business is to shew the connection between things that
seem apart and unrelated in the haphazard order of events in real life"
(III, 49).[67] This design, which is causal, must reveal an action arising out
of purposeful behavior or will, not environment alone. And because
such purpose is willed it is moral. To quote Kenneth Burke again: "When
one talks of the will, one is necessarily in the field of the *moral*; and the
field of the moral is, by definition, the field of action."[68] This field of action, with its moral dimensions, is the staging ground of human behavior; it is the Aristotelian shape of the drama because "the connection
between things" that holds beginning, middle, and end together in dramatic action is the human body in purposeful conduct.

Not surprisingly, Shaw's great success is as a playwright because drama is the representation of human value in terms of moral action. Turning theory into theater (not so difficult since the two words have a common etymology), Shaw dramatizes his ideas about mind, social order, human nature, moral action, and causality, matching speculations to spectators. When Shaw committed himself to playwriting in his mid-thirties, he made his best possible match between metaphysics and aesthetics, which join in the concept of causal action. In fact, so fitting is the correspondence between theory and theater, it is now difficult to ascertain whether his conception of human action and causality derives ultimately from the nature of drama, which he projects onto nature, or from an idea of nature and evolution, which he then dramatizes. We might well ask, did Shaw choose drama or did it choose him?

Whatever the case, Shaw is pledged as both a dramatist and a moralist (since the two are yoked) to deny that action and passion are deterministic. Unlike such writers as Zola, Hardy, Hauptmann, Gorky, Crane, and Dreiser, who in various degrees created a literature of environmental causes for human actions (which in its extreme is all "environ" and no "mental"), Shaw challenges such "scientific realism" because self-control is explained away. In fact, there is something essentially antidramatic and amoral about such literature. It makes situation the shaping power of action, while Shaw wants to place action within a contingent situation.

It is this idea of purpose shaping pattern that underlies Shaw's dramatic action. As he says of his largest (if not his best) work, "the most valuable lesson in *Back to Methuselah* is that things are conditioned not by their origins but by their ends."[69] Shaw's principle of causality is thus teleological. It expresses the Aristotelian idea of final cause, whereby purpose for which a thing is done serves as the measure of its value. This principle requires a sense of direction. It assumes, as Aristotle argues in his *Nicomachean Ethics*, that an action must have an end by which we judge both the agent and the agency of an act. Character thus determines event in Shaw's drama. The characters have aims and values that draw them forward through the action. These causes shape the action and provide the basis for moral judgment—both by the characters within the action and the audience outside of it.

No wonder, therefore, Shaw attacked the drama of the nineteenth century. With its mechanical plots, its characters manipulated by a clockwork design, and its ready-made morality based in romantic illusions, this melodrama is a parody of formal causes and a travesty of final causes. It is the fitting art work for a scientific mechanism, reducing characters to performance in an *a priori* order that may imply necessity but lacks moral justification and psychological validity. Of course, not all plays of the period were well-made in the Scribean mode, but most of these others had another fault just as damaging: a design of arbitrary and accidental action, with sentimentalism and spectacularism, as appeals to the audience, determining character behavior. Sham heroics, tearful resolutions, and insipid morality — what Shaw calls "balderdash" — control the stage. In contrast, Shaw reforms the drama by reestablishing it on valid principles of causality. He replaces the mindless drama of the nineteenth century with a mindful action. On such terms he praises both Ibsen and Shakespeare, for example, because they base action in moral character and avoid either deterministic plot manipulation or mere accident (which he equates with the purposeless "blind chance" of Darwinian natural selection). "In short, pure accidents are not dramatic: they are only anecdotic. They may be sensational, impressive, provocative, ruinous, curious, or a dozen other things; but they have no specifically dramatic interest. . . . As a matter of fact no accident, however sanguinary, can produce a moment of real drama."[70]

And yet William Archer (drama critic, translator of Ibsen, and Shaw's good friend) states that some of Shaw's plays are actually a series of episodes that "might at any moment take a turn in any possible direction without falsifying their antecedents or our expectations. . . . The episodes may grow out of each other plausibly enough, but by no preordained necessity, and with no far-reaching interdependence."[71] And Shaw himself rather disingenuously claimed on occasion that a play develops itself: "I only hold the pen."[72] But all of this is just another way of saying that the plays are not made to fit a mechanical plot — for that is more arbitrary by far than letting the plays develop in terms of their own internal logic of idea and character. Of course Shaw's plays lack a "preordained necessity." We should be surprised if they had such a design, since he purposely wrote to show that no such necessity controls human life absolutely. Human action is the expression of mind in the process of becoming, of will power seeking to define itself not simply as the result of its past (its effective and material causes), but in the light of its future (its formal and final causes). The formal cause is the shaping power of ideas; the final cause is the quality and purpose of those ideas.

Don Juan transforms himself from the rake into the philosopher; Major Barbara removes her Salvation Army uniform and takes up industrialism in order to make war on poverty; St. Joan, the peasant girl, refuses to know her place, becoming instead a soldier, a defense attorney, and, finally, a martyred saint. They all shape the future to meet their purposes. They have definite ends, unlike the characters in *Heartbreak House* whom we judge as defeated people because they have lost direction, giving themselves over to events. In this dark comedy Shaw reveals what happens to a social world when fraud and illusion and self-deception become too pervasive. Everything is false and without will power and moral conscience. Purpose is lost. Only through self-control do we discover direction.

This is the crucial meaning of the Life Force for Shaw. We create ourselves daily. Thus, many of Shaw's characters are without parents or are at least free from parental tyranny. The family bond is a loose one in his plays, as it was in his own family, so individual will comes to the fore. Shaw is determined to show that heredity matters less than will power; or, more specifically, that will power is the most significant aspect of heredity. The genius of strong-willed children is to be self-creating. Vivie in *Mrs Warren's Profession* declares: "I dont want a mother" (I, p. 354). This independence runs throughout his drama.[73] Blanche in *Widowers' Houses* laments: "Oh, if only a girl could have no father, no family, just as I have no mother!" (I, p. 109). Her wish becomes the paradoxical requirement in *Major Barbara*, one of Shaw's greatest achievements, where the central action turns on the proposition that those without parents are most gifted to run the world, to make the future. Cusins, the fiancé of Barbara, "inherits" Undershaft's huge armament business because as an orphan he is an independent soul, like Undershaft before him. Power is transferred on the basis of spiritual qualities, not family ties. By hating poverty Undershaft "became free and great. I was a dangerous man until I had my will: now I am a useful, beneficient, kindly person" (III, p. 173). This will is power. And as Barbara and Cusins learn, ". . . all power is spiritual" (III, p. 181). So, what might have been a mechanical plot device, in the tradition of nineteenth-century drama, becomes in Shaw's drama the basis for an action that causally develops through characterization into thematic design and import.

Likewise, in Shaw's best known play, *Pygmalion*, the independent will is superior to "social destiny" and the manipulative power of others. Eliza goes through two changes, first from flower girl to lady and then from toy woman to free woman. During the play she moves from one kind of independence, free from her father, the properly named Mr. Doolittle, who abandons all claim to controlling her, to a second independence, free from Henry Higgins, who assumes that he does control her. Higgins, the very model of an experimental scientist, has a theory that he can remake Eliza by changing her speech. As a professor of "the science of phonetics" (IV, p. 678), he is confident of his knowledge and power. But while Higgins is successful in changing Eliza's speech patterns, he fails to appreciate that her will is her own to direct. At his moment of apparent triumph he is surprised to discover that "Eliza's bolted" (IV, p. 757). She exits, like Nora in Ibsen's *A Doll's House*, because action arises out of moral purpose and will power. With a future ahead of her, a final cause towards which to aim (even if it is only the job of running her own flower shop), she denies that his power to mold her into a formal lady is her formal cause. He shapes the material with great efficiency, making her tongue sing a new tune, but her actions and values are formed by her deeper character, which he cannot determine. Her act of independence illustrates Shaw's belief in will power and self-control. Given her character, she must walk out. This is why the "happy ending" of marriage in *My Fair Lady* is so wrong: It denies the very life force of the play.

By tracing this life force, drama shows us the causes of human action. "It is the privilege of the drama," Shaw wrote in the 1980s, "to make life intelligible, at least hypothetically, by introducing moral design into it, even if that design be only to shew that moral design is an illusion."[74] Of course, Shaw believed that action can be adequately understood only in terms of moral purpose, but he also believed that both action and purpose are often held together by illusion. Modestly, therefore, he notes that his notion of civilization "is a view like any other view and no more, neither true nor false, but [it is hoped] a way of looking at the subject which throws into familiar order of cause and effect a sufficient body of fact and experience . . ." (II, p. 518). Less modestly, he often argues that his mission is to take the chaotic ways in which we perceive things and to arrange them so as to reveal their essential relationship. The dramatist, he claims, is writing "a genuine science of life and character,"[75] a science of moral consequence not just basic knowledge. In Shaw's worldview the theater is a social institution that at its best rivals any other institution in both expressing and influencing the moral life of culture and society. Such a belief was an article of faith and a justification for him. "The truth is that dramatic invention is the first effort of man to become intellectually conscious. No frontier can be marked between drama and history or religion, or between action and conduct, nor any distinction made between them that is not also the distinction between the masterpieces of the great dramatic poets and the commonplaces of our theatrical seasons" (I, p. 378).

This is a lofty claim, making theater one of the major embodiments of human value and knowledge. We are prepared to grant this to Greek drama, for example, but modern drama, especially comedy, does not usually receive such serious consideration. Shaw challenges us nevertheless to take him at his word. "I am myself a literary artist, and have made larger claims for literature — or, at any rate, put them forward more explicitly — than any writer of my generation as far as I know, claiming a continuous inspiration for modern literature or precisely the same character as that conceded to the ancient Hebrew Scriptures, and maintaining that the man of letters, when he is more than a mere confectioner, is a prophet or nothing."[76] By calling himself a prophet and by making oracular pronouncements, Shaw seems the eminent Victorian, in the manner of Carlyle or Ruskin, Most modern writers (with the exception of such moralists as Jean Paul Sartre or Aleksandr Solzhenitsyn) are diffident about the power and responsibility of literature, especially in the face of a technological culture which seems to dwarf human action, to produce alienation and a sense of powerlessness, and to breed antiheroes (and antiprophets) who live on guilt, cynicism, hate, and self-absorption. Shaw will have none of this; he does not despair. Unlike Carlyle or Ruskin on the one hand and Kafka or Beckett on the other, Shaw remains hopefully committed as a man of letters to a public life of moral criticism and artistic accomplishment that challenges society to improve itself. Without longing for the old ways or hating the new, he criticized both past and present evils and illusions in order to change the way we live.[77] No wonder Bertolt Brecht admired him.

In his most exultant (and possibly wishful) moments Shaw claimed nothing less for the theater than that it is "a temple of the Ascent of Man." (The Darwinian echo and modification are purposeful.) For Shaw, "the apostolic succession from Eschylus to myself is as serious and as continuously inspired as that younger institution, the apostolic succession of the Christian Church."[78] Hyperbole? Only in part, since Shaw was convinced that art, at its best, is the subtlest and most profound moral authority within culture. And the "artist-philosopher" is a prophet who gives metaphysics a palpable shape and order, a manifest destiny. Shaw takes the matters of history — Adam and Eve, the Devil, Caesar, Cleopatra, St. Joan, and contemporary social conditions — and gives them a causal design in dramatic form because ". . . it is [his] business to find some order and meaning in the apparently insane farce of life as it happens higgledy-piggledy off the stage" (III, p. 188). As a comic ironist Shaw delights in the farce, presenting our incongruities, our social disorders, our humors and passions in comic relief. As a "creative evolutionist" he gives history a moral meaning, thus illustrating Aristotle's argument in the *Poetics* that drama and history are two related modes of understanding and representation, with drama supposedly superior because it is more philosophical. It is the work of genius.

Of his own genius Shaw often made substantial claims, not in order to puff himself up (for he was remarkably lacking in vanity, arrogance, envy, or malice), but to acknowledge and champion talent itself because it serves the Life Force. Throughout his life Shaw placed himself in the service of "the drive of evolution, which we call conscience and honor" (IV, p. 567). In these two qualities, which are the supreme topics of great drama—"from Eschylus to myself"—Shaw defines what he expects of himself and of others in all endeavors, and what human life, unlike that posited in the Darwinian world, is driving at. This, in summary, is Shaw's central idea of causality: the moral power and responsibility of mind, of genius, for organizing actions and passions, including the sex drive, into a purpose that is larger than self-centered experience and circumstance. "Life is a constant becoming,"he wrote to Ellen Terry, the great actress whom he admired, "and all stages lead to the beginning of others. . . . I am like the madman in Ibsen's *Peer Gynt* who thought himself a pen and wanted someone to write with him. I want to be used, since use is life."[79] Shaw expresses the Logos itself as a writer dedicated to the Life Force—that is, to the future of human society and justice, to greater self-knowledge (which is the only path to self-preservation), and to the will to create. Thought is for him a passion, expressing the life force. And life is thought turned into action, into moral use. "No person worth a rap makes happiness, or love, or money, or anything but thoroughness of Life itself the criterion of life's value. Beware, beware, beware, beware, beware, BEWARE."[80] Beware of what? Of illusions, of a loss of will, of false values. We must define ourselves and our values by our purposes. In the words of Shakespeare's Prince Hal: "Let the end try the man."[81]

NOTES AND REFERENCES

1. Archibald Henderson, *Bernard Shaw, Playboy and Prophet* (New York: D. Appleton & Co., 1932), p. 50. Richard Ohmann in *Shaw: The Style and the Man* (Middletown, Conn.: Wesleyan University Press, 1951) claims that Shaw's "antiscientism" puts off readers so much that "scientists have pretty generally agreed to ignore Shaw's contributions" to scientific topics (p. 119).
2. Bernard Shaw, "Postscript: After Twenty-five Years," from *Back to Methuselah: A Metabiological Pentateuch* in *Bernard Shaw, Collected Plays with their Prefaces*, vol. 5, ed.

Dan H. Laurence (New York: Dodd, Mead, 1975), p. 693. All subsequent quotations from this seven-volume collection of Shaw's plays and prefaces will be acknowledged parenthetically in the text of the essay by volume number and page. Permission to quote from this collection is granted by The Society of Authors, London, England, on behalf of the Bernard Shaw Estate.

One thing more should be noted here about this collection. To quote its editor, Dan H. Laurence: "Shaw had strong personal opinions about style in printing, many of them highly idiosyncratic, and as he was his own publisher he had no difficulty implementing them. His spellings and contractions were often bizarre (*enterprize* and *wernt*), and sometimes archaic (*shew* for *show*). He had equally strong convictions about the superfluous use of punctuation, noting in *The Author* in April 1902: 'The apostrophes in ain't, don't, haven't, etc., look so ugly that the most careful printing cannot make a page of colloquial dialogue as handsome as a page of classical dialogue. . . . I have written aint, dont, havent, shant, shouldnt and wont for twenty years with perfect impunity, using the apostrophe only where its omission would suggest another word: for example, hell for he'll. . . . I also write thats, whats, lets, for the colloquial forms of that is, what is, let us; and I have not yet been prosecuted" (vol. 1, p. 6). Since Laurence follows this Shavian dictate, I do too.

3. Bernard Shaw, "The Conflict Between Science and Common Sense," *The Human Review*, vol. 1 (April 1900), p. 5. In this article Shaw writes, in his sly, hyperbolic manner: "I have found out the man of science; and in future my attitude towards him will be one of more or less polite incredulity. Imposter for imposter, I prefer the mystic to the scientist — the man who at least has the decency to call his nonsense a mystery, to him who pretends that it is ascertained, weighted, measured, analyzed fact" (p. 10).

4. Shaw writes in his preface to *Back to Methuselah*: "I knew that civilization needs a religion as a matter of life or death; and as the conception of Creative Evolution developed I saw that we were at last within reach of a faith which complied with the first condition of all religions that have ever taken hold of humanity: namely, that it must be, first and fundamentally, a science of metabiology" (V, p. 337).

5. Shaw believed that his ideas were consistent with the essential traits of natural science, if not materialism, so his use of the word "natural" is a calculated extension of the scientific meaning of the word, while also being an ironic challenge of its normal usage among scientists.

6. Bernard Shaw, "Michael and His Lost Angel," *Our Theatres in the Nineties*, vol. 2 (London: Constable, 1932), p. 18.

7. Bernard Shaw, *Collected Letters, 1898-1910*, ed. Dan H. Laurence (New York: Dodd, Mead, 1972), pp. 161 & 163; a letter to H. M. Hyndman, 28 April 1900.

8. In a letter of 18 February 1905 to his biographer Archibald Henderson, Shaw wrote: "Are you going to write a natural history, like a true Shavian, or a romance, like an incorrigible anti-Shavian?" Shaw challenged Henderson to "lay down the lines of a scientific biography" in which an "artificial conscience of morality "is clearly distinguished from "real conscience," based in the will. *Collected Letters, 1898-1910*, pp. 515-16.

9. G. K. Chesterton, *George Bernard Shaw* (New York: Hill & Wang, 1956; first published 1910). p. 51.

10. Bertrand Russell, "George Bernard Shaw," *Virginia Quarterly*, xxvii (Winter 1951), p. 4.

11. See Eric Bentley, *Bernard Shaw, A Reconsideration* (New York: W. W. Norton 1976), esp. pp 79-85.

12. Bernard Shaw, *Everybody's Political What's What* (New York: Dodd, Mead, & Co., 1944), pp. 362-63

13. J. D. Bernal, "Shaw the Scientist," *G. B. S. 90, Aspects of Bernard Shaw's Life and Work*, ed. S. Winston (New York: Dodd, Mead, 1946), p. 120. This is a fine, introductory essay on Shaw and science, both insightful and balanced.

14. For a summary and analysis of the Darwin-Butler controversy see Basil Willey, *Darwin and Butler; Two Versions of Evolution* (New York: Harcourt, Brace, 1960).

15. Letter to E. C. Chapman, 29 April 1891, Bernard Shaw, *Collected Letters, 1874–1897*, ed. Dan H. Laurence (New York: Dodd, Mead, 1965), p. 302.

16. Shaw's doubts about natural selection and the survival of the fittest may seem reactionary or antiscientific to many, but even today the controversy continues. As many scientists now recognize, Darwin's theory is based in part on a central tautology: those who survive are the strongest; the strongest are those who survive. Or: the fittest leave the most offspring; the largest number of offspring are the fittest. The geneticist C. H. Waddington and others have pointed out that this tautology fails to explain evolutionary change. See, for example, R. C. Lewontin, *The Genetic Basis of Evolutionary Change* (1974) and Michael Ruse, *The Philosophy of Biology* (1973).

17. This idea of "striving" is rather distinct from Darwin's idea of a life and death struggle. In chapter three of *The Origin of Species* he writes: "In looking at Nature it is most necessary to keep the foregoing considerations always in mind — never to forget that every single organic being may be said to be striving to the utmost to increase in numbers; that each lives by a struggle at some period of its life; that heavy destruction inevitably falls either on the young or old, during each generation or at recurrent intervals. Lighten any check, mitigate the destruction ever so little, and the number of the species will almost instantaneously increase to any amount." Shaw is describing a mental striving for understanding, not a physical striving for self-preservation.

18. On the one hand, Shaw can state that "our only hope, is in evolution. We must replace the man by the superman"(II, p. 775). Tanner, in *Man and Superman*, calls for selective breeding in order to produce the "superior mind." Yet, on the other hand, Shaw declares, "I have never taught 'the biological gospel of a new race-aristocracy.' My contempt for that disguise of simple snobbery is unbound. . . .There is no such thing as a Superman. There are super mathematicians like Einstein, and super playwrights like myself, but they are bungling amateurs in a dozen other departments of human activity." (Hitlerism and the Nordic Myth," *The Modern Thinker*, IV (Nov. 1933), p. 15 & p. 16.)

19. T. H. Huxley, "Science and Christian Tradition, " *Collected Essays*, Vol. 5 (London: Macmillan, 1893–94), p. 141. See also for a full assessment of Huxley's ideas, James G. Paradis, *T. H. Huxley: Man's Place in Nature* (Lincoln: University of Nebraska Press, 1978).

20. Letter to Mrs. Pakenham Beatty, 1 October 1885, Shaw, *Letters*, vol. one, p. 141.

21. J. D. Bernal, "Shaw the Scientist," p. 131

22. Roger Boxill, *Shaw and the Doctors* (New York: Basic Books, 1969), p. 51.

23. Boxill, p. 52.

24. Shaw, *Everybody's Political What's What?*, p. 190

25. Shaw,[24] pp. 190–91.

26. Letter to E. C. Chapman, 29 July 1891, Shaw, *Letters*, vol. one, p. 303.

27. Shaw, preface to *Three Plays by Brieux* (New York: Brentano's, 1911), pp. xi–xii.

28. Shaw, "On Going to Church," *Shavian Tract No. 5* (London: The Shaw Society, 1957), p. 12. Originally written in 1896 for *The Savoy*.

29. Shaw, *The Quintessence of Ibsenism* in *Shaw and Ibsen*, ed. J. L. Wisenthal (Toronto: University of Toronto Press, 1979), pp. 114–15. This passage is from the 1913 edition of *The Quintessence*, and includes some minor revisions by Shaw of the 1891 edition

30. Shaw, "Discards from Fabian Lecture on Darwin," written 18 March 1906. These "discards" are from unpublished writings of Shaw in the British Library (B.M. Add. MS 50661, Fol. 27, p. 94). Permission to quote this material was granted by The Society of Authors on behalf of the Bernard Shaw Estate. In these same discards Shaw writes: "Will is the motive 355

31. A. N. Whitehead, *Science and the Modern World* (New York: The New American Library, 1948; originally published 1925), pp. 15–16.

32. Shaw, *The Quintessence of Ibsenism* in *Shaw and Ibsen*, p. 112. In this context see John Rodenbeck, "Bernard Shaw's Revolt Against Rationalism," *Victorian Studies*, vol. 15 (June 1972, pp. 409–437. Rodenbeck provides a clear and helpful analysis of this issue.

33. Francis Bacon, *De augmentis scientiarum*, Bk. 2, chap. 4, from *The Works of Francis Bacon*, eds. James Spedding, Robert L. Ellis, and Douglas D. Heath, vol. 8 (New York: Hurd and Houghton, 1869), pp. 484–485. Quoted in William A. Wallace, *Causality and Scientific Explanation*, vol. 2. (Ann Arbor: The University of Michigan Press, 1974), pp. 83–84.

34. Wallace, *Causality and Scientific Explanation*, vol. 2, pp. 13–14.

35. William Blake, "The Marriage of Heaven and Hell," in *Blake Complete Writings*, ed. Geoffrey Keynes (London: Oxford University Press, 1966), p. 157.

36. Emile Zola, "Preface to Thérèse Raquin," in *European Theories of the Drama*, ed. Barrett H. Clark; revised by Henry Popkin (New York: Crown Publishers, 1965), p. 378.

37. Emile Zola, "Naturalism in the Theatre," in *Documents of Modern Literary Realism*, ed. George J. Becker (Princeton, N. J.: Princeton University Press, 1963), pp. 200–01.

38. Zola.,[37] p. 210.

39. Shaw, preface to *Three Plays by Brieux*, p. xii.

40. Shaw, "How William Archer Impressed Bernard Shaw," *Pen Portraits and Reviews*, *The Collected Works of Bernard Shaw*, Ayot St. Lawrence Edition, vol. 29 (New York: Wm. H. Wise, 1932), p. 20.

41. Shaw, "Appendix II: *The Widowers' Houses*," quoted in *Shaw: An Autobiography, 1856–1898*, ed. Stanley Weintraub (New York: Weybright and Talley, 1969), p. 275.

42. Letter to Henry James, 17 January 1909, Shaw, *Letters*, vol. two, p. 828. James wrote back to Shaw: "You simplify too much, by the same token, when you limit the field of interest to what you call the scientific. . . .In the one sense in which The Saloon *could* be scientfic — that is by being done with all the knowledge and intelligence relevant to its motive, I really think it quite supremely so. That is the only sense in which a work of art can be scientific — though in that sense, I admit, it may be so to the point of becoming an everlasting blessing to man" (quoted in Shaw, *Letters*, vol. two, p. 829 from Henry James, *Complete Plays*, ed. Leon Edel, 1949).

43. Shaw, preface to *Three Plays by Brieux*, p.xii.

44 Henry James, "Nana," in *Documents of Modern Literary Realism*, p. 242. This essay originally appeared in *The Parisian*, 26 February 1880. On Zola's lack of humor Shaw wrote: "If Zola had had a sense of humor, or a great artist's delight in playing with his ideas, his materials, and his readers, he would have become either as unreadable to the very people he came to wake up as Anatole France is, or as incredible as Victor Hugo was" (preface to *Three Plays by Brieux*, p. ix). This "artist's delight in playing with his ideas, his materials, and his readers" is of course the quality of mind that Shaw has in abundance, without losing his audience. In a letter to Augustin Hamon, 21 March 1910, Shaw wrote, "Zola was vulgar; Zola had no humor; Zola had no style. Yet he was head and shoulders above contemporaries of his who had refinement, wit, and style to a quite exquisite degree. The truth is that what determines a writer's greatness is neither his accomplishments nor the number

of things he knows by learning or observation; but solely his power of perceiving the relative importance of things" (*Letters*, vol. two, p. 914).

45.   Shaw, "The Illusions of Socialism," *Selected Non-Dramatic Writings of Bernard Shaw*, ed. Dan H. Laurence (Boston: Houghton Mifflin Co., 1965), p. 411

46.   Shaw[45] p.411.

47.   Shaw[45] p. 409

48.   Shaw, "A Dramatic Realist to His Critics," *Selected Non-Dramatic Writings of Bernard Shaw*, p. 327.

49.   Shaw, "Who I Am, and What I Think," *Selected Non-Dramatic Writings of Bernard Shaw*, p. 449.

50.   Shaw, "66 Years Later," typescript preface added March-April 1946 to the manuscript of *The Irrational Knot*, on presentation to the National Library of Ireland. Quoted by Stanley Weintrab in *Shaw: An Autobiography, 1856–1898*, p. 96

51.   Shaw, [49] p. 448. Shaw continues: "Marx's 'Capital' is not a treatise on Socialism; it is a jeremiad against the bourgeoisie, supported by such a mass of evidence and such a relentless Jewish genius for denunciation as had never been brought to bear before. It was supposed to be written for the working classes; but the working man respects the bourgeois, and wants to be a bourgeois; Marx never got hold of him for a moment. It was the revolting sons of the bourgeoisie itself--Lassalle, Marx, Liebknecht, Morris, Hyndman, Bax, all, like myself, bourgeois crossed with squirearchy–that painted the flag red. Bakunin and Kropotkin, of the military and noble caste (like Napoleon) were our extreme left. The middle and upper classes are the revolutionary element in society; the proletariat is the Conservative element, as Disraeli well knew" (pp. 448–49).

52.   Robert L. Heilbroner, "Inescapable Marx," *The New York Review of Books*, vol. 25 (June 29, 1978), p. 35. It is in this same sense that Heilbroner defines Marx's political vision that Shaw writes: "My mind's eye, like my body's, was 'normal': it saw things differently from other people's eyes and saw them better" (I,p. 13).

53.   Shaw, [45] p. 406.

54.   Voltaire, *A Philosophical Dictonary*, in *Works*, vol. 4, (New York: E. R. DuMont and The St. Hubert Guild, 1901), pp. 89–90.

55.   Shaw, [45] p. 444.

56.   Shaw, "Introduction to Charles Dickens' *Hard Times*" (London: Waverly, 1912); reprinted in *Hard Times: An Authoritative Text*, eds. George Ford and Sylvere Monod (New York: Norton, 1966), p.334.

57.   Friedhelm Denninghaus, "Determinism and Voluntarism in Shaw and Shakespeare," trans. John J. Weisert, *The Shaw Review*, 19 (Sept. 1976), pp. 122–23. Denninghaus writes: "The driving force of the dramatic movement in Shaw is *society*. The individual is dramatically significant only so far as he is a social being and embodies social forces through his character and actions. All purely private matters are beyond the field of vision. Dramatic conflicts in Shaw are social conflicts, as they exist in the body of society between various social and political groups. The conflicts within the soul of the individual persons are merely reflexes of those external social-suprapersonal conflicts, into which the concerned person falls through belonging to this or that battling social group" (pp. 120–21).

58.   Letter to an unidentified correspondent, 11 February 1980, Shaw, *Letters*, vol. one, p. 243.

59.   Shaw, *Fabian Essays in Socialism*, ed. G. B. Shaw (London: 1889; reprinted. New York: Doubleday, n.d.), p. 45.

60.   Letter to Henry James, 17 January 1909, Shaw, *Letters*, vol. two, p. 828. In this con-

text Shaw writes: "Would anyone but a buffleheaded idiot of a university professor, half crazy with correcting examination papers, infer that all my plays were written as economic essays, and not as plays of life, character, and human destiny like those of Shakespear or Euripides?" (*Sixteen Self Sketches*, London: Constable, 1949, p. 89).

61. Shaw has no essential argument with the idea that physical things have causes for their actions. Newton's laws are Shaw's. Nor does he doubt that the mechanism of the human body operates on certain causal principles of basic chemistry and biology. In fact, Shaw's vegetarianism and his sexual ideas and habits suggest strongly that he believed in correspondences between body and mind. But Shaw parts company with determinists on the issue of the will. None of the various theories of determinism—physical, psychological, metaphysical (as represented, in turn, by Comte, Spengler, or Hegel)—satisfactorily explain human consciousness and action for Shaw. His metaphysics of the Life Force locates knowledge and power not in an omnipotent and omniscient God or ideal Reason, but in the end towards which human beings are struggling. As he writes to Tolstoy: "God does not yet exist; but there is a creative force constantly struggling to evolve an executive organ of godlike knowledge and power. . . . Every man and woman born is a fresh attempt to achieve this object" (14 February 1910, Letters, vol. two, p. 901). Change is the law of existence and God is the piece-by-piece result of change.

62. Shaw, "The Religion of the British Empire," *The Religious Speeches of Bernard Shaw*, ed. Warren Sylvester Smith (University Park, Penn.: The Pennsylvania State University Press, 1963), p. 6.

63. Theodosius Dobzhansky, *The Biological Basis of Human Freedom* (New York: Columbia University Press, 1956), pp. 132 & 134.

64. Kenneth Burke, *A Grammar of Motives* (Berkeley: University of California Press, 1969), p. 137.

65. Henri Bergson, *Laughter* in *Comedy*, ed. & trans. Wiley Sypher (Garden City, N.Y.: Doubleday Anchor, 1956), p. 79.

66. Letter to E. C. Chapman, 29 July 1891, Shaw, *Letters*, vol. one, p. 301.

67. As Aristotle was the first to argue, dramatic design is causal. It links human destiny to moral (or immoral) actions. It connects characterization to theme by means of plot (Aristotle's *mythos*), which is an orderly arrangement of incidents into a complete and whole action. This design locates consequential actions in character (*ethos*), which is motivational and thus causal, at least in certain essential ways that ask us to evaluate and judge actions. In these terms Shaw writes: "If you look on life as it presents itself to you, it is an extraordinary unmeaning thing. . . .Now what the drama can do, and what it actually does, is to take this unmeaning, haphazard show of life, that means nothing to you, and arrange it in an intelligible order, and arrange it in such a way as to make you think very much more deeply about it than you ever dreamed of thinking about actual incidents that come to your knowledge. That is drama, and that is a very important public service to render. . .'About Actors and Acting," *Shaw on Theatre*, ed. E. J. West, [New York: Hill & Wang, 1958], p. 198). Such an aim is not much different from what Einstein states as the aim of science: "Science is the attempt to make the chaotic diversity of our sense experience correspond to a logically uniform system of thought. In this system single experiences must be correlated with the theoretic structure in such way that the resulting coordination is unique and convincing" ("Considerations Concerning the Fundaments of Theoretical Physics," *Science*, vol. 91, 24 May 1940, p. 487). Drama does not seek one uniform system of thought, but it does put experience into an orderly design of causes.

68. Kenneth Burke, *A Grammar of Motives*, p. 136.

69.   Shaw, "On Clive Bell's Article," in *Shaw on Theatre*, p. 151.

70.   Shaw, *The Quintessence of Ibsenism* in *Shaw and Ibsen*, p. 214, In this Shaw is at one with Aristotle who writes: "Of all plots and actions the episodic are the worst. I call a plot 'episodic' in which the episodes or acts succeed one another without probable or necessary sequence" (*Poetics*, ch. 9). For both Aristotle and Shaw this principle of organization is not merely a matter of dramatic technique but an essential argument about the way drama mirrors nature (or the way we perceive such correspondence, whether illusionary or not).

71.   William Archer, Play-Making (Boston: Small, Maynard & Co., 1912), p. 191.

72.   *Table-Talk of G. B. S.*, ed. Archibald Henderson (London, 1925), p. 75.

73.   Probably no other major dramatist has de-emphasized the family as source and center of action as much as Shaw–not Ibsen, not Aeschylus, not Sophocles, not Euripides (although he comes closer), not Molière, not even Shakespeare, although he does give us daughters of independent will and action in the comedies.

74.   Shaw, *Our Theatre in the Nineties*, vol. 3 (London: Constable, 1932), pp. 18–19.

75.   Shaw, [74] p. 315.

76.   Shaw, "The Artstruck Englishman," *The Nation*, 17 Feb. 1917; reprinted in *Pen Portraits and Reviews*, p. 244.

77.   See Eric Bentley, *Bernard Shaw, A Reconsideration*, for a fine consideration of Shaw's relationship to Carlyle and Ruskin. Also, Bentley's comments on Shaw's public persona of G. B. S. suggest how Shaw trapped himself in a role that limited his influence, if not his renown.

78.   Shaw, "The Author's Apology," *Our Theatre in the Nineties*, vol. one, pp. vi and vii.

79.   Letter to Ellen Terry, 27 January 1897, Shaw, *Letters*, vol. one, pp. 722–23.

80.   Letter to Elizabeth Robins, 13, February 1899, Shaw, *Letters*, Vol. two, p.77.

81.   William Shakespeare, *Henry IV, Part Two* (II, ii, 46).

Index

359